Studies in Computational Intelligence 418

Editor-in-Chief

Prof. Janusz Kacprzyk
Systems Research Institute
Polish Academy of Sciences
ul. Newelska 6
01-447 Warsaw
Poland
E-mail: kacprzyk@ibspan.waw.pl

For further volumes:
http://www.springer.com/series/7092

Ioannis E. Anagnostopoulos, Mária Bieliková,
Phivos Mylonas, and Nicolas Tsapatsoulis (Eds.)

Semantic Hyper/Multimedia Adaptation

Schemes and Applications

Springer

Editors
Ioannis E. Anagnostopoulos
University of Central Greece
Department of Computer Science
and Biomedical Informatics
Lamia
Greece

Mária Bieliková
Slovak University of Technology
in Bratislava
Institute of Informatics and Software
Engineering
Bratislava
Slovakia

Phivos Mylonas
Ionian University
Department of Informatics
Corfu
Greece

Nicolas Tsapatsoulis
Cyprus University of Technology
Department of Communication and Internet
Studies
Limassol
Cyprus

ISSN 1860-949X e-ISSN 1860-9503
ISBN 978-3-642-44155-4 ISBN 978-3-642-28977-4 (eBook)
DOI 10.1007/978-3-642-28977-4
Springer Heidelberg New York Dordrecht London

Printed on acid-free paper

Springer is part of Springer Science+Business Media (www.springer.com)

Editorial on "Semantic Hyper/Multimedia Adaptation: Schemes and Applications"

Nowadays, more and more users are witnessing the impact of Hypermedia/ Multimedia as well as the penetration of social applications in their life. Internet was designed in order maximize user choice and innovation, while Web, as the ultimate service over this multi-layered structure, created a global software environment for millions of users worldwide. Both technological attainments are continuously revolutionizing the way we process, use, exchange and disseminate information. Through this revolution, many real-life applications in the fields of communication, commerce, education, government, and entertainment are redefined.

Parallel to the evolution of Internet and Web, several Hypermedia/Multimedia schemes and technologies bring semantic-based intelligent, personalized and adaptive services to the end users. More and more techniques are applied in media systems in order to be user/group-centric, adapting to different content and context features of a single or a community user. In respect to all the above, researchers need to explore and study the plethora of challenges that emergent personalisation and adaptation technologies bring to the new era.

This edited volume aims to increase the awareness of researchers in this area. It includes thirteen (13) articles authored by researchers from eight (8) different European countries, namely Belgium, Cyprus, Czech Republic, Greece, Italy, Slovakia, Spain, and UK. All accepted contributions provide an in-depth investigation on research and deployment issues, regarding already introduced schemes and applications in Semantic Hyper/Multimedia and Social Media Adaptation. Moreover, the authors provide survey-based articles, so as potential readers can use it for catching up the recent trends and applications in respect to the relevant literature. Finally, the authors discuss and present their approach in the respective field or problem addressed.

For consistency purposes and in order to further highlight the authors' contributions, we divided this edited volume to four (4) separate chapters, which cover most of the topics announced in our open call for papers. The chapter titles are:

- Chapter 1: Semantics Acquisition and Usage,
- Chapter 2: Reasoning for Personalization and Recommendation,
- Chapter 3: Social and Context-aware Adaptation, and
- Chapter 4: Multimedia and Open Standards

The reader can also find analytical prefaces of each chapter, which summarise the aims of each article, and how the work described is related with the chapter topic.

From our part, as Guest-Editors, we would like to thank all authors for their submitted contributions and the opportunity they gave us to edit this volume. We hope that all the contributions that appear in this edited volume will contribute towards a deeper understanding of the key problems in this area, and that they will help researchers and developers to find new solutions to existing problems, opening in parallel new research paths in related topics. We also would like to explicitly acknowledge the help of all referees involved during the review phases. Their valuable comments and suggestions improved the quality of the published works.

Last but not least, we would like to express our gratitude to Prof. Dr. Janusz Kacprzyk, and Dr. Thomas Ditzinger, Editor and Senior Editor of Springer SCI book series respectively, for the all the support and guidance provided to us, as well as, the fruitful cooperation we had.

The Guest Editors

Ioannis E. Anagnostopoulos
Assistant Professor, University of Central Greece
Department of Computer Science and Biomedical Informatics
Papassiopoulou 2-4, 35100, Lamia, Greece
E-mail: janag@ucg.gr

Mária Bieliková
Professor, Slovak University of Technology in Bratislava
Institute of Informatics and Software Engineering
Faculty of Informatics and Information Technologies
Ilkovičova 3, 842 16, Bratislava 4, Slovakia
E-mail: bielik@fiit.stuba.sk

Phivos Mylonas
Assistant Professor, Ionian University
Department of Informatics
7 Tsirigoti Square, 49100, Corfu, Greece
E-mail: fmylonas@ionio.gr

Nicolas Tsapatsoulis
Assistant Professor, Cyprus University of Technology
Department of Communication and Internet Studies,
31 Archbishop Kyprianos Str., CY-3036, Limassol, Cyprus
E-mail: nicolas.tsapatsoulis@cut.ac.cy

Contents

Chapter 1: Semantics Acquisition and Usage

In informatics, every time a meaningful representation of ordinary human activities, observations or tasks is undertaken, a final important step has to be fulfilled, i.e. the interpretation of their semantics in a computationally efficient manner, so as for them to be exploited within our digital world. Given the diversity of languages, symbols, characters, meanings, etc. this effort is still characterized as an open scientific area for anxious researchers. The following articles cover a selection of topics that are central to the study of knowledge and semantics, tackling them from both the acquisition and utilization perspective. Typically the acquisition process follows different paths depending on the specific nature of meaning involved in comparison to the way the meaning is utilized in order to provide meaningful results. For each of these areas, the problems they pose for the semantics acquisition and usage will be examined in general, whereas specific research findings derived from fields like formal computational knowledge representation, embedded systems analysis and synthesis, automatic surveillance systems analysis, as well as exploitation of semantics and mobility within the education and job seeking processes will be presented in detail.

Article 1.1

Title: An Individual Differences Approach in Designing Ontologies for Efficient Personalization

Authors: Nikos Tsianos, Panagiotis Germanakos, Marios Belk, Zacharias Lekkas, George Samaras, and Costas Mourlas

In the search of resourceful personalization methodologies, this contribution discusses the potential role of ontological knowledge structures towards the better handling of their semantics. Its focus lies on possible ways of integrating cognitive individual differences in the representation of users' cognitive characteristics, leading to the development of an proposed ontological approach. By outlining the field of cognitive abilities and by proposing a way for using some of these factors in applied research fields, the interpretation of human behavior in today's digital networking environment is advanced and further explored. Moreover, the enrichment of ontologies with cognitive factors may lead to efficient personalization and measurable benefits for users, thus resulting into very interesting scientific outcomes.

Article 1.2

Title: Formal Methods in High-Level and System Synthesis
Author: Michael F. Dossis

Given the nowadays constantly increasing amount of digital, networked computer devices and their growing complexity, this article takes a step further into researching deeper on their behavior understanding and elucidation. Its discusses relations that exist between software and hardware aspects of portable computing machines and embedded systems, focusing on high-level and system synthesis aspects. It includes a survey and analysis of early and mature formal high-level and system synthesis techniques, as well as the produced hardware implementations. The recently renewed interest of industry and academia in the high-level synthesis topic, motivated research on the underlying knowledge and semantics acquisition, so as to achieve better quality of implementations and shorter specification-to-product times with respect to everyday real-world applications.

Article 1.3

Title: Vision Based Semantic Analysis of Surveillance Videos
Authors: V. Fernandez Arguedas, Q. Zhang, K. Chandramouli, and E. Izquierdo

Among others, a recent hot, cross-domain research topic with respect to the meaningful blending of human activities and perception and computer-guided interpretation is the automatic exploitation of data and knowledge from surveillance systems. As a result, this contribution presents on the one hand an extensive survey of different techniques that have been proposed for surveillance systems, suitably categorized into respective categories, i.e. motion analysis, visual feature extraction and indexing, whereas on the other hand, it describes a practical implementation of an integrated surveillance framework for unsupervised object indexing which closely resembles the generic architectures.

Article 1.4

Title: On the Use of Semantic Technologies to Support Education, Mobility and Employability
Authors: Valentina Gatteschi, Fabrizio Lamberti, and Claudio Demartini

The final article of this book section deals with recent research activities in the field of semantic technologies applied to education and job-seeking contexts. In this framework, issues related to students' and workers' mobility, job-seeking and hiring and the improvement of qualification offers are analyzed and compared, by distinguishing different aspects, such as knowledge base creation, development of integration strategies and definition of meaningful semantic inference rules. The aim of this study is to understand the way technological advances, like the World Wide Web or the Semantic Web, are actively exploited by their users. Towards the production of new knowledge made available to other users.

An Individual Differences Approach in Designing Ontologies for Efficient Personalization

Nikos Tsianos[1], Panagiotis Germanakos[2], Marios Belk[2], Zacharias Lekkas[1], George Samaras[2], and Costas Mourlas[1]

[1] Department of Communication and Media Studies, University of Athens
[2] Department of Computer Science, University of Cyprus

Abstract. This article discusses the potential role of cognitive individual differences in the context of designing ontologies for personalization on users' characteristics. The theoretical framework of the proposed approach is derived from the long tradition of psychometric testing, incorporating implications from the field of differential and cognitive psychology. The current state of the identification and systematization of mental abilities is depicted, with additional emphasis on the constructs of working memory and cognitive style, on which a proposed ontology is based upon. Also, a summary of previously conducted relevant empirical work is presented, providing support to the notion of introducing personalization into educational and commercial websites. To that end, the main argument of this work is that the enrichment of ontologies with cognitive factors may lead to efficient personalization and measurable benefits for users.

1 Introduction

The idea of developing adaptive and personalized systems has been mainly supported by arguments focusing on the drawbacks of the "one-size-fits-all" approach (Brusilovsky and Maybury 2002) and essentially the complexity and vagueness of the ever-expanding World Wide Web (De Bra et al. 2004). In parallel, researchers and practitioners in the field of adaptive hypermedia underline the heterogeneity of the user population, while it is often implied that "static", non-personalized systems fail to satisfy the needs and support the goals of different users (Brusilovsky 2001). A certain degree of recognition towards this approach may be deducted from the fact that web services such as Google, Bing, and Amazon are nowadays offering personalized results and recommendations, by employing rule-based and collaborative filtering techniques. Admittedly, though, these popular services could not be classified as adaptive hypermedia, since they are lacking many important features and functionalities; there is no user model, no varying modes of presentation, and no navigational support, to name but a few of the

I.E. Anagnostopoulos et al. (Eds.): Semantic Hyper/Multimedia Adaptation, SCI 418, pp. 3–21.
springerlink.com

elemental attributes of personalized systems (Brusilovsky 1996). Nevertheless, the notion of personalization has finally found its way in users' everyday interactions in a massive, though rather superficial, manner.

Still, a rather obvious question must be addressed: is it worth to develop extensive personalized services, considering that their technical complexity and requirements far surpass that of static systems? During the incubation period of adaptive hypermedia, Dieterich et al. (1993) identified certain criteria in order to analyze and document the usefulness of adopting a personalized approach: the area of application, the characteristics of the users to be taken into account, the aspects of the system that can be manipulated and adapted, and the goals of the personalized approach. E-learning, for instance, qualifies rather easily according to these criteria, since it is a very wide area, the learner population is much diversified (with different goals, needs, and abilities), while the educational content and the instructional method can be manipulated. However, even in this case, the high cost of designing personalized courses for popular and free to use e-learning platforms has resulted in limited appeal outside the research community (Hauger and Köck 2007). Also, Paramythis and Loidl-Reisinger (2004) stress that in many cases educational adaptive hypermedia are not standard compliant. Therefore, the move towards personalized web applications and services is not expected to be an easy one, especially without support from high profile service providers.

More importantly, there is a need for more research on measuring the actual benefit for users, instead of merely developing elaborated personalization and user modeling techniques, along with corresponding semantic schemes; even if personalization is the key to more efficient interactions and satisfying experience on behalf of the users, there is one undeniable issue to be resolved: how and why would users benefit? Individuals are certainly different from each other, but which would be the underlying theories that could guide research endeavors in producing measurable gains? A first approach would be to identify the levels in which individuals demonstrate considerable differences, such as demographics, social, mental abilities, personality, goals, needs, and experience, and to build a cohesive user model by including characteristics that would be proven to be important in affecting behavior and performance. This could be probably achieved only by conducting extensive empirical work, driven by grounded psychological and sociological theories, and by gradually developing an interdisciplinary framework that would bridge technical possibilities with human factors.

To that end, considering in parallel the main functionalities of adaptive hypermedia, effective personalization of Web content involves two important challenges: i) accurately model and represent user information that is deemed as essential and useful for the adaptation process, and ii) model any hypermedia content in a way that would enable efficient and effective navigation and presentation as a result of the adaptation process. In a more technical view, the challenge is to study and design structures of meta-data (i.e., semantics) at the provider level, aiming to construct a Web-based adaptation mechanism that will serve as an automatic filter adapting the distributed hypertext/hypermedia content based on the user model. Semantics employ specialized approaches and techniques for alleviating difficulties and constraints imposed by the Web and contribute to the whole adaptation process with machine-understandable representation of user profiles and Web content.

In this context, the work that is presented here constitutes an effort to introduce the notion of individual differences as a core element of the abovementioned research directions, focusing mainly on users' cognitive characteristics. Specifically, we discuss the potential role of cognitive abilities in information processing within Web environments, along with possible ways of providing effective personalization processes, with direct implications on the user model, the Web content, and the semantic structures. This approach has been inspired by individual differences research; in the words of Kyllonen and Stevens (1990, p.130), a person may differ from another in "...fundamental cognitive abilities that affect the overall integrity of the individual's cognitive information processing system". It should be noted, though, that such differences are expected to manifest when a certain amount of information processing load is imposed on the user, which consequently involves hypermedia environments and Web systems that present a certain degree of complexity.

Interestingly, a review by DeStefano and Lefevre (2007) suggests that hypertext reading induces higher cognitive load to users, as compared to other forms of reading, and that proper structuring the content and reducing the number of hyperlinks are both beneficial for users with lower cognitive abilities (namely working memory, though experience is also important). Lee and Tedder (2003) also found that such users are facing higher difficulties in information recall in the case of hyperlinked structures, while McDonald and Stevenson (1996) also suggest that hypermedia navigation increases cognitive load. These studies, along with other research that also focuses on cognitive (or information processing) load, have not dealt with heavily demanding learning or training environments, but with rather simple hyperlinked passages of text. Therefore, it seems that accessing hyperlinked information in the Web may be more demanding for users with lower cognitive abilities, and that personalization on such individual differences could be of importance, with measurable gains.

Thus, the following section of this paper discusses the long history of individual differences, especially at the level of cognition, in an effort to identify factors that could provide useful information about users' behavior, preferences, and abilities; utterly, the aim is to provide adequate theoretical support for a proposed user model and semantic scheme, along with ideas for future empirical research. Section 3 focuses on issues that relate to adaptivity implementation considerations, while section 4 presents certain research efforts that we have conducted in order to support the argument of incorporating individual differences in personalized systems for education and commercial Web-sites. Section 6 concludes the paper with discussion and future trends of our work.

2 Individual Differences as User Modeling Factors

The term individual differences is indeed very broad, since it could include from genetics to personality; thus, it should be mentioned that the way that it is used in our research context derives from the field of Differential Psychology. The term (initially in German, Psychologie der individuellen Differenzen) was proposed by William Stern (1900), in order to summarize the research on mental differences, in coordination to a notion of "general psychology". The emergence and proliferation of the individual differences research is not however directly linked to cognitive research; in fact, researchers from the fields of differential and cognitive psychology have often opposed each other, especially on whether psychometrical approaches are truly related to human cognitive structures (Glaser and Pellegrino 1978). Also, it is rather indisputable that, for the most part, individual differences research was based on (or provided a basis for) the study of intelligence (Dillon and Watson 1996).

2.1 Cognitive Abilities

From the beginning of the 20th century, an impressive number of mental abilities tests have been developed, aiming to measure complex and higher cognitive processes that relate to information processing tasks (Cattell 1987). This research led to the emergence of many theories for intelligence, resulting in a somehow fragmented field; a common however focal point of these theories, originating from the work of Thurstone (1938), is that there are certain distinct basic mental abilities (factors), in which people differ at some extent. All this work was at a large extent summarized by Carroll's very influential meta-analysis, which led to the development of his three stratum theory (1993). Carroll identified a general intelligence factor (g), 8 broad abilities (factors), and 69 narrow abilities which are factorial related (and partially explain) to the former. In our research context, these 8 broad abilities (or those identified by similar theories), could provide a basis for personalization on cognitive abilities as user modeling factors; the general intelligence factor (practically a person's IQ), is considered too broad and (by definition) not directly related to specific information processing tasks that would mostly interest in a Web environment.

Specifically, Carroll identified the following abilities: fluid intelligence, crystallized intelligence, general memory and learning, broad visual perception, broad auditory perception, broad retrieval ability, broad cognitive speediness, and processing speed. A recent further development of the three-stratum theory is the Cattell-Horn-Carroll (CHC) Theory (McGrew 1997), which also includes Cattell and Horn's Gf-Gc theory (Horn and Noll 1997). The CHC model of intelligence modifies and raises the number of broad mental abilities to 10, though this number should not be considered as fixed, since the theory is open to new findings and future developments (McGrew 2009). At a descriptive level, we believe that the CHC model provides a very useful understanding on the organization of intelligence factors (narrow to broad), clearly demonstrating the aspects in which people differ at the level of cognitive abilities (see figure 1).

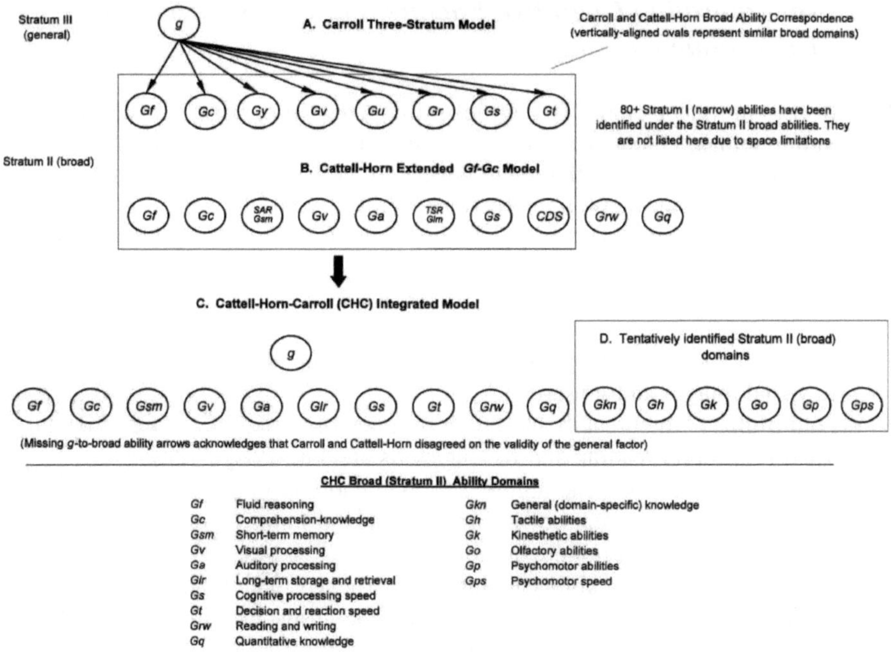

Fig. 1 The Integration of Caroll's Three-Stratum Model and Cattell-Horn's Extended Gf-Gc Model into the CHC Model of Cognitive Abilities (McGrew 2009, p. 4).

As it concerns the 76 narrow abilities that are factorial related to the broad, two examples follow: i) **Comprehension Reasoning** comprises Language Development (LD), Lexical Knowledge (VL), Listening Ability (LS), General (verbal) Information (K0), Information about Culture (K2), Communication Ability (CM), Oral Production and Fluency (OP), Grammatical Sensitivity (MY), Foreign Language Proficiency (KL), Foreign Language Aptitude (LA), and ii) **Short-term Memory** comprises Memory Span (MS) and Working Memory (MW). Obviously, as a result of factor analysis, narrow abilities have different loadings on each broad factor, while they represent performance at corresponding psychometric testing.

It should be clarified at this point that such psychometric theories of individual differences are indeed very elaborate and complex in relation to existing, or even technically viable, approaches in personalization and user modeling. On the other hand, we consider that a broad and thorough understanding of how people differ when required to perform mental tasks is necessary in order to subsequently narrow down the number of possible user attributes that could be used in a personalization scheme. Also, in the context of Web environments, it is very probable that just a few of these factors would suffice in order to predict (at some extent) users' behavior and/or performance; this narrowing down, however, requires both a solid theoretical basis and extensive empirical work. Likewise, our empirical work (see section 4) involves assessing the role of a limited number of factors at a time, with the purpose of gradually enriching and optimizing a user model of cognitive abilities and preferences.

2.2 Incorporating Cognitive Factors into Personalized Web Environments

The effort to introduce cognitive abilities into the design of adaptive Web systems by creating semantic schemes (such as ontologies) and filtering mechanisms is mainly hampered by the fact that there is very limited experience on which factors are the most important in Web interactions. Still, it would be very useful to keep in mind that according to Deary's (2001) review on the relatively recent state of research on intelligence, individuals predominantly differ in the following factors (the definitions are cited from McGrew, 2009, p.5-6):

- **Visual** *(and spatial)* **ability**: "the ability to generate, store, retrieve, and transform visual images and sensations".
- **Verbal ability**: "the breadth and depth of a person's acquired store of declarative and procedural reading and writing skills and knowledge".
- **Memory** *(short-term)*: "The ability to apprehend and maintain awareness of a limited number of elements of information in the immediate situation (events that occurred in the last minute or so)".
- **Processing speed**: "The ability to automatically and fluently perform relatively easy or over-learned elementary cognitive tasks, especially when high mental efficiency (i.e., attention and focused concentration) is required".

Thus, in our opinion, these areas of cognition are a good starting point for the study of their respective effects in Web environments. Deary also points that intelligence factors are highly stable throughout a person's lifetime, though fluid reasoning, memory, and speed tend to deteriorate with high age. Most importantly, according to this survey, performance in psychometric tests is a rather strong predictor of educational and occupational achievement, along off course with numerous other parameters (i.e., social).

The abovementioned factors, with the exception of processing speed, could easily be related to the use of the Web, albeit tentatively; the Web requires both visual and verbal (reading) processing of information, while maintaining awareness of different elements (i.e., hyperlinks) is essential, and as depicted in the introductory section persons with lower levels of working memory may face increased difficulties in hypertext environments. These assumptions do not necessarily imply that more intelligent persons excel at Web interactions – each individual may have different strengths and/or weaknesses, and perhaps the employment of personalization techniques could result in providing tailor-suited environments. Such an approach would be very relevant to the proposed theories of cognitive style, a construct that refers to habitual or preferred modes of problem solving, thinking, perceiving, and remembering (Tennant 1988), or to consistent individual differences in preferred modes of organizing and processing information and experience (Messick 1984).

In fact, Riding and Cheema (1991) identified two independent dimensions of cognitive style, by integrating a large volume of preexisting style research into their theory: Verbalizer – Imager, and Wholist – Analyst. The first dichotomy represents individuals' preference for receiving and processing information in either visual or verbal mode, while the second refers to a corresponding preference for information in whole or in parts; individuals without preferences are classified in each scale as "intermediates". The implications of the verbal – imager dimension are rather clear (imagery vs. textual and auditory information); the wholist – analyst dimension, however, is derived from Witkin's construct of "psychological differentiation" (Witkin et al 1971; Witkin et al 1977), and its implications are somehow more complex. In a nutshell, analysts are better at active analysis and perception differentiation, tend to act independently, are self-oriented and self-reinforced, and develop their own strategies. Wholists prefer social interaction and collaboration, while they require external direction, reinforcement, feedback, defined goals and specific structures. It should be noted that this research has been conducted primarily in the field of learning, but cognitive style has been developed as a broader concept than learning style, which on the contrary focuses on educational strategies and measurement tools in schooling environments (Rayner and Riding 1997).

According to our theoretical analysis, so far, it seems that there is a possibility to link the abovementioned individual differences to certain aspects of Web environments: individuals may excel at or prefer either visual or verbal information (i.e., correspondingly tagged content), while there may be preference for either loosely or highly structured environments (i.e., navigational freedom and level of control). Due to the fact that Riding and Cheema's theory of style (Cognitive Style Analysis – CSA) provides certain rather clear guidelines on how to distribute information to individuals, we opted for adopting CSA as an indicative measure of users' information processing preferences at a preliminary level of exploring individual differences.

Still, the potential role of (short-term/working) memory has not been discussed yet; the following subsection aims to provide some insight on this dominant in the field of cognitive psychology construct. As it concerns processing speed, it could perhaps be related to Web systems that impose time constraints, but in our opinion it would be nearly impossible to increase individuals' performance with current personalization techniques. Also, the role of processing speed may be more important in lower level mental tasks, which is not the case with most Web environments.

2.2.1 The Significance of Working Memory

Working Memory is a central construct in the field of cognitive psychology and neuroscience, as the main mechanism for maintaining information that is required for the performance of mental tasks (Shah and Miyake 1999). Working memory is also closely related to individual differences psychology, since many aspects of this construct have emerged through psychometric procedures (Baddeley 1992). In fact, measuring working (or "immediate") memory has been an important part of intelligence testing since the beginning of the century (Ackerman et al. 2005). Baddeley (1992, p.1) also provides a relevant description: working memory refers to a brain system that provides temporary storage and manipulation of information that is necessary for complex cognitive tasks such as verbal comprehension, learning, and reasoning. It should be mentioned that in relation to the (unitary) concept of short-term memory, working memory usually refers to system that consists of distinct components, including a subsystem for the control of attention.

Moreover, besides being such an important construct for the study of information processing, it has been supported that working memory shares considerable variance with general and/or fluid intelligence (g/Gf) (Kyllonen and Christal 1990; Conway et al. 2003; Heitz et al. 2005). In general, working memory is highly correlated to intelligence, with substantial loadings on the g factor in psychometric testing; thus, it could be hypothesized that the construct of working memory is not only important per se, but also quite (in terms of broadly measuring) representative of an individual's mental abilities. On the other hand, working memory is not isomorphic to intelligence, nor it relates to every dimension of mental abilities.

Operationally, in the context of personalization, it is also very important that this construct has been extensively described by a number of theories, thus allowing a better understanding of individuals' specific capabilities and limitations. In our approach, mostly based on the well-known models of Baddeley (Baddeley and Hitch 1974; Baddeley 2000), Cowan (Cowan 1999; Cowan 2005), and Kane and Engle (Kane and Engle 2000; Kane et al. 2001), the main differences among healthy individuals with varying degrees of working memory span are: a) the ability to control attention, and b) the amount of information that can be manipulated simultaneously (memory span); still, the latter is probably influenced or even caused by the former. The issue of attention control is also central in the rather popular in educational and training hypermedia Cognitive Load Theory (Sweller 2003; Paas et al. 2003), in relation to the form and complexity of learning material.

Accordingly, we consider that Web environments can be manipulated in a way that could compensate for certain individuals' lower levels of attention control and memory span, mainly: i) by restructuring the content, ii) reducing the number of simultaneously presented stimuli, and iii) providing information at a slower pace. These methods are essentially personalization techniques that could be employed in almost every (complex enough) Web system, though their efficiency can only be validated through empirical research. The following section presents a possible way of integrating such functionalities in personalized systems, with cognitive style and working memory as the first areas of individual differences to be explored.

3 Technical and Design Considerations

3.1 Ontology-Based Web Content Annotations

Apart from building a theoretical framework for identifying important information processing parameters, it is also necessary to study and design the structure of meta-data (semantics) coming from the providers' side, aiming to construct a Web-based personalization mechanism that will serve as an automatic filter adapting the hypertext/hypermedia content based on users' profiles. The functionalities of the personalization mechanism are directly derived from the theoretical framework; the assumptions on which methods may assist users' interactions according to their abilities/preferences are in fact the filtering rules. As it concerns the profiling procedure, a large number of corresponding psychometric tools are readily available in either electronic form (i.e., CSA), or could be "ported" online (for instance, we developed an online version (http://adaptiveweb.cs.ucy.ac.cy/profileconstruction) of a common reading span working memory test); however, the presentation of such tools is out of the scope of this paper.

Still, as it was depicted in the introductory section, a main prerequisite for the proliferation of personalized Web services is the establishment of a set of standards that will be supported by high profile providers. Currently, at the level of semantics and ontologies, there have not been proposed any schemes including cognitive individual differences, at least to the authors' knowledge (excluding perhaps certain strictly educational hypermedia research approaches based on learning style). On the other hand, existing systems from other areas could easily be modified in order to map the Web content on users' cognitive profiles.

One such system is OntoSeek (Guarino et al. 1999), which is designed for content-based information retrieval from online yellow pages and product catalogues. OntoSeek uses simple conceptual graphs to represent queries and resource descriptions. The system uses the Sensus ontology (Knight and Luk 1999), which comprises a simple taxonomic structure of about 50,000 nodes. Another similar system developed by Labrou and Finin (1999) uses Yahoo! as an ontology. The system semantically annotates Web-pages via the use of Yahoo! categories as descriptors of their content. The system uses Telltale (Chowder and Nicholas 1996; Chowder and Nicholas 1997; Pearce and Miller 1997) as its classifier. Telltale computes the similarity between documents using n-grams as index terms. The ontologies used in the above examples use simple structured links between concepts.

A richer and more powerful representation is provided by SHOE (Heflin et al. 1999; Luke et al. 1997). SHOE is a set of Simple HTML Ontology Extensions that allow Web authors annotate their pages with semantics expressed in terms of ontologies. SHOE provides the ability to define ontologies, create new ontologies which extend existing ontologies, and classify entities under an "is a" classification scheme. Most importantly, Google has also recently announced that their search engine is going to support enhanced searching in Web-pages, by using RDFa and Microformats embedded in XHTML. Google states that the extra (structured) data will be used in order to get results for Product Reviews (e.g., CNET Reviews), Products (e.g., Amazon product pages), People (e.g., LinkedIn profiles) and any other types of resources will be made public through the data-vocabulary.org.

In our research, these approaches were a starting point for enriching the adaptation process with machine understandable representation of user profiles and Web content; specifically, the corresponding ontologies were taken into consideration for the design and development of our ontological adaptation mechanism.

3.2 A First Step in Individualizing Ontologies

In the theoretical section of this paper, a large number of cognitive factors were presented and discussed; however, as mentioned above, after a narrowing down process we opted for using working memory and cognitive style as personalization parameters. In terms of implications on the information space, these two constructs are somehow more relevant and far clearer in comparison to factors such as fluid reasoning or crystallized knowledge, which is rather important at this initial and explorative level of research. For a better understanding of these two adaptivity parameters' implications and their relation with the information space, figure 2 shows the possible content transformations/enhancements, based on the theoretical assumptions.

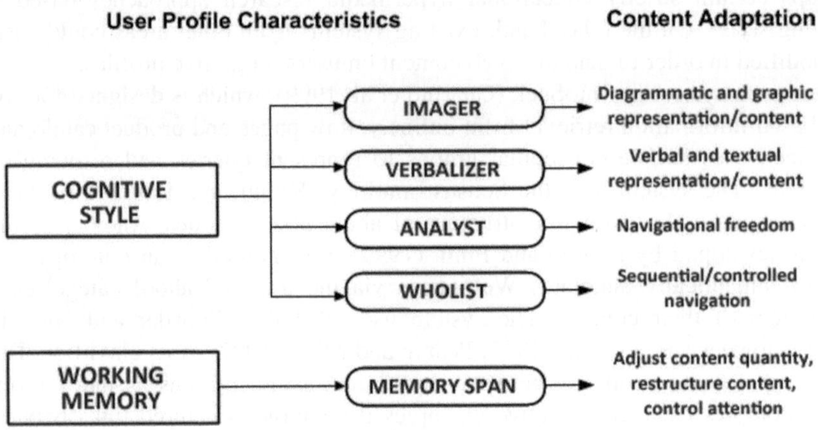

Fig. 2 Web design transformations/enhancements.

Specifically, the cognitive meta-characteristics of a user profile are: imager or verbalizer, analyst or wholist, and working memory span (low/medium/high), and have a particular impact on specific characteristics of the information space: images, text, information quantity and structure, links, control, and navigation support. These transformations represent groups of data affected during the mapping process on the selected cognitive factors. The main rationale is to process and/or alter the same content in different ways (according to a specific user's profile each time), without degrading in any way the message conveyed.

This approach led us to an Ontological Cognitive User Model (OCUM) (Germanakos et al. 2010); an RDFa vocabulary has been designed based on the theoretical framework and can be found online (http://adaptiveweb.cs.ucy.ac.cy/resources/rdf.xml). This vocabulary (User Model) consists of a number of classes and properties which describe a user's profile (table 1). The main class of this vocabulary is

Person, which represents a living or fictional person. The Person class has the following basic properties: i) "name" property; the Person's name, ii) "title" property; the Person's title (i.e., Prof. or Managing Director), iii) "affiliation" property; the Person's affiliation. A Person class has also the following properties with regards to cognitive style parameters: i) "imagerverbal" property; imager or verbal, ii) "wholistanalyst" property; wholist or analyst, and iii) "workingmemory" property; the Person's working memory capacity (i.e., low/medium/high). In this respect, the Person class, for example, in the RDFa instance (table 1) is the main entity. Specializations of the Person entity are the Cognitive Styles and Working Memory entities. Furthermore, there are three implicitly defined entities: the person's cognitive style, working memory capacity, and his personal details.

Table 1 RDFa Instance of a User's OCUM

```
<div xmlns:v="http://adaptiveweb.cs.ucy.ac.cy/resources/rdf/#" typeof="v:Person">
   <div><span property="v:name">John Smith</span>
      <span property="v:title">Managing Director.</span>
      <span property="v:affiliation">AWeb Solutions</span>
   </div>
   <div>Cognitive Style
      <span rel="v:imagerverbal"> Imager</span><br />
      <span rel="v:wholistanalyst">Analyst</span>
   </div>
   <div>Working Memory
        <span rel="v:workingmemory">Low</span>
   </div>
</div>
```

A practical example of this conceptualization would be the following: A user may be identified as verbalizer(V)/wholist(W), in respect to his/her style, with lower levels of working memory capacity (usually one standard deviation below the mean of a group). The content affected for this particular instance, according to the filtering rules, is: images (fewer images displayed), text (predominant mode of information delivery), navigation support (activated and/or enhanced), and info quantity (less quantity, split of competing for attention resources, clear structure with tags and guidelines, or further support with the use of a temporary buffer/notepad holding active particular information during task execution).

This process is rendered possible with the use of a proposed Ontological Adaptation Mechanism (OAM) that is composed of three main layers (figure 3): i) User Profile Layer, ii) Adaptation Mapping Layer, and iii) Web Content Layer.

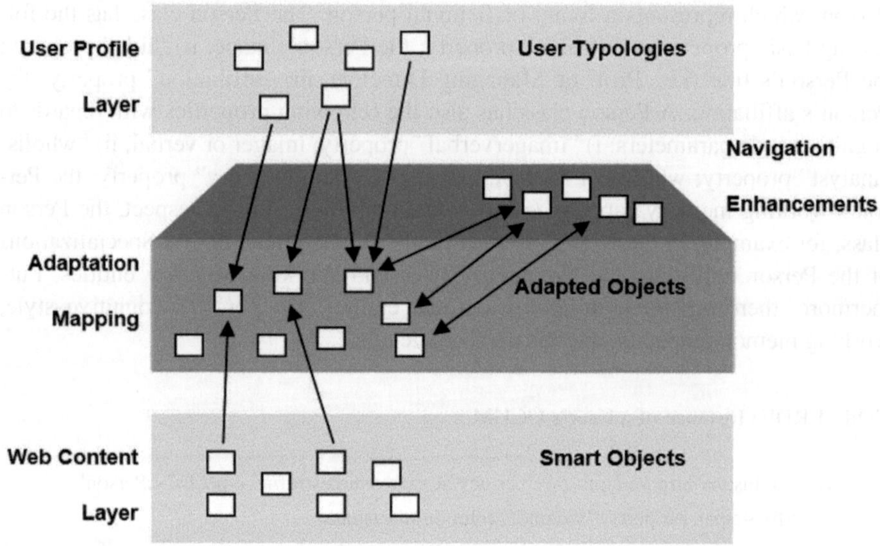

Fig. 3 Ontological Adaptation Mechanism.

The first layer of the OAM is the User Profile Layer, on which users' cognitive parameters are modeled; from a high level point of view, each user profile is a semantically defined object (RDFa object) that contains users' intrinsic characteristics. On the other end of the OAM, the Web Content Layer models hypermedia Web content with specific meta-characteristics, using again an RDFa vocabulary to annotate specific areas of an XHTML document as Adaptable Objects (Belk et al. 2010). In the middle layer, the Adaptation Mapping Layer is responsible for mapping User Profiles with the Adaptable Objects of the Web Content Layer. Based on this mapping, all Adaptable Objects of an XHTML document are manipulated (i.e., show more images and content in a diagrammatical form in case of an Imager), and extra navigation enhancements may be provided to the end-user.

Both the OCUM and OAM can be integrated in various Web-environments, regardless of orientation; still, the usefulness of proposing such an approach lies exclusively on the possibility of providing users with considerable benefits. Since the focus is on information processing, it is anticipated that the field of Web education would provide easily measurable data. However, personalization nowadays is emerging at a broader level, involving everyday interactions; thus, it is of high importance to explore whether an ontological conceptualization of individual differences (admittedly limited at this phase of research) is any useful in non-educational Web-sites. The following section briefly presents our empirical findings in both educational and commercial fields of application, in order to substantiate the usefulness of using OCUM and OAM.

4 Empirical Findings on Mapping Web-Content on Individual Differences

During the process of designing OCUM and developing OAM we conducted a series of empirical experiments in order to assess the positive effects, if any, of personalization on users' performance and/or quality of interactions. In this section, we briefly present the findings of three experiments that have been elsewhere published, elucidating our notion of "benefits for the user". The construct of cognitive style has been used throughout the entirety of this empirical work; on the contrary, the measure of visual short-term memory was replaced in the third experiment by a more robust working memory span and executive control measurement tool.

4.1 Experiment I

The first experiment was conducted in the field of e-learning (Tsianos et al. 2009), and included the following factors as personalization parameters: i) cognitive style, ii) visual short-term memory, iii) processing speed, and iv) self-reported anxiety (as a non-cognitive measure). The experiment was conducted in two phases: in the first phase the effect of personalization on cognitive style was explored (n=138), while the second (and later) phase involved the remaining parameters (n=81), controlling for the effect of style by personalizing the environment on users' style preference. The participants were social sciences university students who volunteered (convenience sampling) to take an online lesson on computer science; they generally considered the course as an additional aid on an academic subject on which they have minimal experience and usually perform poorly.

The procedure, which lasted for about an hour, in both phases was the same: users created their profiles through a series of psychometric tests, logged into the system, took an online course on algorithms and flow charts, and afterwards participated in an on-line exam assessing their level of comprehension of the lesson. In the case of cognitive style (phase I), the filtering process was based on the aforementioned personalization techniques: i) selection of content based on preference for imagery or verbal information, and ii) navigational freedom, level of control, and access to information based on analytic or wholistic preference (with the use of a navigation panel and the manipulation of links). The design of the experiment was between participants, and a match/mismatch methodology was employed: half of the participants were instructed in a matched to their preferences way, while the other half were instructed in a mismatched way. According to the findings of phase I, personalization on cognitive style increased learners' performance by 8.74 points (the mean score was 66.53 out of 100 in the personalized condition, and 57.79 in the mismatched condition, while intermediates scored 58.58): $F(2,137)=4.395$, $p=0.014$. In fact, the difference is statistically significant only between the matched and mismatched condition (post hoc analysis); intermediates perform somewhere in between.

In the second phase of the experiment, the filtering rules were the following: i) visual short-term memory: users that had low levels received segmented content that unfolded gradually, ii) processing speed: different time limits were set for each category of learners, and iii) anxiety: aesthetical enhancement of the environment (font size, colours, and annotations). The match/mismatch methodology was again applied; according to the analysis of the data (2x2x3 Anova), we found a significant effect of matching the instructional style to: a) users' visual short-term memory ($F=_{(1,80)}=4.501$, $p=0.037$), and b) to their levels of anxiety ($F=_{(2,80)}=3.128$, $p=0.05$). The increase in performance was more than 10 points, albeit including the positive effect of matching style. On the other hand, processing speed was not found to have any effect on score or interaction with the other parameters.

4.2 Experiment II

The second experiment involved the commercial Web-site of laptop manufacturer (Belk et al. 2010). In this case, the profiling procedure was the same, but the design was within-participants: each user navigated in both the original and personalized version of the Web-site (n=89 university students). Users were asked to answer certain questions in relation to the available information (tasks), and they were afterwards given a satisfaction questionnaire. Thus, the dependent variables were: i) task accuracy, ii) task completion time, and iii) user satisfaction.

The personalization factors were the following two: i) cognitive style, and ii) visual short-term memory. In the case of style, the filtering rules were the same as in experiment I. In the case of visual memory span, however, an additional support tool was employed: a readily available "notepad", serving as temporary external memory (there was no segmentation of information). According to the results, task accuracy was significantly higher in the personalized condition (1.9 vs. 1 out of three correct answers): Wilcoxon Signed Ranks $Z=-4.755$, $p=0.000$). Also, users required less time to fulfill the tasks: 412 vs. 512 seconds, paired samples $t(88)=4.668$, $p=0.000$. As it concerns the satisfaction questionnaire, 31 users leaned towards the personalized environment, 38 had no preference, and 20 preferred the original version. These differences should probably be attributed to style, since the use of the external memory tool was found to be limited.

4.3 Experiment III

The third experiment was focused on the construct of working memory in an e-learning environment, for the reasons that were depicted in section 2 (Tsianos et al. 2010). The experimental design, the educational course, and the sampling procedure were the same as in experiment I, albeit exploring specifically the effect of personalization on working memory; cognitive style preferences were manipulated as a control and not personalization variable (all preferences were matched), and the sample consisted of 230 university students (voluntary participation). In this case, there were two main personalization techniques that were employed: i) split of attention-demanding objects by segmentation and gradual unfolding of the content, and ii) annotation of certain key pieces of information and concepts. According to the findings, users in the personalized condition outperformed their counterparts by approximately 9 points ($F_{(1,226)}$=8.380, p=0.004).

Generally, in all three experiments we observed a consistent positive effect of personalization, confirming most of our research hypotheses on the role of individual differences in information processing within Web environments (see table 2).

Table 2 Summary of Experimental Results

	Personalization Variables (match/mismatch)	**User Benefit**
Experiment I *(e-Learning Course)*	Cognitive Style (phase I), Visual Short term Memory and Anxiety (phase II)	Increase of learner performance by 8.74 (phase I) and 10 to 13.43 (phase II) points in post course retention test
Experiment II *(Commercial Website)*	Cognitive Style and Visual Short-term Memory	Better accuracy (1.9/3 vs. 1/3 correct answers) and task completion time (412 vs. 512 sec.), slightly increased user-reported satisfaction
Experiment III *(e-Learning Course)*	Working Memory	Increase of learner performance by 9.18 points in post course retention test

It should be noted though that this work involved only a small part of the domain of individual differences, and in a very limited number of Web-environments (namely an educational and a commercial). Therefore, in spite of these positive findings, the benefits of personalization still need to be further researched and consistently demonstrated in various websites and human-computer interactions.

5 Discussion

In this article we made an effort to sum up the theoretical background and the empirical work that has led us to the development of a proposed ontological approach. As such, the focus is not actually on the OCUM and the OAM, but on possible ways of integrating cognitive individual differences in the representation of users' cognitive characteristics. We would expect that by outlining the field of cognitive abilities and by proposing a way for using certain of these factors in applied research, some new ideas could perhaps emerge in exploring elements of human behavior in the Web.

It is evident that it is hard to operationalize cognitive factors in terms of personalization techniques, since such an endeavor would require a continuous trial and error process. However, the significance of individual differences in many areas of cognition and the existing well validated psychometric tools are perhaps too concrete to ignore; thus, the formation of a corresponding ontology and semantic scheme would be very useful in conducting large scale studies on Web behavior, information processing, and personalization. We actually believe that only the bridging of the field of differential and experimental psychology with the field of semantics and Web technologies could produce robust results, for two main reasons: i) the extremely vague area of Web resources and their relation with the numerous cognitive factors would be explored far more efficiently through automated filtering processes, using extendable ontologies, rather than single experimental studies such as ours, and ii) the establishment of standards seems to be a key prerequisite for the adoption of personalization by high profile Web providers.

As it concerns the specific contribution of our theoretical and empirical work, the following arguments may be supported: i) cognitive style preferences seem to relate to how individuals perceive and process information in hypermedia environments, ii) working (and short-term) memory span represents individuals' information processing ability at some extent, and iii) it is possible to produce measurable benefits for users with the use of personalization techniques. Therefore, individual differences seem to have an effect on users' performance in both educational and commercial hypermedia.

It is certainly possible that these factors would not be equally important in different settings, and there is undoubtedly room for more accurate measurements and elaborated personalization techniques. This is actually the reason why we focused on mental abilities and on the current state of research on intelligence in the theoretical section of this paper; we seek new ideas for enriching OCUM (and user ontologies in general), new fields of application, and innovative methods for personalizing Web interactions. More specifically, as it concerns our future research directions, we aim to explore i) the role of verbal abilities in processing textual information in the Web (i.e. news and online encyclopedias), and ii) the role of fluid reasoning in online education, and the relation of the Gf factor with the pace of instruction.

As a closing remark, it should be noted that mental abilities certainly do not define or predict an individuals' behavior; motivation, personality factors, and emotion, to name but a few, are probably equally or even more important. Nevertheless, the long tradition of intelligence testing has produced accurate methods of measurement, and has systemized a large part of human individual differences in a quantitative manner that is quite compatible with the design and development of Web structures and functionalities. In the introductory section we posed the obvious question: is it worth it to personalize the Web, and in particular on the basis of cognitive abilities? Our experience has shown that there may be something of importance in linking abilities and preferences with the perceived by users aspects of the Web; thus, in an era of seeking added value in services, the answer would be positive.

References

Ackerman, P.L., Beier, M.E., Boyle, M.O.: Working Memory and Intelligence: The Same or Different Constructs? Psychological Bulletin 131(1), 30–60 (2005)

Baddeley, A.: Working Memory. Science 255, 556–559 (1992)

Baddeley, A.D.: The episodic buffer: A new component of working memory? Trends in Cognitive Sciences 4(11), 417–423 (2000)

Baddeley, A.D., Hitch, G.: Working memory. In: Bower, G.H. (ed.) The Psychology of Learning and Motivation. Academic Press, New York (1974)

Belk, M., Germanakos, P., Tsianos, N., Lekkas, Z., Mourlas, C., Samaras, G.: Adapting Generic Web Structures with Semantic Web Technologies: A Cognitive Approach. In: Proceedings of the 4th International Workshop on Personalized Access, Profile Management, and Context Awareness in Databases, in conjunction with VLDB 2010, pp. 35–40 (2010)

Brusilovsky, P.: Methods and techniques of adaptive hypermedia. User Modeling and User Adapted Interaction 6(2-3), 87–129 (1996)

Brusilovsky, P.: Adaptive Hypermedia. User Modeling and User-Adapted Interaction 11(1,2), 87–110 (2001)

Brusilovsky, P., Maybury, M.T.: From adaptive hypermedia to the adaptive web. Communications of the ACM 45(5), 30–33 (2002)

Caroll, J.B.: Human Cognitive Abilities: A Survey of Factor Analytical Studies. Cambridge University Press, Cambridge (1993)

Cattell, R.B.: Intelligence: Its structure, growth and action. North Holland, Amsterdam (1987)

Chowder, G., Nicholas, C.: Resource Selection in Café: an Architecture for Networked Information retrieval. In: Proceedings of Workshop on Networked Information Retrieval, pp. 1343–1355 (1996)

Chowder, G., Nicholas, C.: Meta-Data for Distributed Text Retrieval. In: Proceedings of Workshop on Networked Information Retrieval, pp. 317–332 (1997)

Conway, A.R.A., Kane, M.J., Engle, R.W.: Working memory capacity and its relation to general intelligence. Trends in Cognitive Sciences 7(12), 547–552 (2003)

Cowan, N.: An embedded-processes model of working memory. In: Miyake, A., Shah, P. (eds.) Models of Working Memory: Mechanisms of Active Maintenance and Executive Control. Cambridge University Press, Cambridge (1999)

Cowan, N.: Working memory capacity. Psychology Press, New York (2005)

De Bra, P., Aroyo, L., Chepegin, V.: The Next Big Thing: Adaptive Web-Based Systems. Journal of Digital Information 5(1) (2004)

Deary, I.J.: Human intelligence differences: a recent history. Trends in Cognitive Sciences 5(3), 127–130 (2001)

De Stefano, D., Lefevre, J.: Cognitive load in hypertext reading: A review. Computers in Human Behavior 23(3), 1616–1641 (2007)

Dieterich, H., Malinowski, U., Kühme, T., Schneider-Hufschmidt, M.: State of the art in adaptive user interfaces. In: Schneider-Hufschmidt, M., Kühme, T., Malinowski, U. (eds.) Adaptive user Interfaces: Principles and Practice. North-Holland, Amsterdam (1993)

Dillon, A., Watson, C.: User analysis in HCI-the historical lessons from individual differences research. International Journal of Human-Computer Studies 45(6), 619–637 (1996)

Germanakos, P., Belk, M., Tsianos, N., Lekkas, Z., Mourlas, C., Samaras, G.: An Ontological User Model for Adapting Generic Web Structures. In: Proceedings of the IEEE Computer Society 5th International Workshop on Semantic Media Adaptation and Personalization, pp. 7–12 (2010)

Glaser, R., Pellegrino, J.W.: Uniting cognitive process theory and differential psychology: Back home from the wars. Intelligence 2(3), 305–319 (1978)

Guarino, N., Masolo, C., Vetere, G.: OntoSeek: Content-Based Access to the Web. IEEE Intelligent Systems 14(3), 70–80 (1999)

Hauger, D., Köck, M.: State of the Art of Adaptivity in E-Learning Platforms. In: Proceedings of the 15th Workshop on Adaptivity and User Modeling in Interactive Systems, pp. 355–360 (2007)

Heflin, J., Hendler, J., Luke, S.: SHOE, A Knowledge Representation Language for internet Applications, Technical Report CS-TR-4078. Institute for Advanced Computer Studies: University of Maryland, College Park (1999)

Heitz, R.P., Unsworth, N., Engle, R.W.: Working memory capacity, attentional control, and fluid intelligence. In: Wilhelm, O., Engle, R.W. (eds.) Handbook of Understanding and Measuring Intelligence. Sage Publications, London (2005)

Horn, J.L., Noll, J.: Human cognitive capabilities: Gf-Gc theory. In: Flanagan, D.P., Genshaft, J.L., Harrison, P.L. (eds.) Contemporary Intellectual Assessment: Theories, Tests, and Issues, Guildord, New York (1997)

Kane, M.J., Bleckley, K.M., Conway, A.R.A., Engle, R.W.: A controlled-attention view of working-memory capacity. Journal of Experimental Psychology: General 130(2), 169–183 (2001)

Kane, M.J., Engle, R.W.: Working-Memory Capacity, Proactive Interference, and Divided Attention: Limits on Long-Term Memory Retrieval. Journal of Experimental Psychology: Learning, Memory, and Cognition 26(2), 336–358 (2000)

Knight, K., Luk, S.: Building a Large Knowledge Base for Machine Translation. In: Proceedings of the Twelfth National Conference on Artificial Intelligence, vol. 1, pp. 773–778 (1999)

Kyllonen, P.C., Christal, R.E.: Reasoning ability is (little more than) working-memory capacity?! Intelligence 14(4), 389–433 (1990)

Kyllonen, P.C., Stevens, D.L.: Cognitive Abilities as Determinants of Success in Acquiring Logic Skill. Learning and Individual Differences 2(2), 129–160 (1990)

Labrou, Y., Finin, T.: Yahoo! As an Ontology-Using Yahoo! Categories to Describe Documents. In: Proceedings of the 1999 ACM Conference on Information and Knowledge Management, pp. 180–187 (1999)

Lee, M.J., Tedder, M.C.: The effects of three different computer texts on readers' recall: based on working memory capacity. Computers in Human Behavior 19(6), 767–783 (2003)

Luke, S., Spector, L., Rager, D., Hendler, J.: Ontology-Based Web Agents. In: Proceedings of the first International Conference on Autonomous Agents, pp. 59–66 (1997)

McDonald, S., Stevenson, R.J.: Disorientation in hypertext: the effects of three text structures on navigation performance. Applied Ergonomics 27(1), 61–68 (1996)

McGrew, K.S.: Analysis of the major intelligence batteries according to a proposed comprehensive Gf-Gc framework. In: Flanagan, D.P., Genshaft, J.L., Harrison, P.L. (eds.) Contemporary Intellectual Assessment: Theories, Tests, and Issues, Guildord, New York (1997)

McGrew, K.S.: CHC theory and the human cognitive abilities project: Standing on the shoulders of the giants of psychometric intelligence research. Intelligence 37(1), 1–10 (2009)

Messick, S.: The nature of cognitive styles: problems and promises in educational research. Educational Psychologist 19(2), 59–74 (1984)

Miyake, A., Shah, P.: Models of working memory: Mechanisms of active maintenance and executive control. Cambridge University Press, New York (1999)

Paas, F., Tuovinen, J.E., Tabbers, H., Van Gerven, P.W.M.: Cognitive Load Measurement as a Means to Advance Cognitive Load. Educational Psychologist 38(1), 63–71 (2003)

Paramythis, A., Loidl-Reisinger, S.: Adaptive Learning Environments and eLearning Standards. Electronic Journal of e-Learning 2(1), 181–194 (2004)

Pearce, C., Miller, E.: The Telltale Dynamic Hypertext Environment: Approaches to Scalability. In: Nicholas, C. (ed.) Intelligent Hypertext. LNCS, vol. 1326, pp. 109–130. Springer, Heidelberg (1997)

Rayner, S., Riding, R.: Towards a Categorisation of Cognitive Styles and Learning Style. Educational Psychology 17(1-2), 5–27 (1997)

Riding, R.J., Cheema, I.: Cognitive Styles-An Overview and Integration. Educational Psychology 11(3-4), 193–215 (1991)

Stern, L.W.: Uber Psychologie der individuellen Differenzen Ideen zu einer Differentiellen Psychologie, Leipzig, Barth (1900)

Sweller, J.: Evolution of human cognitive architecture. Psychology of Learning and Motivation 43, 215–266 (2003)

Tennant, M.: Psychology and Adult Learning. Routledge, London (1988)

Thurstone, L.: Primary Mental Capabilities. University of Chicago Press, Chicago (1938)

Tsianos, N., Germanakos, P., Lekkas, Z., Mourlas, C., Samaras, G.: Working Memory Span and E-Learning: The Effect of Personalization Techniques on Learners' Performance. In: Proceedings of the 2nd and 18th International Conference on User Modeling, Adaptation, and Personalization, pp. 64–74 (2010)

Tsianos, N., Lekkas, Z., Germanakos, P., Mourlas, C., Samaras, G.: An Experimental Assessment of the Use of Cognitive and Affective Factors in Adaptive Educational Hypermedia. IEEE Transactions on Learning Technologies 2(3), 249–258 (2009)

Witkin, H.A., Moore, C., Goodenough, D., Cox, P.: Field-Dependent And Field-Independent Cognitive Styles and Their Educational Implications. Review of Educational Research 47, 1–64 (1977)

Witkin, H.A., Oltman, P., Raskin, E., Karp, S.: A Manual for Embedded Figures Test. Consulting Psychologists Press, Palo Alto (1971)

Formal Methods in High-Level and System Synthesis

Michael F. Dossis

Department of Informatics and Computer Technology,
Higher TEI of Western Macedonia, School of Kastoria

Abstract. The extreme complexity of today's embedded systems and portable computing machines has motivated research in the area of high-level and system synthesis (HLS). HLS is the process of modeling the behavior of the system in a high, algorithmic level using a regular imperative programming language such as C or ADA, and automatically transforming this behavioral model into hardware implementations. If the HLS transformations that are applied on the source code are formal, then the generated system implementation is correct-by-construction. The HLS tasks include scheduling of operations, allocation and binding to library resources, and techniques for these tasks are explained here. This work surveys and reviews past research in the area of formal methods for high-level and system synthesis. Since this area of research has been active for almost three decades, first the early HLS attempts are reviewed and then the more mature toolsets and HLS algorithms are analyzed. A prototype HLS compiler tool that has been developed by the author is presented here which uses compiler-generators, RDF rules and logic programming in combination with XML validation interfaces for the internal state of the compiler. All these make the compilation process formal. The embedded scheduler PARCS and the formal compilation of the system are tested with a number of benchmarks and real-world applications.

1 Introduction

The extremely high complexity of today's electronic systems and integrated digital systems that are found in computing machinery such as e.g. in embedded and portable systems has renewed the interest of industry and academia in high-level synthesis, so as to achieve better quality of implementations and shorter specification-to-product times. Currently, the established industrial approach in designing complex digital systems is the use of Register-Transfer Level (RTL) coding in languages such as VHDL and Verilog. However, for designs that exceed an area of a hundred thousand logic gates, the use of RTL models for specification and design can result into years of design flow loops and simulation-based verifications. With the current short lifetime of computing products in the market, the industry just doesn't have anymore the luxury to wait for so long until to launch

I.E. Anagnostopoulos et al. (Eds.): Semantic Hyper/Multimedia Adaptation, SCI 418, pp. 23–81.
springerlink.com

such products in the market (from the time the specification is captured). Therefore, higher abstraction level and executable types of specifications are required to help industries to become competitive. The first time that HLS became popular was in the 80s where some rather simple HLS tools attempted to synthesize mostly linear (dataflow-oriented) applications into hardware netlists. However the broad acceptance of HLS by the engineering community was prevented for a long time by the poor synthesis results from specifications that coded complete systems with hierarchy and complex (e.g. nested) control flow constructs in the specification program. In particular the programming style of the source algorithmic code has an unpredictable impact on the quality of the synthesized system. This is deteriorated by applications with hierarchical blocks, subprogram calls as well as nested control constructs (e.g. if-then-else and while loops), where the complexity of the transformations that are required for the synthesis tasks (compilation, algorithmic transformations, scheduling, allocation and binding) increases at an exponential rate, for a linear increase of the design size.

Usually, the input code (such as ANSI-C or ADA) to the HLS tool, is first transformed into a control/data flow graph (CDFG) by a front-end compilation phare. This involves a number of compiler-like optimizations such as code motion, dead code elimination, constant propagation, common sub-expression elimination, loop unrolling, hardware-oriented optimizations such as minimization of syntactic variances, retiming, and code transformations which are based on the associativity and commutativity properties of some operators, in order to deliver simpler expressions. Then various synthesis transformations are applied on the CDFG to generate the final implementation. The most important HLS tasks of this process are scheduling, allocation and binding. Scheduling makes an as-much-as-possible optimal order of the operations in a number of control steps or states. Optimization at this stage includes making as many operations as possible parallel (or in other words assigned onto the same control step) so as to achieve shorter execution times of the hardware implementation. Allocation and binding assign operations onto known functional units and variables and data structures onto registers, wires or memory positions, which are available from an implementation library.

A number of commercial HLS tools exist nowadays, which often impose their own extensions or restrictions on the programming language code that they accept as input, as well as various shortcuts and heuristics on the HLS tasks that they execute. Such tools are the CatapultC by Mentor Graphics, the Cynthesizer by Forte Design Systems, the Impulse CoDeveloper by Impulse Accelerated Technologies, the Synfony HLS by Synopsys, the C-to-silicon by Cadence, the C to Verilog Compiler by C-to-Verilog, the AutoPilot by AutoESL, the PICO by Synfora, and the CyberWorkBench by NEC System Technologies Ltd. The analysis of these tools is not the purpose of this work. Nevertheless, most of them are suitable for linear, dataflow dominated (e.g. stream-based) applications, such as pipelined DSP and image filtering.

An important aspect of the HLS tools is whether or not their transformation tasks (e.g. within the scheduler) are based on formal techniques. The latter would guarantee that the produced hardware implementations are correct-by-construction. This means that by definition of the formal process, the functionality of the implementation matches the functionality of the behavioral specification model (the source code). In this way the design will need to be verified only at the behavioral level, without spending hours or days (or even weeks for complex designs) of lengthy simulations at the generated RTL. The behavioral code can be verified by building a module that produces test vectors and reads the results and this verification can be realized with simple compilation and execution with the host compiler of the language (e.g. GNU C compiler and linker). This type of behavioral verification is orders of magnitude faster than long RTL or even more than gate-netlist simulations.

2 The Scheduling Task

In general there are two major categories of the scheduling problem: time-constrained scheduling and resource-constrained scheduling. Time-constrained scheduling attempts to deliver the lowest hardware cost (e.g. area or number of functional units) when the total number of control steps (states) is fixed (time constraint). Resource-constrained scheduling attempts to deliver the fastest schedule (the fewest control states) when the amount of hardware resources or area is fixed (resource constraint). Integer linear programming (ILP) formulations for the scheduling problem have been proposed but their execution time grows exponentially with the number of variables and inequalities. Therefore, ILP is practical and suitable only for very small designs. Thus, heuristic methods have been introduced to deal with larger designs and to provide sub-optimal but practical solutions. Heuristic scheduling uses in general two techniques: constructive solutions and iterative refinement. Two constructive methods are the as-soon-as-possible (ASAP) and the as-late-as-possible (ALAP) algorithms. During both of them the hardware operations are placed at a precedence-based list. With the ASAP method, one operation is taken from the list at a time and the algorithm tries to position the operation at the earliest possible control step. With the ALAP method, each operation from the list is moved at the latest possible control step. The operations that are found in the same control steps of both ASAP and ALAP schedules of the design, define the critical path of the design.

In ASAP and ALAP scheduling methods, the operations on the critical path of the design are not given any special priority, and therefore, excessive delay may be imposed on them, if the resource constraints are too hard and so only a few operations can be assigned on similar functional units in each control cycle. On the contrary, list scheduling uses a global priority function to select the next operation to be scheduled. This global priority function can be either the mobility [1] of the operation, or its urgency [2]. For example, the mobility of an operation can be calculated as the absolute difference between its ASAP and ALAP control step. Force-directed scheduling [3] calculates the range of control steps for each operation between the operation's ASAP and ALAP state assignment. It then attempts to reduce the total number of functional units of the design's implementation, by using distribution graphs of operations to the same state, in order to evenly distribute the operations of the same type into all of the available states of the range.

Constructive scheduling doesn't lookahead into future assignment of operations into the same control step, so the generated design solution may be sub-optimal. In the iterative scheduling, after an initial schedule is delivered by any scheduling algorithm, new schedules are produced by iteratively re-schedule sequences of operations that maximally reduce the cost functions [4]. When no further improvement can be achieved, the scheduling process stops. The above scheduling techniques are usually applied on linear dataflow dominated sequences of operations. For control-intensive problems with constructs such as loops, loop pipelining [5] and loop folding [6] techniques have been reported in the bibliography.

3 Allocation and Binding Tasks

The allocation task determines the type of resource units from a library of components. This library may contain one component for each operation (e.g. multiplier) and data object (e.g. register) of the input program. It also calculates the number of resources of each type that are needed to implement every operation or data variable. Binding assigns operations, data objects (e.g. variables or data structures) and data transfers onto functional units, storage elements (registers or memory blocks) and interconnections respectively, and makes sure that the design functionality is preserved using the selected library components. As mentioned above, the three interdependent tasks of binding are: functional-unit binding, storage-element binding and interconnection binding. Functional-unit binding assigns operations onto functional units and operators (e.g. adders, subtractors, multipliers, ALUs) from the datapath library. Storage binding maps data objects such as variables, constants, and data structures (e.g. arrays, or records) onto hardware elements such as registers, wires (hardwired on the power or ground lines) and RAMs/ROMs, respectively, which are placed on the datapath implementation. Interconnection binding maps data transfers onto sets of interconnection units, along with the necessary multiplexing to implement the required data routing in the design.

In general, there are three types of solutions to the allocation problem: constructive techniques, decomposition techniques and iterative approaches. The constructive techniques start with an empty implementation and progressively build the datapath and control unit by adding more functional, storage and interconnection elements while they traverse the CDFG or other type of internal graph formats. Constructive techniques are fairly simple but the implementations they deliver are far from optimal. Decomposition techniques divide the allocation problem into a sequence of well-defined independent sub-tasks. Each such sub-task is a graph-based theoretical problem which is solved with well known graph methods. Three well known graph-based methods are the clique partitioning, the left-edge technique and the weighted bipartite matching technique. The three sub-tasks of functional unit, storage and interconnection allocation are mapped onto a well-known problem of graph clique partitioning [7]. The nodes of the graph are operations, values and interconnection elements. Finding the minimum cliques in the graph, which is the solution for the sub-tasks, is a NP-hard problem, therefore heuristic approaches are utilized [7].

Because the sub-task of storage allocation, totally ignores the side-effects between the storage and interconnections allocation, using the clique partitioning technique, graph edges were enhanced with weights that represent the effect on interconnection complexity, due to sharing registers among different variables [3]. The left-edge algorithm is applied on the storage allocation problem, and it allocates the minimum number of registers [8]. In contrary to the clique partitioning complexity which is NP-complete, the left-edge algorithm has a polynomial complexity. However, the left-edge algorithm doesn't take into account the interdependence with the interconnect cost (which is considered in the weighted graph edges of the clique partitioning solution).

A weighted, bipartite-matching algorithm can be used to solve both the storage and functional unit allocation problems. First, a bipartite graph is generated which contains two disjoint sets, e.g. one for variables and one for registers, or one for operations and one for functional units [9]. An edge between one node of the one of the sets and one node of the other represents an allocation of e.g. a variable to a register. The bipartite-matching algorithm has a polynomial complexity. In addition to this it allocates the minimum number of registers. Moreover, it considers the effect of register allocation on the design's interconnection elements, since the edges of the two sets of the graph can be weighted [9]. The datapaths that are generated by either constructive or decomposition allocation techniques can be further improved iteratively, either by a simple assignment exchange, using the pairwise exchange of the simulated annealing, or by using a branch-and-bound approach which reallocates groups of elements of different types in order to refine the datapath [10].

4 Early High-Level Synthesis

HLS has been an active research field for more than two decades now. Early approaches of experimental synthesis tools that synthesized small subsets of programming constructs or proprietary modeling formats have emerged since the late 80's. An early tool that generated hardware structures from algorithmic code, written in the PASCAL-like, Digital System Specification language (DSL) is found in [11]. In this work, the three domains of integrated circuit designs were defined as the behavioral, structural and geometrical. Behavioral synthesis is the transformation of behavioral descriptions (e.g. program code) into circuit structures. This can be done at different levels such as e.g. at the register-transfer, at the gate or logic level, at the transistor or electric level, etc. The geometrical domain involves the generation of the integrated circuit geometrical features such as the circuit layout on the silicon array. What the authors in [11] named as behavioral synthesis is of course defined in our days in much more detail and it comes with the name High-Level Synthesis. The problem of HLS is extremely complex, but is much more understood in our days than in the early days of the first synthesis systems. The main tasks that were identified in [11] were compilation, datapath & control synthesis from imperative specifications (e.g. in DSL), optimization (area & speed) and circuit assembly. The circuit structure generated by the tool in [11] is coded in the Structure Description Language

(STRUDEL), and this in turn is ported to the Carlsruhe Digital System CADDY to generate the geometrical description of the circuit. Examples of other behavioral circuit specification languages of that time, apart from DSL, were DAISY [12], ISPS [13], and MIMOLA [14]. The synthesis system in [11] performs the circuit compilation in two steps: first step is datapath synthesis which is followed by control synthesis. The authors claimed the lack of need for circuit verification since it is correct by construction due to automated circuit synthesis methods.

The PARSIFAL DSP synthesis system from GE Corporate and Development division was one of the earliest synthesizers that targeted DSP applications [15]. PARSIFAL can describe the intended circuit in a combination of algorithmic and structural level and it is synthesized in a bit-serial DSP circuit implementation. PARSIFAL is part of a larger E-CAD system called FACE and which included the FACE design representation and design manager core. FACE includes interfaces to synthesis tools, analysis tools, physical assembly tools, libraries, and external tool interfaces such as parsers and formatters. The synthesis sub-system of FACE is interactive and it utilizes the FACE core functions. FACE focuses on design transformations and optimizations suitable for pipeline architectures, which are also suitable for non-pipelined structures. FACE synthesis process includes the following algorithms: minimize the execution time of expressions, maximize hardware resource sharing, insert multiplexers, and schedule operations in pipeline stages. It is thus obvious that FACE and PARSIFAL were particularly suitable for DSP pipelined implementations, and it wasn't suitable for a generic hardware synthesis system.

The synthesis optimization algorithms which include scheduling of operations and allocation of registers and busses, all in the light of timing and hardware resource – constraints are analyzed in [16]. According to [16] scheduling consists of determining the propagation delay of each operation and then assigning all operations into control steps, or states of a finite state machine (FSM). Different types of scheduling approaches are explained. List-scheduling attempts to minimize the total execution time of the state machine while satisfying resource constraints. List-scheduling uses a local priority function to postpone the assignment of operations into states, when resource constraints are violated. On the contrary, force-directed scheduling (FDS) tries to satisfy a global execution deadline (time constraint) while minimizing the utilized hardware resources (functional units, registers and busses). The way FDS does this is by positioning similar operations in different control states, so that the concurrency of operations is balanced without increasing the total execution time of the circuit. Thus each structural unit retains a high utilization which results into reducing the total number of units that are required to implement the design. This is done in three basic steps: determine the time-frame of each operation, generate a distribution graph and calculate the force associated with each operation assignment. In [16], other important problems are addressed as well, such as minimizing the cost of storage units and interconnections. Moreover, the force-directed list scheduling (FDLS) algorithm attempts to implement the fastest schedule while satisfying fixed hardware resource constraints. FDLS is similar to the list scheduling approach, however in FDLS the force is the priority function rather than the mobility or the urgency of operations. Another implementation exploration approach is also outlined in [16] by combining FDS and FDLS. In this approach, first the FDS technique is applied to find the near-optimal allocation by satisfying a fixed maximum time constraint. Then the designer runs FDLS on the resulting (by FDS) allocation in order to seek for an even faster implementation. After scheduling, the following problems are addressed: bind operations to functional units, bind storage actions to registers and bind data-transfer operations to interconnections. Other optimization actions are explored such as merging registers, merging multiplexers, and good design partitioning. However, there are no measures or indexes as to how fast synthesis runs using the above algorithms.

The authors in [17] redefined the tasks involved in HLS. According to them, the main problem in HLS is the mapping of a behavioral description into a RTL circuit description which contains a datapath and a control unit. In our days the latter can be implemented with a FSM which controls a datapath of operators, storage elements and a number of data-steering multiplexers. According to [17] the main tasks in HLS include allocation, scheduling and binding. According to [18] scheduling is finding the appropriate sequence of operations to execute in order to produce a schedule of control steps with allocated operations in each step of the schedule; allocation is setting aside the required number of functional, storage and interconnect units; binding is assigning operations to functional units, variables and values to storage elements and forming the interconnections amongst them to form a complete working circuit that executes the functionality of the source behavioral model. First the input behavioral description is transformed into a CDFG. Then various optimization algorithms run on the CDFG, in order to drive the implementation of the final circuit implementation. The CDFG captures the algorithmic characteristics of the input behavioral program (e.g. in VHDL or Verilog) as well as the data and control dependency between the operations to be scheduled. Apparently two operations that have a read-after-write dependency from one another, they cannot be scheduled in the same control step (or state). The authors in [18] introduce various problems that are encountered within various scheduling approaches: the unconstrained scheduling (UCS) problem, the time-constrained scheduling (TCS) problem, the resource-constrained scheduling (RCS) problem and mixed approaches such as the time and resource – constrained scheduling (TRCS) problem. Also advanced synthesis issues such as chaining (concatenating a number of different operations within the same control step in a chain) and multicycling (spreading the execution of an operation over more than one control step), handling in a special way control structures such as nested if-then-else and loop constructs, and various issues of constraining the global execution time and the latency of operations by the user of the synthesis tool, are addressed in [18]. Moreover, the tutorial in [18] defines and analyses the most common scheduling algorithms and approaches which include: ASAP scheduling, ALAP scheduling, list scheduling, force-directed scheduling, and ILP.

The V compiler [19] translates sequential descriptions into RTL models using parsing, scheduling and resource allocation. The source sequential descriptions are written in the V language which includes queues, asynchronous calls and cycle blocks and it is tuned to a particular type of parallel hardware RTL implementations. The utilized parser is built from a LALR (Look-Ahead LR) grammar [20], and the parse tree with leaves representing syntactic tokens and vertices (nodes) representing syntactic units of the source code. The V compiler marks the statements in the generated RTL and simulation code so that the user can trace the statements back in the V code by observing the token number. It also treats the hardware state machine as a directed, (possibly) cyclic control graph. Each vertex of the graph represents a state of the state machine and a set of operators to execute on the particular cycle. Each edge between vertices represents a state transition which can be conditioned on a Boolean variable. Thus, if a vertex has multiple transitions to other vertices, then the conditions of these multiple edges must be mutually exclusive. The inputs and outputs of operations and the conditions on the state transitions and operators are initially treated all as variables. Later on, during RTL implementation, these variables are implemented with wires or with registers. The V compiler utilizes percolation scheduling [21] to compress the state machine so as to achieve the required degree of parallelism and thus meet time constraints. Furthermore, the tool generates simulation code in PL/I so that the generated hardware can be verified with any compatible simulator.

A timing network is generated once from every particular behavioral design in [22] which is annotated with parameters for every different scheduling approach. The timing network is based solely on CDFGs, derived from the input specification, before scheduling and allocation are executed. The scheduling optimization algorithm in [22] attempts to satisfy a given design cycle for a given set of resource constraints, using the timing model parameters. An integrated approach for scheduling, allocation and binding in datapath synthesis is explained in [23]. This approach utilizes an ILP that minimizes a weighted sum of area and execution time using highly generalized modules. These modules can execute an arbitrary number of different operations, using e.g. different numbers of control steps for different operations. Moreover, the same operation can be executed on a variety of modules involving different number of control steps. The approach in [23] attempts to minimize the execution time and the hardware area of an initial data-flow graph (DFG) by using two types of constraints: the data dependency constraints (DD-constraints) and an operation ordering based on the sharing of functional units by the operations. This sharing is related with the fact that operations that share the same unit, have to be ordered, in a way that none of them can start its execution before its predecessor (which shares the same unit) has finished its execution on the same unit. This operation ordering is called unit-use ordering which is determined by unit-use constraints (UU-constraints). The work presented in [23] includes extensions of the ILP approach for pipelined functional units and for operation chaining. A prototype synthesizer called Symphony is presented in [23], in combination with 3 benchmarks that were executed through the Symphony system, namely: a fifth order elliptical wave filter, a differential equation, and a bandpass filter. For these benchmarks and according to the authors of [23] the Symphony tool delivers better area and speed than ADPS [24]. It seems from the type of scheduling approach as well as the used testcases, that the approach in [23] is more suitable for data-flow dominated designs such as DSP blocks.

The CALLAS synthesis framework [25], transforms algorithmic, behavioral VHDL models into VHDL RTL and gate netlists, which can be further implemented with available commercial logic synthesis tools. If the timing constraints are too tight for the scheduler, then CALLAS produces an ASAP schedule and issues a corresponding error message. The EXPANDER tool is connected to the backend of CALLAS in order to support low-level synthesis of the design in respect with specific delay, area and library components. CALLAS produces the final implementation via a number of iterative high-level and RTL transformations upon an initial structure which is derived from the algorithmic VHDL input. The user of CALLAS can drive these transformations using synthesis scripts. Compilation of the algorithmic code (from a subset of the VHDL language) delivers initial data flow and control flow graphs, and the initial allocation creates a starting ALAP schedule without resource constraints. Then the control flow graph is reduced so that the fixed I/O operation schedule is satisfied. The initial structure is refined with a number of high-level and RTL transformations in order to be optimized. Then the produced structure is processed by logic optimization and technology mapping (EXPANDER tool), and thus a VHDL or EDIF [26], [27] netlist is generated. The generated circuit is modeled using a Moore-type FSM, which is consistent with the semantics of the VHDL subset which is used for the specification code. The synthesis transformations in CALLAS include removal of superfluous edges in the control flow graph, removal of unnecessary data transfers between registers, control flow graph reduction (scheduling) so that the specified I/O-timing constraints are met. Other optimizations include lifetime analysis, register sharing, operator sharing, multiplexer optimization, arithmetic and logic transformations, optimizing of the datapath/controller interface, flattening of complex functional units, partitioning and logic minimization. The above utilize techniques such as clique partitioning, path analysis and symbolic simulation. The formal verification methods that are used in the CALLAS framework comprise the equivalence checking (in terms of FSM equivalence) between the original VHDL FSM and the synthesized FSM by using the symbolic verifier of the Circuit Verification Environment (CVE) system [28]. A number of benchmarks and industrial designs were executed within the CALLAS framework and confirmed its usability.

The Ptolemy framework [29] allows for an integrated hardware-software co-design methodology to be implemented from the specification through to synthesis (of hardware and software components), simulation and evaluation of the implementation. Ptolemy is a tool-set that allows the simulation and rapid prototyping of heterogeneous hardware+software systems. The basic unit of modularity inside Ptolemy is the block. Blocks communicate with each other and with their computing environment through portholes. The tools of Ptolemy can synthesize assembly code for a programmable DSP core (e.g. DSP processor), for a synthesis-oriented application. A domain in Ptolemy consists of a set of blocks, targets and associated schedulers that conform to the operational semantics which determine how blocks interact. Some of the simulation domains supported by Ptolemy include the synchronous dataflow (SDF), dynamic dataflow (DDF) and digital hardware modeling (Thor). E.g. for every commercial DSP processor there are corresponding models and a simulator which is invoked when the user wants to verify a design that contains this processor. Mixed digital and analog components such as e.g. A/D and D/A converters and filters can be represented as components with their functional models in the SDF domain. The engineers of Ptolemy have supported the generation of C and C++ code for a variety of processors. In Ptolemy, an initial algorithm that models the entire system is partitioned manually into the software and hardware parts which are synthesized in combination with their interface synthesis. The partitioning of software parts amongst multiple available processors is done automatically by invoking a parallel scheduler. Then the hardware, software and interface implementation models can be co-simulated and the overall system prototype can be evaluated. This unified representation of hardware and software components allows the migration of functions between the two implementations with their interfaces being automatically synthesized as well. This process is not fully-automatic but provides interoperability of tools to the users of Ptolemy.

An iterative partitioning process, based on a hardware extraction algorithm driven by a cost function, is implemented in the Cosyma framework [30]. The primary target in this work is to minimize customized hardware in microcontrollers but the same time allow for exploration of a large design space. The hardware-software co-synthesis approach in Cosyma is based on a processor core, memory and custom coprocessing hardware. In this approach, system implementation is focused on the generation of machine code for the embedded microprocessor. However, custom hardware is preferred only when timing constraints are violated by the generalized processor implementation, or when basic, available and cheap I/O peripherals are required for the completion of the embedded system. The custom coprocessors of the embedded system are usually synthesized using HLS tools. The hardware/software partitioning process in Cosyma is automatic. Initially the whole system is implemented in a set of hardware components. Then gradually as many as possible of these hardware components are moved into software components, as soon as timing constraints and system synchronization are satisfied. The specification language contains the following C extensions: timing (minimum and maximum delays), tasks, and task intercommunication. Partitioning occurs at different levels of system granularity: task, function, basic block, and single statement. Parallelism in Cosyma C language is explicit and it is defined by the user's coding. The extended syntax (ES) graph is used as the internal representation of the design in Cosyma which is extended by a symbol table as well as data and control dependencies. The ES graph is used for both partitioning and cost estimation as well as for software and hardware C generation. The hardware description is in turn ported to the HLS Olympus tool [31]. Cosyma utilizes its ES internal format to estimate possible speed-ups of the hardware implementations of critical loops in the design and therefore aid towards the required software-hardware partitioning. This partitioning process is based on hardware extractions which use a partitioning cost function, to drive the hardware implementation of the system components that can be implemented well in hardware. Such a cost function embodies knowledge about synthesis, compilers and libraries. An example is a specific cost function for the extraction of coprocessors that implement computation-time-intensive parts of the application such as nested loops. In [30] tests were run on a configuration of an embedded system, which is built around the Sparc microprocessor.

A methodology which exercises co-simulation and co-synthesis of mixed hardware-software specifications is analyzed in [32]. During co-synthesis, hardware-software partitioning is run in combination with control parallelism transformations. The hardware-software partition is defined by a set of application-oriented functions which are implemented with application-specific hardware. The control parallelism is defined by the interaction of the processes of the functional behavior of the specified system. Finding the appropriate control concurrency involves splitting of merging processes, or moving functionality from one process to another. The co-simulation environment developed in [32] produces a mixed system model that is functionally correct but it might not meet design goals. The co-synthesis tools are then used to modify the hardware-software partition and the control concurrency so that design goals are satisfied. Then the software part is implemented with standard compilation into system memory and the hardware part is synthesized with HLS tools and implemented with a programmable platform (three Xilinx FPGAs) and interconnection modules (two field-programmable interconnect chips from Aptix) that are plugged in the backplane of the host computer, suitable for implementation measurements. There are three abstractions of hardware-software interaction: send/receive/wait transactions between application program and custom application hardware, register reads/writes between the I/O driver running in the host computer and the bus interface of the custom hardware, and bus transactions between the two I/O bus sides. The system behavior is modeled by a set of communicating sequential processes [33]. Each process can be assigned to either hardware or software implementations. The following types of inter-process communication primitives exist in the system: synchronized data transfer, unsyncronized (unbuffered) data transfer, synchronization without data transfer, and communication with a shared memory space. Co-simulation is implemented in [32] using a Verilog simulator and its programming language interface (Verilog PLI). Two example applications tested the co-synthesis and co-simulation environment: the Sphinx speech phoneme recognition system and a data compression/encryption application.

A hardware-software co-design methodology is presented in [34], which synthesizes heterogeneous systems. The synthesis process is driven by timing constraints which drive the mapping of tasks onto hardware or software parts so that the performance requirements of the designed system are met. This method is based on using modeling and synthesis of programs written in HardwareC, which enables the use of the Olympus chip synthesis system for prototyping [31]. The HardwareC model consists of a set of interacting processes which are instantiated in blocks, using declarative semantics. Every process restarts itself upon completion of its tasks and all processes can execute concurrently in the system model. Hierarchically-related sequencing graphs are produced from the input HardwareC specification. Within each graph, vertices represent input program operations and edges represent dependencies between operations. Two vertices, namely the source (beginning) and sink (end) represent no operations. The graph model allows for operations in different graphs to pass messages to each other, such as send and receive. This is a very important feature in modeling heterogeneous systems, because the processor (which implements the software part of specification) and the custom hardware (which implements the hardware part of specification) may run on different clocks and speeds. Timing constraints are important and they are used to select the specific system implementation which satisfies performance requirements. Timing constraints are of two types: min/max delay constraints and execution rate constraints. For example min delay constraints are captured by providing weights at the edges of the graph, to indicate delay to the corresponding source operations of each edge. Performance measurement is done on the basis of operation delays. These delays are estimated separately for the hardware and software parts of the system, based on the type of hardware technology, which is used to implement the hardware part of the system, and the processor that is used to run the software. The assignment of an operation to software or hardware determines the delay of the operation. Moreover, moving operations from the hardware to software parts and vise versa involves additional delays due to inter-communication delays. All these delays are used to determine the hardware/software partitioning of the final system implementation. A testcase for the methodology in [34] was an Ethernet-based network coprocessor. The authors concluded that the use of their proposed hardware-software co-design methodology aids the development of embedded, real-time systems which have a simple configuration as compared to that of a general purpose computing system.

5 Mature High-Level Synthesis

More mature methodologies and tools started appearing from the late 90s and continue with improved input language subsets as well as scheduling and other optimization algorithms. Also system level synthesis matured in the last decade by using more specialized and platform-oriented methodologies. The CoWare hardware-software co-design environment [35] is based on a data model which is used to specify, simulate and synthesize heterogeneous architectures from heterogeneous specification models. The choice of implementing real-time telecommunications DSP applications on programmable DSP processors or application-specific hardware is usually driven by trade-offs between cost, power, performance and flexibility. The work in [35] focuses on designing telecommunication systems that contain DSP, control loops and user interfaces. The SDF type of algorithms can be found in a category of DSP applications and they can easily synthesized into hardware from languages such as SILAGE [36], DFL [37], and LUSTRE [38]. The advantage of this type of implementations is that they can be scheduled at compile time and the execution of the compiled code can be two orders of magnitude faster than event-driven VHDL simulations. In contrast to the above, DDF algorithms consume and produce tokens that are data-dependent, and thus they allow for complex if-then-else and while loop control constructs. One way to deal with the data-dependent DDF algorithms is to map them onto the worse case SDF and schedule them at compile time. Another way is to partition the DDF into partial SDFs that are triggered by internal or external Boolean conditions. Then these partial SDFs need to be scheduled at run time using the I/O timing constraints of the DSP signals and external events. CAD systems that allow for specifying both SDF and DDF algorithms and perform as much as possible static scheduling are the DSP-station from Mentor Graphics [39], PTOLEMY [40], GRAPE-II [41], COSSAP from Synopsys and SPW from the Alta group [42]. Modularity in the specification models is realized in CoWare [35] by means of processes. A behavioral interface using ports implements the communication between processes. Process ports that communicate are connected through a channel. The data model is hierarchical and allows for gradual refinement of channels, ports and protocols into lower levels of objects, by adding detail. The most abstract object is the primitive object. In contrast, a hierarchical object contains implementation detail. A thread is a single flow of control within a process. There are slave threads and autonomous threads [35]. Communication between threads in different processes is called interprocess communication. Shared variables or signals, are declared within the context of a process, and they are used for intraprocess communication. Channels and ports can be refined via adding more detail onto them, through the CoWare's design flow. The CoWare data model is suitable for merging of processes and for design for reuse and reuse of designs. Software-to-hardware communication is implemented in CoWare by means of memory-mapped I/O, instruction-programmed I/O, and interrupt control. The CoWare methodology was implemented in [35] using a design example, which is a pager, based on spread-spectrum techniques. One important conclusion in [35] was that there is a pressing need for bottom-up formal verification tools, which can evaluate both functionality and timing of the design.

The authors in [43] present an approach for mapping C models into hardware. The C models include dynamic memory allocation, pointers and the functions malloc and free. The solution which is proposed in [43] instantiates a custom (to the application) hardware memory allocator, which is coupled with the specific memory architecture. This work also supports the resolution of pointers without any restriction on the underlying data structures. Many networking and multimedia applications are implemented in hardware or mixed hardware/software platforms and they feature heavy use of complex data structures which are sometimes stored in one or multiple memory banks. An immediate result of this is that some features of C/C++ which were originally designed for software development are now strong candidates for hardware design as well. The SpC tool which was developed through this work [43] resolves pointer variables at compile time and thus C functional models are synthesized into hardware efficiently. In a hardware implementation of pointers, memory allocation may be distributed onto multiple memories, and the data which are referenced by the pointers may be stored in memories, registers or wires. Therefore the hardware synthesis tool needs to automatically generate the appropriate circuit to allocate, access (read/write) and deallocate data. Pointer analysis is a compiler technique that identifies at compile-time the potential values of all pointers of an application program. In order to implement dynamic memory allocation in hardware there is a need to synthesize circuits to access, modify or deallocate the location which is referenced by each pointer. For this purpose the aliasing information [43] must be both safe and accurate. In [43] it is assumed that the computational complexity of flow-sensitive and context-sensitive analysis is not high because of the small size and simplicity of the programs and function calls which are used for hardware synthesis. This of course is not guaranteed since modern system descriptions could easily contain some thousands of lines of hierarchical code to describe complex hardware architectures. The subset of C which is accepted by the methodology in [43] includes malloc/free and all types of pointers and type casting. However, pointers that point to data outside the scope of a process (e.g. global variables) are not allowed. The synthesis of functions in C, and therefore the resolution of pointers and malloc/free inside these functions, is not included in this work. In order for the C code with the pointers to be efficiently mapped onto hardware, first the memory is partitioned into sets which can include memories, registers or wires, and which can also represent pointers. Pointers are resolved by encoding their value and generating branching statements for loads and stores. Dynamic memory allocation and deallocation are executed by custom hardware memory allocators. The SpC tool [43] takes a C function with complex data structures and generates a Verilog module. The different techniques and optimizations described above have been implemented using the SUIF compiler environment [44]. The memory model consists of distinct location sets, and it is used to map memory locations onto variables and arrays in Verilog. Then, the generated Verilog module can then be synthesized using commercial synthesis tools such as the Behavioral Compiler of Synopsys. The case studies that tested this methodology included a video algorithm and an asynchronous transfer mode (ATM) segmentation engine.

An efficient heuristic for scheduling behavioral specifications with heavy use of conditional control flow, within HLS tools, is presented in [45]. This heuristic is based on a specific intermediate design representation which apart from established techniques such as chaining and multicycling, it allows to apply more advanced techniques, suitable for scheduling conditional behaviors, such as conditional resource sharing and speculative execution. This work intended to bridge the gap in design implementation quality between HLS results from dataflow-dominated descriptions, and those from conditional controlflow-dominated specifications. Generally, although HLS was accepted by the engineering community earlier for dataflow oriented applications, it took some time before it became adopted, and it is still not widely accepted for designs that contain complex conditional control flow, such as nested if-then-else and loop constructs. The intermediate design representation, which is used by the HLS flow, is called hierarchical conditional dependency graph (HCDG), and the invented heuristics for HLS tasks that are based on the HCDG, have been developed to deal with complex control flow and with control hierarchy. HCDGs introduced two new concepts: a hierarchical control representation and the explicit representation of both data and control dependencies in the design. This explicit representation of control dependencies is suitable to explore maximum parallelism in the implementation by rearranging these control dependencies. Being able to express maximum parallelism at the intermediate form level of a hardware design is essential, since exploiting parallelism is easier for custom hardware designs than for software ones. The HCDG can be successful in avoiding the negative effects of syntactic variance in the specification code of a system.

The hierarchical control representation of HCDG allows to safely and efficiently perform the HLS tasks such as scheduling, allocation, binding, etc. In the context of the work in [45], symbolic names are given to the Boolean conditions under which the various operations are executed and values are assigned to variables. Those symbolic names are called guards. In a HCDG there are two types of nodes: guard nodes and operation nodes. Guard nodes represent the symbolic names of the various conditions under which operations are executed. Operation nodes represent I/Os, computations, data multiplexing, and storage elements. In a HCDG there are two types of edges: data dependencies and control dependencies. Data dependencies are precedence constraints from one operation node to another. This defines the dataflow-dependent order of operation execution order. Control dependencies indicate which conditions (guards) must evaluate to true so that the data values are computed and considered as valid. Each operation node has its control dependency edge from its guard. Guards can be also hierarchical, which results into a graphical representation of nested control constructs (e.g. and if-then-else nested inside another if-then-else, and so on). Therefore there is a guard hierarchy graph for every design in [45]. Deriving HCDGs from conditional behaviors is being exercised in [45], but deriving them from loop constructs is reported in the particular work [45] as being the subject of future work. In order to schedule conditional behaviors efficiently the mutual exclusiveness of the conditions has to be exploited. This means being able to conditionally share resources and schedule operations effectively. In order to do this, complete lists of mutually exclusive guards have to be constructed. For large and complex designs this means that a very large number of mutual exclusiveness tests have to be performed on potential pairs of guards. Nevertheless, this number of tests can be drastically reduced in [45] by using the inclusion relations represented by the guard hierarchy graph. Using the above techniques the following HLS transformations are facilitated: lazy execution, node duplication, speculative execution, false-path elimination, and conditional resource sharing. Moreover, considerations are taken for operation chaining and multicycle operations. A special priority function based on guard hierarchy and graph node mobility, is utilized in order to obtain the node priorities to perform the scheduling task. Mutual exclusiveness information is very useful for applying register allocation and for other types of resource sharing such as those applied to interconnects. The HLS techniques presented in [45] were implemented in a prototype graphical interactive tool called CODESIS which used HCDG as its internal design representation. The tool can generate VHDL or C code from the HCDG, but no reports about translating a standard programming language into HCDG are known so far.

An advanced HLS approach is presented in [46], which applies a coordinated set of coarse-grain and fine-grain parallelizing transformations on the input design model, so as to deliver synthesis results that are free from the negative effects of complex control constructs in the specification. The transformations are applied both during a presynthesis phase and during scheduling, in order to improve the quality of synthesis. During presynthesis the following transformations are applied: common sub-expression elimination (CSE), copy propagation, dead code elimination, loop-invariant code motion, as well as restructuring transformations such as loop unrolling and loop fusion. Then during scheduling, aggressive speculative code motions (transformations) are used to re-order, speculate and some times duplicate operations in the design. In this way, maximum parallelizing is applied on the synthesis results. A technique called dynamic CSE, dynamically coordinates CSE, speculation and conditional speculation during scheduling. The code motions that are enabled during scheduling, move operations through, beyond and into conditional blocks with the purpose of maximizing parallelism and therefore increase design performance. The compilation and scheduling tasks include the scheduling heuristic, the code motion heuristic, dynamic transformations and loop pipelining, and all of these use functions from a tool library, which includes percolation and trailblazing, speculative code motions, chaining across conditions, CSE and copy propagation. Then, during the binding and control synthesis steps, the operation and variable binding as well as FSM generation and optimization are executed. All these techniques were implemented in the SPARK HLS tool, which transforms specifications in a small subset of C into RTL VHDL hardware models. The scheduler in SPARK is a resource-constrained scheduler and it is essentially a priority-based global list scheduling heuristic. The user provides SPARK with a library of resources, which include amongst others the type and number of each resource. This user library is used by the HLS tool, to allocate operations and registers onto library components. In terms of intermediate design representations, SPARK utilizes both CDFGs as well as an encapsulation of basic design blocks inside hierarchical task graphs (HTGs) [46]. HTGs allow for coarse-grain code restructuring such as loop transformations and an efficient way to move operations across large pieces of specification code. This is why the combination of CDFGs and HTGs in SPARK is so successful. Nevertheless, there are serious restrictions on the subset of the C language that SPARK accepts as input, and limitations such as lack of design hierarchy (e.g. subprograms) and of "while" type of loops. SPARK is validated in [46] by synthesizing three large examples: MPEG-1, MPEG-2 and the GIMP image processing tool.

The typical HLS tasks such as scheduling, resource allocation, module binding, module selection, register binding and clock selection are performed simultaneously in [47] so as to achieve better optimization in design energy, power and area. The scheduling algorithm utilized in the HLS system in [47] applies concurrent loop optimization and multicycling and it is driven by resource constraints. The state transition graph (STG) of the design is simulated in order to generate switched capacitance matrices. These matrices are used to estimate power/energy consumption of the design's datapath. The initial schedule is optimized by multiple sequences of module selection, module sharing and register sharing tasks. Nevertheless, the input to the HLS tool which was developed in [47] is not program code in a popular language, but a proprietary format representing an enhanced CDFG as well as a RTL design library and resource constraints. Special nodes and edges were added to this proprietary CDFG to aid the capturing of control constructs such as if-then-else and loops, as well as memory access sequences. The scheduler takes the CDFG and resource constraints as input and produces a form of an optimized STG. In the synthesis algorithm the cost function (for optimization) can be area, power, or energy. The synthesis process is iterative and it continuously improves the cost function until all of the constraints and data dependencies are met. The iterative improvement algorithm is executed in multiple passes until there is no potential improvement on the cost functions. In every pass, a sequence of the following moves is created; the moves can be: module selection, module sharing and register sharing. After each move, the behavior of the system is re-scheduled and the cost is estimated again. If the move generates the best reduction of cost then this move is saved, otherwise different moves are selected. If the cost is reduced in the current pass, then a new pass is generated and the scheduling continues. This iterative process runs until there is no potential improvement in the cost functions. The output from the tool is RTL Verilog. The developed HLS system is targeted at control-intensive applications but it is also applicable to dataflow dominated designs. The system was tested using a number of control-intensive benchmarks, such as for loop, concurrent loops, nested loops, greatest common divisor, a fifth-order Elliptic wave filter, and a popular dataflow dominated benchmark. The synthesis results focused more on power reduction with a rate up to 70% rather than area or speed results. Most of the benchmarks took a number of minutes to execute on a conventional Pentium III PC.

The work in [48] describes an incremental floorplanner that is used in order to combine an incremental behavioral and physical optimization into high-level synthesis. These techniques were integrated into an existing interconnect-aware HLS tool called ISCALP [49]. The new combination was named IFP-HLS (incremental floorplanner high-level synthesis tool), and it attempts to concurrently improve the design's schedule, resource binding and floorplan, by integrating high-level and physical design algorithms. Important factor considered in this work was the impact of interconnect on area and power consumption of integrated circuits. To define the problem this method is based on the following equation:

$$T_{clock} = T_s / \text{csteps}$$

where T_{clock} is the system clock period, T_s is the constraint on the input data rate (sample period), and csteps is the number of clock cycles required to process an input sample. Given csteps, an ASAP schedule is then generated for an initial solution to determine whether it meets timing. An iterative improvement task is then applied on this initial solution, in order to reduce the switched capacitance while it still satisfies the sample period constraint. From the way the problem and the solution are defined in this HLS approach, it seems that the latter is suitable for dataflow-dominated designs and not for control-intensive applications. IFP-HLS generates one initial solution, at the maximum number of csteps and then it applies incremental floorplanning and it eliminates redundant operations. In this way, the solution is improving as the figure csteps decreases. If a solution meets its timing requirement after re-scheduling, then re-binding is not necessary. In any other case, it re-binds some tasks and uses parallel execution to improve performance. Possible pairs of tasks that are initially assigned to the same functional unit, are split onto separate functional units [48].

For a given csteps, the floorplan is incrementally modified to see if this improves the solution quality. If it does, then this change is saved. Otherwise the floorplan change is rejected and other modifications are attempted to determine whether they improve the solution, and so on. In order to guide these changes, the tool extracts physical information from the current, incrementally generated floorplan. Therefore, IFP-HLS incrementally performs scheduling, allocation and binding by modifying iteratively csteps, and determines which operations need to be re-scheduled or re-bound (split) in order to meet the timing constraints, and in each step the floorplanner is updated. An incremental simulated annealing floorplanner is embedded into the IFP-HLS tool which was designed for design quality and not for speed. The floorplanner handles blocks with different aspect ratios and generates non-slicing floorplans. The synthesis moves either remove a single module or they split a module into two. Therefore, most of the modifications are small and their effects on the floorplan are local, rather than global. In this way, an existing floorplan can be used as the base for each new floorplan. In practice, the authors found this approach to deliver quality-of-results and performance improvements, even compared with a very fast constructive floorplanner.

Fifteen different benchmarks were used to test the utility of the approach on this work [48]. The average improvements of IFP-HLS over ISCALP, for implementations with non-unity aspect ratio functional units, are 14% in area, 4% in power consumption, 172% in reduction in the number of merge operations and 369% in CPU time. The average improvements of IFP-HLS over ISCALP, for implementations with unity aspect ratio functional units, are 12% in area, 7% in power consumption, 100% in reduction in the number of merge operations, and for some benchmarks the IFP-HLS CPU time was 6 times less than that of the ISCALP method.

A HLS methodology which is suitable for the design of distributed logic and memory architectures is presented in [50]. Beginning with a behavioral description of the system in C, the methodology starts with behavioral profiling in order to extract simulation statistics of computations and references of array data. This allows the generation of footprints which contain the accessed array locations and the frequency of their occurrence. Using these footprints, array data reference operations with similar access patterns are grouped together into a computation partition. A method to assign each such partition onto a different subsystem is used via minimizing a cost function that includes balancing the workloads, synchronization overheads and locality of data accesses. Then array data are distributed into different partitions so that their accesses will be as much as possible local to each subsystem, based on the clustering of their reference operations. In order to implement correct communication between different partitions, synchronization code is inserted into the implementation's behavior. This results into a distributed logic/memory micro-architecture RTL model, which is synthesizable with existing RTL synthesizers, and which consists of two or more partitions, depending on the clustering of operations that was applied earlier. These techniques are implemented into an industrial tool called Cyber [51] and several benchmark applications were run on the tool to produce distributed logic/memory implementations. The results were encouraging with up to twice performance increase and reduction up to 2.7 times of the delay X energy product over single-memory and homogeneously partitioned designs.

The methodology presented in [52] implements communicating processes which are part of a system specification. In contrast to the conventional HLS approach, which synthesizes each concurrent process of the system individually, the approach in [52] considers the impact of the operation scheduling globally, in the system critical path (as opposed to the individual process critical path). First the system is scheduled by assuming that there are unlimited resources for each process. Next, the scheduled design behavior is simulated, and using the simulation's execution traces, system performance is analyzed and the critical path(s) of the behavior is (are) extracted. Using this information about the system, the criticality of operations is calculated based upon whether they belong to the critical path(s) or the near-critical path(s). Then, depending on the type and number of critical operations that a process contains, the relative resource requirement of each process is calculated. With this information for each process, the resources for the overall system are budgeted. This resource budget is then used as a constraint to reschedule the whole system. The rescheduled behavior is simulated again and the critical paths are yet one more time extracted from the traces. If the critical path changes, then the above process is repeated again and again until the critical path remains the same after each resource re-allocation. When the critical path remains stable, and using the last resource budget, the behavior model is passed to the rest of the HLS tasks, such as resource sharing and generation of the controller and datapath. In this way, the RTL hardware description of the multiple processes is built. It is argued by the authors in [52] that this methodology allocates the resources where they are mostly needed in the system, which is in the critical paths, and in this way it improves the overall multi-process system performance.

The synthesis approach in [53] contributes towards incorporating memory access management within a HLS design flow. It mainly targets digital signal processing (DSP) applications but also more general streaming systems with specific performance constraints. A memory sequencer architecture is analyzed in the publication and utilized by the presented methodology which can pipeline both static and dynamic memory access sequences. In order to take advantage of the memory sequencer, specific enhancements of the typical HLS flow are proposed. The targeted architecture template for the signal processors includes the processing unit (PU) which contains the datapath and a controller, the memory unit (MemU) which executes pipeline accesses to memories, and the communication unit (ComU) which handles communication from/to the rest of the design's computing environment. The synthesis process is performed on the extended data-flow graph (EDFG) which is based on the signal flow graph. The EDFG handles the access and data computations, the transfers of data, and the condition statements for addressing, computation and data transfers respectively. Mutually exclusive scheduling methods [54], [55] are implemented with the EDFG, since it allows both data and conditional semantics to be handled in the same way, and in this way, the exploitation of potential design parallelism can be maximized.

Special structure nodes are defined for the EDFG, so as to represent the arrays and their components access in the application. In order to handle memory access dependencies, the write after write, and read after write dependencies are taken into account and the structure nodes are renamed after e.g. a write access, in order to remove ambiguous dependency accesses for scalar load and store operations. Conditional nodes are defined in order to handle conditioned operations and memorizations. There are dependencies between the calculation of the condition's value and all the following conditioned operations (inside the conditional structure). The function t(u), for operation u, annotates the EDFG edge, in order to capture the delay (time) that the operation takes from the change of its inputs to propagate the result at the outputs (see following paragraphs for HLS internal format descriptions). This is the transfer time from the predecessor of the operation, to its successor. In a first annotation step, all operations including the dynamic address calculations, are assumed to be implemented in the datapath unit of PU. Also, using the available memory mapping data, the data nodes are also annotated. In order to transform an annotated graph into a coherent graph [53], the location of the graph nodes is checked. If the location of all the predecessors and successors of a node is not the same, then a transfer node is inserted. Based on a set of criteria [53], dynamic address computation operations are moved from the datapath unit onto the sequencer, which is called address computation balancing. This is done in order to increase the overall system performance. The resulting graph is then given to the GAUT HLS tool [56] to perform operator selection and allocation, scheduling and binding. A static list scheduling algorithm is selected by GAUT to maximally parallelize the initial schedule. The above methodology is suitable for dataflow dominated applications such as video streaming and linear DSP algorithms.

In [57] a combined execution of decomposition and pattern-matching techniques is applied in HLS problems, in order to reduce the total circuit area. This is done via reducing the datapath area, which is implemented by decomposing multicycle operations, so that they are executed on monocycle functional units (FUs that take one clock cycle to execute and deliver their results). Moreover, other techniques such as regularity exploitation can deliver high quality circuits, when they are used to guide operator decompositions, so that operations extract their most common operation pattern, which is usually repeated in many clock cycles. In this way, the circuit that is needed to execute the selected operation pattern is shared among many operations in many cycles, and therefore the total hardware area is drastically reduced. The HLS algorithm presented in [57] takes as input a behavioral design description and time constraints, and it selectively decomposes complex operations into smaller ones, in order to schedule in every clock cycle a similar number of decomposed fragments of operators, with the same pattern. The method considers only operation decompositions that meet the time constraints, and some of decompositions cause a reduction in the length of the clock cycle, which results into increasing the system's performance. The output from the HLS process is a complete datapath with FUs, multiplexers, registers and some glue logic, as well as a controller. The number, type and width of the resources used in the produced datapath, are generally independent from the input behavioral hardware description, due to the operation decompositions which are applied through the synthesis process.

The research approach in [58] discusses a simple formal model that relies on a FSM-based formalism for describing and synthesizing on-chip communication protocols and protocol converters between different bus-based protocols. The discussed formalism allows for detailed modeling of existing commercial protocols, and the analysis of protocol compatibility. Moreover, the most important is that it allows for automated and correct-by-construction synthesis of protocol converters between existing popular communication protocols. The FSM-based format is at an abstraction level which is low enough for its automatic translation into HDL descriptions which are synthesizable with commercial tools. Typically a system-on-a-chip (SoC) includes intellectual property (IP) blocks that are connected together either directly on a bus or via specialized wrappers. The wrappers play the role of converters from the IP's interface into the bus protocol, so that all the SoC parts collaborate with each other. Usually these wrappers are manually built, based on a non-formal knowledge about the bus protocol and up to the publishing of this work, there were no automated converter synthesis techniques employed in industrial or academic practice. This work contributes towards three aspects of converter synthesis: a formal, FSM-based model for protocol definition, a precise definition of protocol compatibility and a definition of converters and converter correctness (for a given pair of existing protocols).

In [58] protocols are modeled as synchronous FSMs with bounded counters that communicate via channels. Channels are attributed with a type and a direction. A channel type can be control or data, and direction can be input or output. A channel action can be write, read or a value test on a channel. The bounded counters are used in the model in order to keep a data item valid on a channel for a number of clock cycles. Between two changes in the counter value, any read or write action indicates repetition of data values. Bounded counters allow for smaller and precise models of data bursts on channels. Protocols executing concurrently are described by the parallel composition of the protocols. The parallel composition of two protocols describes all the possible control states that the protocols may be in, when they run concurrently. In order to make sure that data flows between these protocols, the following constraints must be satisfied: data is read by one protocol only when it is written by the other; a specific data item can be read as distinct exactly once; no deadlocks can occur and livelocks can always be avoided. The third condition makes sure that every data transfer can terminate in a finite number of steps. In order to satisfy the second constraint of correct data flow, the data actions, along a path between two protocols, need to be correct. This means that every written data item is read as new before it is read as a repeated item. Also a new data item can be written only if the previous one has been read. Moreover, there should be no read repetition between a new write and a new read action. The formal model used, allows for analyzing and checking compatibility between two existing protocols. The model is validated with an example of communication protocol pairs which included AMBA APB and ASB, which are checked regarding their compatibility, and using the formal model.

In order to make sure that correct data flow is happening through a protocol converter the following constraints have to be checked by the formal techniques: data is read by a protocol(/converter) only when the data is written by the converter(/a protocol); a data item can be read as distinct exactly once; no deadlocks and livelocks can occur; every data that is sent (written) from P1 (P2) to the converter, will be sent (written) by the converter to P2 (P1); every data written to the converter was previously written by a protocol. A converter itself is a FSM with bounded counters and a finite buffer for each output data channel. The input to the converter synthesizer consists of a pair of protocols, data channel mapping, and buffer sizes. The converter synthesizer generates the most general (correct) protocol converter. Amongst the protocol converter examples that were used in the context of the work in [58] there were an ASB to APB converter and a set of converters between the Open Core Protocol (OCP) and the AMBA family of bus protocols. The existing synthesis framework is limited to protocols that can be defined by a single FSM. Future work by the authors include more than one FSM per protocol description capabilities.

SystemCoDesigner [59] uses an actor-oriented approach in order to integrate HLS into electronic system level (ESL) design space exploration. Its main aim is to automate the design and building of correct-by-construction System on a chip (SoC) implementations from a behavioral model. The design starts with an executable SystemC system model. Then, commercial synthesizers such as Forte's Cynthesizer are used in order to produce hardware implementations of actors from the behavioral model. The produced actor implementations are characterized on the basis of their area (number of lookup tables, and block memories) and their latency. This facilitates the design space exploration in finding the best candidate architectures (mixtures of hardware and software modules). After deciding on the chosen solution, the appropriate target platform is synthesized with the implementations of the hardware and software parts.

Every module or process in [59] is modeled as an actor which communicates with other actors via a number of communication channels. This is the starting point for modeling a system in [59]. The specification language of an actor is a subset of SystemC which is defined in SysteMoC library [61]. SystemC actors communicate via SystemC FIFO channels and their functionality is implemented in a single SystemC thread. Each SysteMoC actor is defined by a FSM, which defines the communication methods which are controlled by the FSM. Each such actor can be transformed into hardware (using Cynthesizer) or software implementations. Furthermore the use of a commercial tool for hardware synthesis allows for arriving at design solution evaluations, so as to decide about the most suited solution, in terms of hardware resources and throughput. The inputs for design space exploration include the performance information, the executable actor system specification and an architecture template. The architecture template is represented by a graph, which represents all the possible hardware modules, the processors, and the communication infrastructure. From this graph the designer can select the solutions that satisfy the user requirements and which produce tradeoffs between hardware size and performance.

The final step is to generate the FPGA-based SoC implementation from the chosen hardware/software architecture solution. This is done by connecting existing IP blocks and processor cores with the communication elements from an appropriate library. Also the program code for each processor is generated, so as to achieve rapid prototyping. The final FPGA bitstream is generated in [59] using the Xilinx EDK (Embedded Development Kit) tools. A motion-JPEG test application was used to validate the proposed methodology in [59]. The architecture template used an embedded Microblaze processor core, 224 FIFOs, and 19 modules generated by the HLS tool. The complete system specification used 319 actor mapping edges and the design space exploration produced 5 x 10^{33} different alternative solutions. During platform synthesis, for each processor used, a Microblaze core including memory and bus resources were instantiated. The rest of the hardware modules were inserted in the form of Verilog netlists which were generated by Cynthesizer (HLS) and Synplify by Synplicity (RTL synthesis) tools. Moreover, FIFO primitives were inserted for communication between the system's blocks. For the particular test (JPEG), the objectives taken into account during design space exploration, included throughput, latency, number of flip-flops, number of look-up tables, block-RAMS and multipliers. Based on the proposed methodology, the formal underlying mechanisms and the used examples, it was concluded that the SystemCoDesigner method is suitable for stream-based applications, found in areas such as DSP, image filtering and communications. Up to now, there are no indications on how well this methodology would perform in applications with complex control flow.

In [60] a formal approach is followed in order to prove that every HLS translation of a source code model produces a RTL model that is functionally-equivalent to the one in the behavioral input to the HLS tools. This approach is called translation validation and it has been maturing via its use in the optimizing software compilers. The HLS process is approached in [60] from a viewpoint of a sequence of refinements of the original specification (behavioral) code down to the final RTL model of the design implementation. It is argued in [60] that formally proving that these refinement steps maintain the behavioral properties of the original design, is very important since it eliminates the need to re-verify (e.g. with simulations) the correctness of the produced RTL model. The validating system in [60] is called SURYA and it is using the Symplify theorem prover to implement the validation algorithms. The HLS system that was validated with SURYA was the SPARK system [46], and SURYA managed to catch two bugs in the SPARK compilations, which were unknown before.

The translation validation [60] methodology includes two algorithm components: the checking algorithm and the inference algorithm. Given a simulation relation, the checking algorithm determines whether or not this relation is a correct refinement simulation relation. The inference algorithm uses the specification and the implementation programs to infer a correct simulation relation between them. The inference algorithm establishes a simulation relation that defines which points in the specification program are related to the corresponding points in the implementation program. In order to check that a program is a relation of another program the inference algorithm is applied to infer a simulation relation and then the checking algorithm is used to verify that the produced relation is the required one. The translation validation algorithm models the environment of the design as a set of processes that are executed in parallel with the processes of the specification and the implementation. The simulation relation includes a set of entries of the form: (gl1, gl2, ø) where gl1 and gl2 are locations in the specification and implementation programs respectively and ø is a predicate over variables in the specification and the implementation. The pair gl1, gl2 captures how control flow points (control states) relate in the specification and implementation programs and ø captures how data are related between the two programs. The checking algorithm establishes correctness of a relation, if having an entry (gl1, gl2, ø), and if the specification and implementation programs start executing in parallel from states gl1 and gl2 respectively, where ø holds, and they reach another pair of states gl1' and gl2', then in the resulting simulation relation entry (gl1', gl2', ø'), ø' holds in the resulting states. If there are multiple paths from an entry the checking algorithm in [60] checks all of them.

The inference algorithm in [60] begins by finding the points in the specification and simulation programs that need to be related in the simulation. Then it moves further down in the control flow, in both the specification and implementation programs until it finds a branch or a read/write operation on a visible channel. Also, the algorithm co-relates the branches in the specification and the implementation, and it finds the local conditions which must hold in order for the visible instructions to match. When instructions write to visible output channels, the written values must be the same. When instructions read from externally visible input channels, the local conditions state that the specification and implementation programs read from the same point in the conceptual stream of input values. Once the related pairs of locations gl1, gl2 are all collected, a constraint variable is defined to represent the state-relating formula that will be used in the relation for this pair. Then, a set of constraints are applied on these constraint variables to make sure that the relation is indeed a simulation. The first kind of constraints makes sure that the computed simulation relation is strong enough, so that the visible instructions have the same behavior in the specification and the implementation programs. A second kind of constraints is used to state the relationship between a pair of related locations with other pairs of related locations. When all the constraints are generated, then an algorithm sets all the constraint variables to true and strengthens the constraint variables until a theorem prover can show that all constraints are satisfied.

The formal model of refinement in [60] assumes that the specification and the implementation are single entry and single exit programs. Every process in these programs is represented by a transition diagram, which uses generalized program locations and program transitions. A program location represents a point in the control flow of the program and it is either a node identifier, or a pair of two locations which refer to the state of two processes that are running in parallel. A program transition is represented by instructions and it defines how the program state changes from one location to another. In this frame of models, two execution sequences of programs are equivalent, if the two sequences contain visible instructions that are pairwise equivalent. In case the inference algorithm cannot find an appropriate relation, then the user can provide the simulation relation by hand, and use the checking algorithm in order to verify that the relation is a correct one. In a practical context of [60], the SURYA validation system was applied on the SPARK HLS compilation to evaluate the equivalence between the intermediate representation (IR) of SPARK and the scheduled IR of the same translation process.

The replacement of flip-flop registers with latches is proposed in [62] in order to yield better timing in the implemented designs, since latches are inherently more tolerant to process variations than flip-flops. The latch replacement in [62] is considered not only during the register allocation task, but in all steps of HLS, including scheduling, allocation and control synthesis. A certain constraint for latch replacement is to avoid having latches being read and written at the same time. The authors in [62] introduce the concept of the p-step which is the period during which the clock is high or low. Using the p-step as the elementary time unit instead of the conventional clock-cycle, makes scheduling more flexible and it becomes easier to reduce the latency. In [62] list scheduling is extended with the p-step concept, and a method is proposed to reduce the latency by determining the duty cycle of the clock, and therefore the p-step. In order to control the p-step schedule, dual-edge-triggered flip-flops are used in controller synthesis, since both clock edges are used to define the boundaries of the p-steps. These methods were integrated into a tool called HLS-1. HLS-1 accepts behavioral VHDL and produces a synthesized netlist. The design flow was tested with a number of behavioral benchmarks as well as an industrial H.264 video coding application.

The execution time of operators (e.g. multipliers and adders) is formulated mathematically in [62], in terms of clock period, the maximum delay of a functional unit (FU) executing the specific operator, the clock-to-Q delay, the setup time for the utilized flip-flops, and the delay of multiplexers. The transparent phase is the period of time that the clock is high, and the non-transparent phase is the remainder of the clock period. The mathematical analysis of timing using latches, is formulated in [62] using the clock period, the time of the transparent phase and the residual delay (the remainder of the delay time from modulo-dividing with the clock period). Using these formulations, and assuming a 30% duty cycle, it is shown in [62] that an example multiplier will need four p-steps, if it is scheduled to start at the first transparent phase, and will need three p-steps if it is scheduled to start at the second (non-transparent) phase. If flip-flops are used instead, the same multiplier will need two clock steps (equivalent to four p-steps) to execute. In a similar way, and by modifying the duty cycle of the clock, the whole schedule can be optimized by using the p-step as the basic scheduling unit. P-steps enable scheduling at both edges of the clock, so tighter schedules can be produced. Also, register allocation is facilitated since operations can complete in a non-transparent p-step, and therefore read/write conflicts that are inherent in latch registers can be resolved. This method assumes that the delay of the controller is negligible, as compared to the transparent and non-transparent phase times. Nevertheless, implementing registers with latches instead of edge-triggered flip-flops is considered to be difficult due to the complicated timing behavior of latches.

The MORPHEUS approach [63] is introducing a new (abstract) platform for heterogeneous components that allows coupling of dynamic reconfigurable devices and general-purpose processors (as well as SoCs). The methodology is exploiting the best characteristics of each component and the design framework is based on a series of tools and languages, such as C (mainly ANSI C but also GriffyC is used), FlexEOS Macros, SPEAR, CASCADE (STEP standard format), MADEO etc. The MORPHEUS approach is suitable for control-oriented and telecommunication applications, however multimedia and security applications can also be implemented, selecting the appropriate processing core.

5.1 Considering Interconnect Area and Delay

Conventional HLS tools are based on crude estimates of area and timing from an aggregate sum of area and timing models of their functional unit, storage and interconnect elements taken from the component libraries that they use. However, by moving into the deep submicron era the area and timing of the chip's wires become significant. Therefore, new models and optimization approaches are needed. More mature HLS tools adopt various approaches to consider more accurate models of the impact of interconnect on area and timing of the implementation. Given a DFG, a set of resources and resource delays, as well as the clock cycle, a performance estimation tool calculates a lower-bound completion delay for non-pipelined, resource constrained scheduling problems [64]. An algorithm which computes lower bounds on the number of functional units of each type which are required to schedule a DFG in a given number of control steps is presented in [65]. The lower bounds are found by relaxing either the precedence constraints or the integrity constraints, and the method estimates functional-unit area in order to generate resource constraints and thus reduce the search space, or in combination with exact formulation for design space exploration.

A high-level area estimation approach is proposed in [66] which is suitable for standard cell implementation and it focuses on predicting the interconnect area. Simultaneous functional unit binding and floorplanning during synthesis is proposed in [67]. An analytical technique which includes the placement and binding problems in a single, mixed, ILP model, is proposed in [68] for a linear, bit-slice architecture, with the model being able to minimize the overall interconnections of the resulting datapath. In [69] a datapath synthesis is presented which is based on multiple-width shared bus architecture. This methodology uses models of circuit area, delay, power consumption, and output noise which are related to functional unit grouping, binding, allocation and different word-lengths. The functional unit grouping and multiple-width bus partitioning are executed during allocation but before scheduling which, according to the authors, increases the synthesis flexibility and the possibility for better synthesis. The aim is to reduce the delay, as well as the interconnection cost and power consumption of the implementation.

5.2 Synthesis for Testability

Testability for HLS can be achieved by reducing the number of self-looped registers, while considering the tradeoff between testability improvement and area increase. In [70] the switching property of multiplexors and buses is used to reduce the area and test generation costs, by analyzing the location of switches during the selection of partial test registers, and by using these switches to transfer test data through them. In [71] the testability of the design at the behavioral level is analyzed, by considering loops and other control structures, in order to improve the testability by including test constraints, during allocation of registers and generation of interconnections. Simultaneous scheduling and allocation of testable functional units and registers under testability, area and performance constraints, are performed in [72] using a problem-space genetic algorithm. Binding for testability is implemented in two stages in [73]. First a binder with test cost function generates a design almost without any loops. Then the remaining register self-loops are broken by alternating the module and register binding.

A two-stage objective function is used in [74] to transform the behavioral code so that synthesis requires less area and test cost, by estimating the area and testability as well as the effects of every transformation. As a second step a randomized branch-and-bound decent algorithm is used to identify that particular sequence of transformations which offers the best results (in terms of area and test cost). A high-level synthesis-for-testability approach was followed in [75] where the testability of the hardware was increased by improving the controllability of the circuit's controller (control flow). This constitutes a testability increase via better controller design of the circuit.

5.3 Synthesis for Low Energy

Because of the low energy requirements of a number of portable and embedded computing systems and applications such as mobile (smart) phones, PDAs, etc, low power consumption is becoming very important in all of the VLSI and embedded design applications. In the last decade significant research was invested on regarding VLSI techniques and HLS for low power design. In order to incorporate low energy achievements in the results of HLS and system design, new techniques for estimating power consumption at the high-level description, are needed. Switching activity and power consumption are estimated at the RTL in [76], taking also into account the glitching activity in a number of datapath and controller signals. Some important properties from the algorithm represented in the behavioral description are identified in [77] such as the spatial locality, the regularity, the operation count and the ratio of critical path to available time, with the aim to reduce the power consumption of the interconnections. The scheduling, allocation and binding tasks of the relevant HLS system, take into account such algorithmic statistics and properties so as to reduce the fanins and fanouts of the interconnect wires, and thus reduce the complexity and the power consumed on the capacitance of these buses [78].

The synthesis approach in [79] takes into account the effect of the controller on the power consumption of the datapath and the authors proposed a special datapath allocation technique that achieves low power. Pipelining and module selection was proposed in [80] for low power consumption. The activity of the functional units was reduced in [81] by minimizing the transitions of the unit's inputs which were utilized in a scheduling and resource binding algorithm, in order to reduce power consumption. The DFG in [82] is simulated with profiling stimuli, provided by the user, in order to measure the activity of operations and data carriers. Then the switching activity is reduced, by carefully selecting a special module set and schedule. Reducing supply voltage, disabling the clock of idle elements, and architectural tradeoffs were utilized in [83] in order to minimize power consumption within a HLS environment.

The work in [84] addresses the energy consumption of memory subsystem and the communication lines within a multiprocessor system-on-a-chip (MPSoC). This work targets streaming applications such as image and video processing that have regular memory access patterns, which can be statically analyzed and predicted at compile time. This allows the replacement of conventional cache memories with scratch-pad memories (SPMs), which are more efficient than cache, since they don't use tag memory blocks. The key to realize optimal solutions for MPSoCs is to execute the memory architecture definition and the connectivity synthesis in the same step. In [84], memory and communication are co-synthesized, based on multiprocessor data reuse analysis. The latter determines a number of buffers which contain the frequently used data in the main memory. Then, the buffer mapping onto physical memory blocks is done simultaneously with communication synthesis for minimal energy consumption, while satisfying time constraints.

5.4 Issues of Controller (FSM) vs Datapath Tradeoffs

Typically, using the conventional HLS approaches the controller is implemented after the datapath is completed. However, there are tradeoffs between datapath and controller in terms of area and performance. For example, excessive loop unrolling, particularly on many levels of nested loops in the behavioral model, can produce hundreds of states which generate a very large state encoder and thus increase the achievable clock period. The latter will definitely reduce the speed of the design. On the other hand a very simple controller with straight command wires towards the datapath, will end up in heavy use of data routing multiplexers which will have again a negative impact on the minimum clock cycle which can be reached. The controller overhead is taken into account at every level of the hierarchy before the datapath is fully implemented in [85], an approach which is suitable for regular behavioral model algorithms. The system clock period is minimized during allocation in [86], by an allocation technique which considers the effect of the controller on the critical path delay.

6 HLS Internal Formats

One of the most important aspects of compilers and therefore HLS systems is the internal representation of the algorithmic information that they use. The internal format used in [1] consists of three graphs which all share the same vertices. If P is a program in DSL, after decomposing complex expressions into simple, three address operations, W is the set of variables and V the set of operations then the internal form S of the program contains the three directed graphs as follows:

$S = (G_S, G_D, G_C)$ where
$G_S = (V, E_S)$
$G_D = (V \cup W, E_D)$
$G_C = (V, E_C)$

and where the edges E correspond to: E_S = sequence, E_D = dataflow and E_C = constraints. The Sequences set of edges are related to the dependency arcs from one operation to the other. The dataflow edges represent the data processing structure (with connected inputs and output of every operation) inside the specified circuit, and the constraints edges represent timing deadlines from one operation to another.

The HCDG [45] is an intermediate design representation that consists of operation nodes and guard nodes. The HCDG also contains edges which are precedence constraints for data which determine the order of operations, and for control (guards), which determine which Boolean expressions must evaluate to true, in order for the dependent operations to execute. Guards are a special type of node and they represent Boolean conditions that guard the execution of nodes which operate on data. When there are multiple but mutually exclusive assignments to the same variable (e.g. within different if-then-else true/false branches), then this is represented in HCDG with a static single assignment, taking the results of the mutually exclusive operations through a multiplexing logic block (operation node in HCDG). The HCDG itself and the mutual exclusiveness of pairs of guard are utilized in [45] to aid the efficient HLS scheduling transformations of the method developed there.

The HTG is a hierarchical intermediate representation for control-intensive designs which is used by the SPARK HLS tools [46]. A hierarchical task graph is a hierarchy of directed acyclic graphs, that are defined by: $G_{HTG}(V_{HTG}, E_{HTG})$, where the vertices V_{HTG} are nodes of one of the three following types:

1. Single nodes that have no sub-nodes and they capture basic blocks. A basic block is a sequence of operations that have no control flow branches between them.
2. (Hierarchical) Compound nodes that are HTGs which contain other HTG nodes. They capture if-then-else or switch case blocks or a series of other HTGs.
3. Loop nodes that capture loops. Loop nodes contain the loop head and loop tail, which are simple nodes, as well as the loop body which is a compound node.

The set of edges E_{HTG} represents flow of control between HTG nodes. An edge (htg_i, htg_j) in E_{HTG}, where htg_i, $htg_j \in V_{HTG}$, and where htg_i is the start node and htg_j is the end node of the edge, denotes that htg_j executes after htg_i has finished its execution. Each node *htgi* in V_{HTG} has two special nodes, *htgStart(i)* and *htgStop(i)*, which also belong to V_{HTG} . The *htgStart* and *htgStop* nodes for all compound and loop HTG nodes are always single nodes. The *htgStart* and *htgStop* nodes of a loop HTG node are the loop head and loop tail respectively and those of a single node are the node itself [46]. HTGs are providing the means for coarse-grain parallelizing transformations in HLS, since they enable the movement of operations through and across large chunks of structured high-level code blocks, such as complex conditional structures in the C language. Therefore, the HTGs are very useful in optimizing transformations of a HLS scheduler. Their drawback though is their high complexity in translating from regular programming languages' code into a combination of CDFGs and HTGs, such as it happens in the SPARK HLS tool [46].

The authors in [47] have developed their own version of CDFG, which serves two purposes: to be the input design format for their HLS tool, and to facilitate HLS optimizations and in particular to preserve and exploit further the parallelism inherent in control-dominated designs. In order to achieve this, the following special nodes were introduced in the CDFG:

1. The SLP node represents the start of a loop and it selects between the initial value of the loop variable and the value, which is calculated from the last iteration of the executed loop.
2. The EIF node represents the end of an if-then-else branch and it is used to select the correct value from the if and else branches. Both SLP and EIF nodes are implemented with multiplexers in the corresponding datapath.
3. The BLP node sends the values of the variables which are computed by the current iteration of a loop, back to the SLP node which will select the value for the next iteration.
4. The ELP represents the end of the loop. All the loop variables have to be passed through the ELP node, before they are fed to operations which use them, outside of the loop.

Thus, the HLS tool can recognize the beginning, the end and the range of a loop using the SLP, BLP and ELP nodes of the graph. The CDFG model in [47] includes support for memory operations such as load and store. Memory access sequences are defined with a special edge of the CDFG between the corresponding load and store nodes.

The extended data-flow graph (EDFG) which is used in [53] is a finite, directed and weighted graph

$$G = (V, E, t)$$

where V is the set of vertices (nodes) of computation and memorization, E is set of edges representing precedence rules between the nodes, and the function f(u) is the computation time (delay) of node u. A path in G is a connected sequence of nodes and edges of the graph G. The EDFG is used in a synthesis system which optimizes the computing part and the sequencer part of a custom DSP subsystem [53].

6.1 The Intermediate Predicate Format

The Intermediate Predicate Format (IPF)[1] was developed by the author of this paper as a tool and media for the design encapsulation and the HLS transformations in the CCC hardware compilation tool[2]. A near-complete analysis of IPF syntax and semantics can be found in [87]. Here, the basic features and declarative semantics of IPF are discussed along with some example IPF data. The IPF file consists of a number of tables (lists) of homogeneous Prolog predicate facts. These facts are logic relations between a number of objects that are sited as a list of positional symbols or reference numbers of other facts of the same, or other tables inside the IPF database. Therefore, the IPF style allows for both declarative (Prolog) and sequential (list-based) processing of the logic facts of IPF by the CCC HLS tool transformations in a formal way.

The formal methodology discussed here is motivated by the benefits of using predicate logic to describe the intermediate representations of compilation steps. Another way to use the logic predicate facts of IPF is to use the resolution of a set of transformation Horn clauses [88], as the building blocks of a HLS compiler with formal transformations and a state machine optimization engine. A logic inference engine constitutes the most critical (HLS - oriented back-end) phase of the CCC hardware compiler. The inference engine allows for an efficient implementation of the hardware compilation tasks, such as the mapping of complex data and control structures into equivalent hardware architectures in an optimal way, as well as scheduling and grouping the abstract data operations of the source programs into hardware machine (FSM) control states. The choice of logic predicates turns the HLS process into a formal one. Moreover, there have been a number of other successful uses of Prolog in safety-critical areas, such as e.g. in solving the scheduling problem of air-flight control and aircraft landing, expert decision systems etc.

The IPF database is produced by the front-end phase of the CCC compiler (see following paragraphs), and it captures the complete algorithmic, structural and data typing information of the source programs. The source programs model the system's behavior in a number of ADA subroutines. The IPF facts are formal relations between values, types, objects and other programming entities of the source program, as well as other IPF facts. The general syntax of the IPF facts follows the following format:

fact_id(object1, object2, ..., objectN) (form 1)

---

[1] The intermediate predicate format is patented with patent number: 1006354, 15/4/2009, from the Greek Industrial Property Organization.

[2] This hardware compiler method is patented with patent number: 1005308, 5/10/2006, from the Greek Industrial Property Organization.

The predicate name fact_id relates in this fact the objects object1 to objectN in a logical way. IPF facts can represent an algorithmic feature such as a program operation, a data object description, a data type, an operator, a subprogram call, etc. An example of a plus (+) operation which is an addition is described by the following, program table fact:

prog_stmt("subprogram2", 3,0,63,3,9,10,5) (form 2)

The predicate fact prog_stmt of form 2, describes an addition (operator reference = 63), which is the 3rd operation of the ADA subprogram subprogram2. This operation adds two operands with reference numbers 3 and 9 and produces a result on a variable with reference number 10. These operands and result data descriptions are part of the data table and for this example they are given in the following examples (form 3, form 4 and form 5).

data_stmt("subprogram2","dx", 3, 2,"var",sym("node")) (form 3)
…
data_stmt("subprogram2","dy", 9, 2,"var",sym("node")) (form 4)
data_stmt("subprogram2","xc", 10, 2,"var",sym("node")) (form 5)

In the above data table facts we see their reference numbers 3, 9 and 10 which are used in the program table fact of form 2, and their names (variable ids) are dx, dy, and xc respectively. Apart from their host subprogram, these facts describe variables (see the "var" object) and they are of type 2 (integer). This type is described in the type table of the same IPF database with the type fact as it is below:

type_def(2,"integer",32,"standard",0,"single_t",0,0,0) (form 6)

The type definition of the integer type is given in form 6, and there are various objects related under this type fact such as the kind of this type ("single_t" which means with no components or with a single component), the name ("integer") of the type and its size in bits (32). A complete description of the IPF fact objects and structure is not the purpose of this work. Nevertheless, a near-complete description can be found in [87].

From the above, it can be derived that the IPF facts are logical relations of objects, and this formal description is used by the back-end phase of the CCC compiler in order to implement the HLS transformations of the tool. The CCC HLS transformations use the IPF facts of the source design, along with other logical rules in order to "conclude" and generate the RTL hardware models at the output of the CCC compiler. The generated RTL models (coded in VHDL) can be used in turn, by commercial and research RTL synthesizers, along with other, technology-specific and EDA vendor-specific ECAD back-end tools, in order to produce the final hardware implementations of the described system.

It can be derived from the above forms (IPF excerpts) that IPF is suitable for declarative processing which is done by Prolog code, as well as linear, sequential processing. The latter is enabled from the way that some IPF facts (e.g. data table facts) are referenced with their linear entry numbers in other IPF facts (e.g. program table facts). In this way, lists can be built and processed by prolog predicates that utilize list editing and with recursive processing to easily implement the generation and processing of lists, such as e.g. the initial schedule (state list) which includes the FSM states and their components (e.g. operations and conditions-guards). Also both these types of processing in Prolog as well as the nature of the IPF predicate facts are formal, which makes the whole transformation from the source code down to the generated FSM-controlled hardware, a formal one. This eliminates the need to re-verify the functionality of the generated hardware with traditional and lengthy RTL simulations, since due to the formal nature of the transformations the generated hardware is correct-by-construction. Therefore, valuable development time is saved in order to take important decisions about the high-level architecture template options such as the positioning of large data objects such as arrays on embedded scratch pad or external shared memories.

7 A Formal HLS Compiler

Due to the extreme complexity of today's digital systems and applications, methods to shorten the specification-to-product development and verification times are needed. Therefore, robust HLS tools and methodologies are becoming invaluable. These methods become more successful if their transformation processes are formal, which shortens significantly development time, since it eliminates the need for RTL and gate-netlist simulations that are very slow and difficult (or sometimes practically impossible) to achieve 100% test coverage. Here, a formal HLS compiler is discussed, which is called the CCC (custom coprocessor compilations) hardware compiler. The CCC tool is based on compiler-compiler generation techniques and logic programming transformations which make the whole compilation process formal. Therefore, the user of the CCC flow can rely on the high-level ADA code compilation and fast verification in order to make sure that the functionality of the system to be synthesized has been captured, in the ADA source code. Nevertheless, since RTL VHDL implementable models are generated by the CCC tool, and although not necessary, RTL verification can be executed as well. Also, the CCC process is enhanced with validation of the internal state of the compiler, by using a robust XML interface and XML schema implementations are automatically generated by the tool. Moreover, and in combination with a web site that executes the CCC compiler (and thus makes it available to the public) RDF technology is exploited to make the CCC execution formal in its web interface state.

7.1 The CCC Design Flow

The hardware compilation flow consists of two major steps: the front-end phase and the back-end phase. These two compilation phases are connected with and exchange data through a tabular intermediate format. The Intermediate Predicate

Format constitutes the link between the two phases and the tool also generates a functionally-equivalent XML schema in order to validate the compilation process as shown in Figure 1.

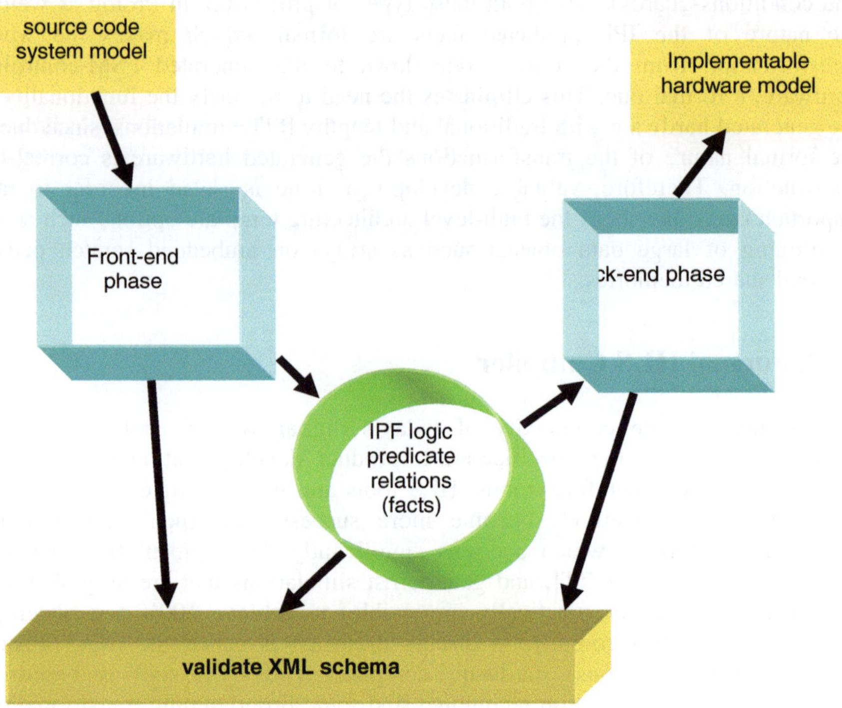

Fig. 1 Hardware compilation and XML validation flow

IPF utilizes the elegance, expressiveness and formal features of the Horn clauses syntax [88], to describe the complete algorithmic and hierarchical information of the source programs, and also to allow for the HLS transformations of the back-end phase to take place. In this way, and because of the alternative formal view of the IPF model, the XML schema view, which is also produced automatically by the front-end phase to formally validate the translation, IPF constitutes a formal link between the two formal phases of the hardware compilation.

The front-end phase compiles source code system-models into the IPF (logic clause) database. The latter contains the typing, structural and algorithmic (control & data flow) information of the source code programs. During the front-end phase, the complex statements of the source code model are analyzed and translated into a set of 3-address operations (with 2 source operands and 1 target operand), which are modeled with the IPF logic predicate facts.

The back-end phase of the hardware compiler utilizes formal transformation techniques, based on logic rule relations between the predicate facts of the IPF model, after the latter is validated in its XML schema view. The generated IPF database is formally transformed into a set of standalone hardware FSMs. Each FSM is functionally equivalent to that of the corresponding source code subprogram. The initial FSM state schedules, which are directly produced from analyzing the IPF data, as well as additional user parameters (such as e.g. the use of shared memories for large data objects), are optimized using a parallelizing transformation algorithm called PARCS (Parallel Abstract Resoure – Constrained Scheduling). PARCS attempts to absorb as many as possible, parallel operations in each state, in an aggressive manner, without violating data dependencies and abstract resource (operator) user-constraints. These user-constraints, can be passed as options to the back-end compiler, in order to determine the type and max number of available parallel hardware resources (operators). The generated RTL hardware description code is then implemented into silicon using commercial RTL HDL synthesis E-CAD tools.

The set of available data types and operators can be extended in a library module which is written in a IPF syntax subset, and integrated into the flow of Figure 1 via the use of a library compiler.

The front-end compiler performs the typical tasks of a program compiler, such as lexical and syntactic analysis, generation of syntactic and semantic error messages, production and processing of abstract syntax trees, type-checking of data declarations and program statements, optimization of auxiliary variables and constants etc. Moreover, the front-end compiler transforms the source programs into the Prolog and the XML schema views of the IPF model. The XML schema view can be used to validate the program transformation into the IPF predicate facts. The validation of the XML view can be done either at the XML code level or at the graphical representation of its schema structures, using regular XML schema visualization tools (e.g. XML editors) which are commercially available or license-free.

7.2 Back-End Compiler Transformations

The back-end compiler consists of a very large number of logic rules. The back-end compiler logic rules are coded with logic programming techniques, which are used to incorporate in this compilation phase a number of HLS transformations and optimization algorithms. As an example, one of these algorithms reads and incorporates the IPF tables facts' data into the compiler's internal inference engine of logic predicates and rules. The backend compiler rules are given as a great number of definite clauses [88] of the following form:

$$A_0 \leftarrow A_1 \wedge \ldots \wedge A_n \text{ (where } n \geq 0) \qquad \text{(form 7)}$$

where $\leftarrow$ is the logical implication symbol ($A \leftarrow B$ means that if B applies then A applies), $\wedge$ is the logical conjunction symbol, and $A_0, \ldots, A_n$ are atomic formulas (logic facts) of the form:

$$\text{predicate\_symbol(Var\_1, Var\_2, ..., Var\_N)} \qquad \text{(form 8)}$$

where the positional parameters Var_1,…,Var_N of the above predicate "predicate_symbol" are either variable names (in the case of the back-end compiler logic rules), or constants (in the case of the IPF table statements) [87]. The predicate syntax in form 8 is typical of the way of the IPF facts (and other facts used internally in the inference engine) are organized. Therefore, the input programs are formally transformed, by the back-end phase, into a set of functionally-equivalent hardware modules. This turns the overall transformation into a provably-correct formal process and the generated hardware accelerators are correct-by-construction. In essence, the IPF file consists of a number of such atomic formulas, which are grouped in the IPF tables. Each such table contains a list of homogeneous facts which describe a certain aspect of the compiled program. E.g. all prog_stmt facts for a given subprogram are grouped together in the listing of the program statements table.

In this way, IPF can be considered as a means to relate objects (the positional parameters and the predicate symbols) to each other and to predicate symbols. Every such relation denotes an algorithmic and programming semantic or syntactic property of the source programs.

7.3 *Inference Logic and High-Level Synthesis Transformations Validated with XML*

The inference engine of the back-end phase of the hardware compiler includes formal logic rules (logic relations) such as the one in form 7. A new design to be synthesized is loaded via its IPF logic view into the back-end inference engine. Hence, the IPF's statements "drive" the logic rules of the back-end compiler which generate provably-correct hardware architectures. The hardware compiler generates technology-independent, standard template – free, and custom (V)HDL RTL implementable hardware models. These VHDL models can be either single-FSM-controlled machines, or pairs of FSM+datapath microarchitectures. This selection is done by option parameters passed to the back-end compilation by the user of the CCC flow.

The following is an example inference rule (logical relation) from the back-end phase code. Its purpose is to build a new (optimized) PARCS state from the so far absorbed parallel operations and states from the initial schedule.

```
build_parcs_state(Module, Existing_schedule, Incomp_ops, Incomp_states, Parcs_state) ←
  str_int(State_num_str, Parcs_state),
  concat("state_", State_num_str, State_name),
  Provisional_next_state = Parcs_state + 1,
  assertz(state(Module, "parcs", Parcs_state, State_name, Provisional_next_state,
    0, Incomp_ops, [])),
  build_rescheduled(Module, Existing_schedule, Incomp_states, "parcs", Parcs_state),
    set_last_parcs_state(Module, Parcs_state).                                (form 9)
```

In this predicate clause (logic rule), the comma (,) is used in place of the conjunction symbol (∧) of form 7. The above example maintains the syntax of form 7, with the only difference being that the last (rightmost) atomic formula of the rule is terminated with the full-stop symbol (.) to indicate the end of the inference rule's conditions.

The above logic rule, denotes that the atomic formula with predicate symbol build_parcs_state (at the left side of the ←) is valid (is true) when all of the five atomic formulas at the right side of the logical implication symbol ← apply as well. The rule executes when the execution of the five predicates at the right side of the logical implication (←) completes (when they are all true). Desirable side effects of this rule execution include the increment of the PARCS state counter by one, as well as the assertion of a state(…) fact (atomic formula like in form 8) for the newly created optimized PARCS state. Another, desirable side effect of the execution of the rule is the assertion of the last PARCS state, global variable (the last predicate of the right side of the rule). This last assertion is implemented with the predicate:

set_last_parcs_state(Module, Parcs_state)

As mentioned above, the PARCS optimization is applied either on the initial schedule or on the one, which is enhanced with additional communication hardware protocols, such as automatically generated external memory ports and interfaces that the user may want to implement (for external, shared memories).

Therefore, the design optimization is applied on the complete set of source model operations including the interface communication functions. This in turn improves the quality of the generated hardware logic. As mentioned already, XML schema is used to validate the hardware compilation intermediate representations and processes, at the various stages of the hardware compiler. The following code is an excerpt from the XML schema validation of the state(…) predicate of the above inference rule:

```
<complexType name="state">
  <sequence>
    <element name="Module_1"/>
      <element name="parcs"/>
      <element name="parcs_state_number"/>
      <element name="parcs_state_name"/>
      <element name="parcs_next_state"/>
      <element name="no_conditional_transition"/>
      <element name="scheduled_operation_list">
        <complexType>
          <sequence>
            <element name="Operation_1"/>
             <element name="Operation_2"/>
             <element name="Operation_3"/>
          </sequence>
```

```
      </complexType>
    </element>
  <element name="no_conditional_operations"/>
 </sequence>
</complexType>
```

for the case of a PARCS state of design Module_1, with three scheduled operations in parallel and with no conditional operations or transitions. The graphical validation of this XML schema view of the PARCS state is shown in Figure 2. From the above XML validation of the state(…) predicate it can be argued that XML schema is another formal way to relate objects and predicates. Therefore, one of the two above views can validate the other and vise-versa.

The IPF file consists of a number of atomic formulas, or predicate facts (in Prolog's terminology) like the one in form 8. These facts are grouped into a number of IPF tables. Every table contains a number of homogeneous facts which describe a certain aspect (semantic group) of the compiled program. Examples are the prog_stmt facts for a given subprogram, which are grouped together to describe the 3 address operations of that subprogram, and the data_stmt table facts which provide the semantics of the source code data objects. The IPF syntax is formally defined, but its complete description is outside the purposes of this paper [87].

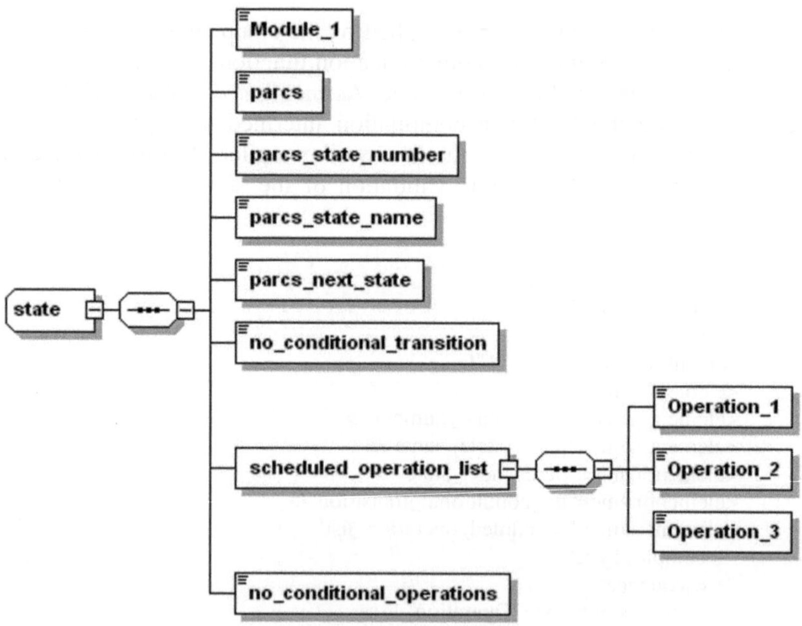

Fig. 2 XML schema validation of PARCS state

Loading the IPF file into the inference engine of the back-end compilation phase works through simple predicate relations that define the functionality of the compiler. The rules of the inference engine, such as the previous example in this paragraph are object relations between the IPF, as well as various additional inference predicates. In this way, both the logic and the XML views of the IPF, which are produced automatically by the front-end compilation phase as well as the back-end inference rules can be modelled (and validated) in both logic programming and XML code. Moreover, the relations between the program semantics (e.g. data, types, operations, etc.) are implemented in the form of pointers from any IPF facts to other facts or tables of the IPF file. The characteristic cause and result/effect relation between these predicates at the left side of symbol $\leftarrow$ and the right side of it denotes the method by which the back-end inference engine rules work.

Experience showed that coding a compiler with predicate relations makes it easier to implement the high-level synthesis transformations of a hardware compiler. This applies because only the relations between the compiler and source program semantics are needed, in contrast to the traditional sequential programming which requires the detailed implementation of every step of the algorithm. The result of this approach was the drastic reduction of numbers of code lines.

As already mentioned the XML view contains relations between predicate symbols, e.g. transform1, and algorithmic objects, e.g. operation1, left_operand_A, right_operand_B, result_data_C. This is an example transformation that is implemented with a logic predicate, which logically relates the two operands and the result of a source code operation. This can be validated with the logic view (predicate) of the relation, with the XML view of the relation between the 3 operands or with the graphical validation of the XML view of the predicate transform1. The logic programming view of this relation is:

```
transform1(operation1,left_operand_A,right_operand_B, result_data_C).
```

The XML schema view is shown in the following code excerpt:

```
<complexType name=" transform1">
  <sequence>
    <element name=" operation1"/>
    <element name=" left_operand_A "/>
    <element name=" right_operand_B "/>
    <element name=" result_data_C "/>
  </sequence>
</complexType>
```

The graphical validation of the XML schema is shown in Figure 3.

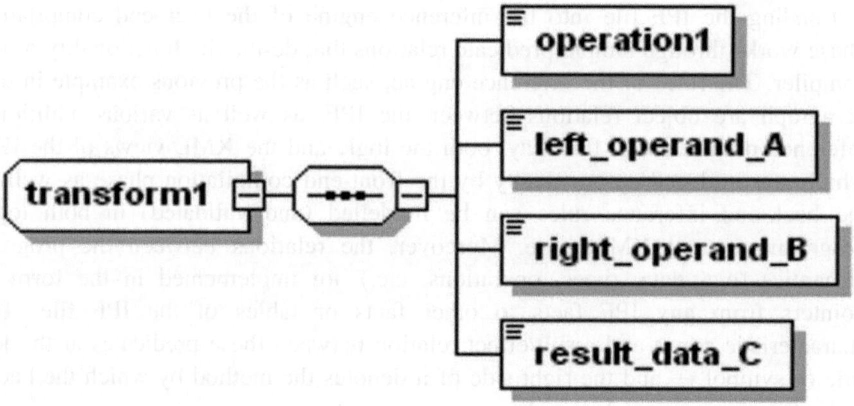

Fig. 3 Graphical validation of the predicate's XML schema

7.4 The Web Interface

A web page interface (http://ccchlstool.tk) was built that accommodates the CCC tools and makes them available to the scientific and engineering community. This web page includes buttons for uploading the ADA source code, the user constraint files and the memory interface files, for executing the front-end and the back-end compilers, for passing user constraints and other user parameters to the compilation, as well as for downloading the result files of the flow (VHDL output and log files).

There is a certain (imposed by the page) execution order from one step of the flow to the following to help the user identify easily the order of his actions on the client web page. The processes which are executed on the server each time the user activates a button are guided by equivalent number of PERL scripts. The main web page which is accessed by remote users is coded in HTML and PSP. The configuration of the machine that the web server is running on is OS Windows XP-64-bit SP3 and Web Server Apache 2.2, on an Intel Xeon 3 GHz processor.

In terms of an intelligent web service, the discussed web application is comprised of a RDF-based schema, with a large set of subject-predicate-object statements which contain the IPF facts (predicates) as well as the inference engine rules (cause-effect relations). Another way to think of the back-end compiler inference engine rules is to analyze them as RDF-like subject-predicate-object relations. For example the following back-end compiler rule:

dont_schedule(Operation1, Operation2) ← predecessor(Operation1,Operation2).
(form 10)

is a logic relation between two predicates. The "cause" predicate predecessor is in the form subject-predicate-object with Operation1 as subject and Operation2 as object ("operation1 is a dataflow-predecessor of operation2" in linguistic terms). The "result" predicate dont_schedule is also in the form subject-predicate-object, with Operation1 as subject and Operation2 as object ("operation1 must not be scheduled with operation2" in linguistic terms). This cause and result/effect relation between these two predicates is characteristic in the way that the back-end compiler's inference engine rules work.

As already analyzed the two predicate relations (inference rule) example of form 10, is depicted with the RDF graph which is shown in Figure 4. The two predicates of form 10 with the subject-object connection and the relation arrow between the cause and effect (result) predicates can be seen in Figure 4.

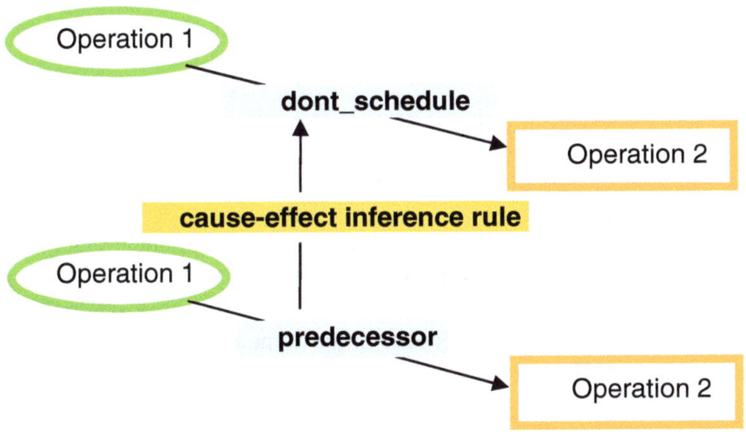

Fig. 4 The RDF graph of the 2 predicates rule example

It is clear from Figure 4 that the cause subject-predicate-object is transformed into the effect (result) subject-predicate-object relation at the top of the diagram. These inference rules are invoked in the back-end compiler's Prolog code, which in turn is activated from the respective buttons of the web-client of our applications.

The front-end compiler, activated by the respective button of the client page, converts in a systematic and provably-correct way (compiler-generator techniques) the source high-level ADA program statements into functionally equivalent IPF predicate facts, which are also RDF subject-predicate-object relations.

The back-end compiler, activated as well by the respective button of the client page, loads the generated IPF predicates into the inference engine. The inference engine is comprised of a large number of predicate rules, similar to the example in Figure 4. The inference engine, activated by the web-application, schedules the states of the IPF in an optimum way and amongst other actions, it generates the custom (V)HDL hardware code of the coprocessors. Therefore, the web interface, provides an intelligent web service, which uses subject-predicate-object relation rules to provide rapid hardware compilations from high-level program code to custom hardware coprocessor definition code.

7.5 Generation of Hardware Accelerators

The inference engine of the back-end compiler consists of a great number of logic rules which make conclusions based on a number of input logic predicate facts, and produce another set of logic facts and so on. Eventually, the inference logic rules produce the logic predicates that encapsulate the writing of RTL VHDL hardware accelerator models. These hardware models are directly implementable to any hardware (e.g. ASIC or FPGA) technology, since they are technology and platform – independent. For example, accelerator RTL models produced in this way from the prototype compiler were synthesized successfully into hardware implementations using the Synopsys DC Ultra, the Xilinx ISE and the Mentor Graphics Precision E-CAD software without the need of any manual alterations of the produced RTL VHDL code. In the following form 11 an example of such an inference rule is shown:

```
dont_schedule(Operation1, Operation2) ←
    examine(Operation1, Operation2),
    predecessor(Operation1, Operation2).          (form 11)
```

The meaning of this rule, that combines two input logic predicate facts in order to produce another logic relation (dont_schedule) is that when two operations (Operation1 and Operation2) are examined and the first is a predecessor of the second, then they will not be scheduled in the same control step. This rule is part of a parallelizing optimizer which is embedded in the back-end compiler and it aggressively "compresses" the execution time of the generated hardware accelerator, by making a number of operations parallel in the same control step, as soon as of course the data and control dependencies are still satisfied.

The way that the inference engine rules (predicates relations & productions) work is depicted in Figure 5. The last produced (from its rule) predicate fact is the VHDL RTL writing predicate at the top of the diagram. Right bellow level 0 of predicate production rule there is rule at the -1 level, then level -2 and so on. The first predicates that are fed into this engine of production rules, belong to level –K, as shown in this figure. Level –K predicate facts include of the IPF facts that are loaded into the inference engine along with the rest predicates of this level. It is clear from this that the back-end compiler works with inference logic on the basis of predicate relation rules and therefore, this process is a formal transformation of the IPF source program definitions into the hardware accelerator (implementable) models. There is no traditional (imperative) software programming involved in this compilation phase and the whole implementation of the backend compiler is done using logic programming techniques. Of course in the case of the CCC compiler, there are a very large number of predicates and their relation rules that are defined inside the implementation code of the back-end compiler, but the ultimate concept of implementing this phase is as shown in Figure 5.

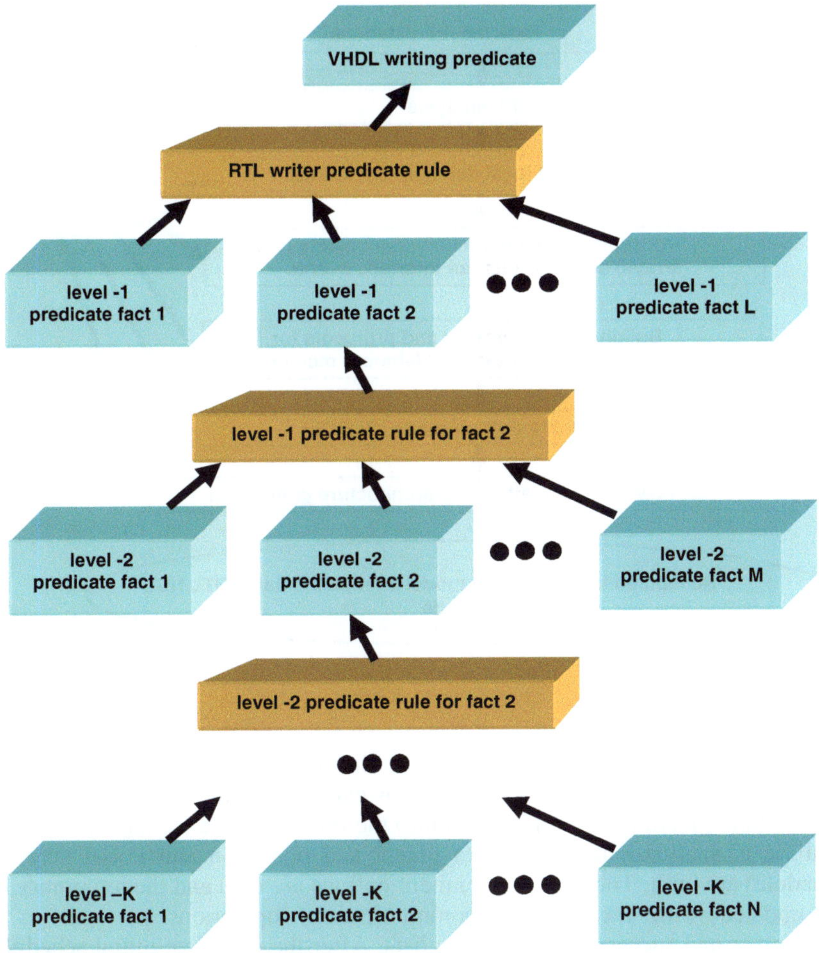

Fig. 5 The back-end compiler inference rules mechanism

The above fact production rules include the loading of IPF, the generation of list of data objects (virtual registers) and operators (including the special operators of communication with memories and the external computing environment), the initial and the processed list of states, the list of (prede-)cessor relations between operations, the PARCS optimizer, the datapath and FSM architecture generators etc. The user of the back-end compiler can select certain environment command list options, via an external memory port parameter file, as well as drive the compiler's optimizer with certain resource constraints for the available hardware operators.

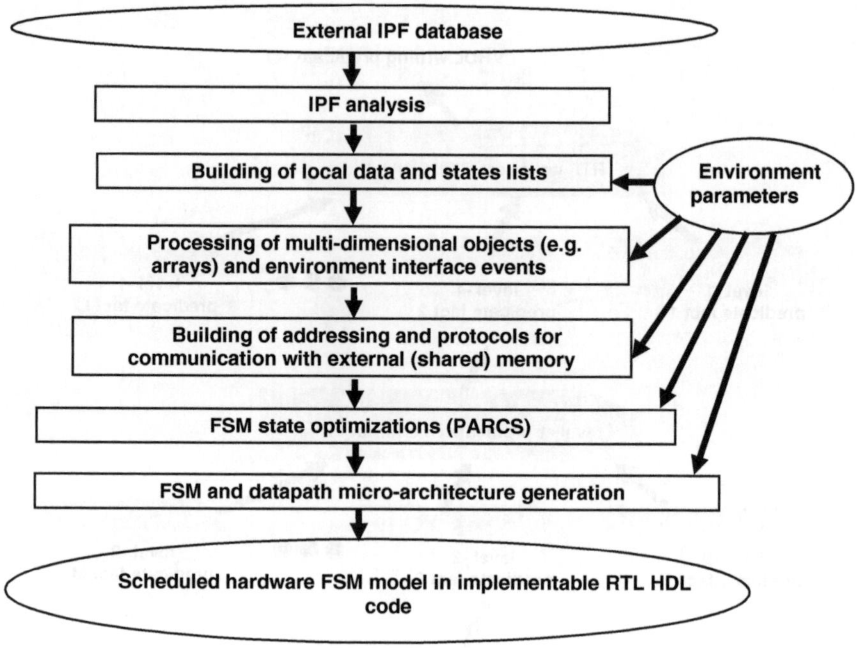

Fig. 6 The various processing stages of the back-end compiler

The above predicate rules implement the various stages of the back-end compiler. The most important of these steps can be seen in Figure 6. The process starts with the loading of the IPF facts into the inference rule engine. After the IPF database is analyzed, the local data object and operation and initial state lists (schedule) are built. Then the environment options are read and the temporary lists are updated with the special (communication) operations as well as the predecessor and successor dependency lists. After the complete initial schedule is built and concluded, the PARCS optimizer is run and the optimized schedule is delivered to the micro-architecture generator. The HLS transformations are concluded with the formation of the FSM and datapath implementation, and the writing of the RTL VHDL model for each accelerator that is defined in each subprogram of the source code program.

As mentioned already, from each subprogram, which is coded in the source program, a separate hardware accelerator (or coprocessor) model is generated. All these hardware models are directly implementable into hardware using commercial E-CAD RTL synthesizers. Also the hierarchy of the source program modules (subprograms) is maintained and the generated accelerators may be hierarchical. This means that an accelerator can invoke the services of another accelerator, from within its processing states, and that other accelerators may use the services of yet another accelerator and so on. In this way, a subprogram call in the source code is translated into an external coprocessor interface event of the corresponding hardware accelerator. The two hierarchical coprocessors are synchronized via a synchronous handshake interface, similar to this that is generated for the start and completion of its operation.

7.6 The PARCS Optimizer

As already mentioned, the various source program operations are scheduled using an aggressive algorithm called PARCS. PARCS schedules as many as possible operations in the same control step, respecting the data and control dependencies and the specific resource (operator) constraints if there are any, that may have been provided, via the run-time, environment options and/or a constraint file.

1. Start with the initial schedule (including the special external port operations)
2. Current PARCS state <- 1
3. Get the 1st state and make it the current state
4. Get the next state
5. Examine the next state's operations to find out if there are any dependencies with the current state
6. If no dependencies then absorb the next state's operations into the current PARCS state; if there are dependencies then finalize the so far absorbed operations into the current PARCS state, store the current PARCS state, PARCS state <- PARCS state + 1; make next state the current state; store the new state's operations into the current PARCS state
7. If next state is of conditional type (it is enabled by guarding conditions) then call the conditional (true/false branch) processing predicates, else continue
8. If there are more states to process then goto step 4, otherwise finalize the so far operations of the current PARCS state and terminate

Fig. 1 Pseudo-code of the PARCS scheduling algorithm

The pseudo-code for the main parts of the PARCS scheduler is shown in Figure 7. The PARCS scheduler consists of a great number of predicate logic rules. All these are part of the inference engine of the back-end compiler. The various used predicates are produced with formal logic rules such as the one in form 7. Every new design to be synthesized is loaded via its IPF database into the back-end compiler's inference engine. Hence, the IPF's facts, as well as the newly created predicate facts, from the so far inference logic processing, "drive" the logic rules of the back-end compiler which generate provably-correct hardware architectures.

For example the following PARCS scheduler rule, examines whether there are no dependencies between the current state's scheduled operations and a new operation. If there are no dependencies then it absorbs the new operation list into the currently processed (optimized) state:

process_state_ops(Current_state, Current_list_of_operations, New_states_list_of_operations) ←
no_dependencies(Current_list_of_operations, New_states_list_of_operations),
absorb(Current_state, New_states_list_of_operations). (form 12)

This rule is a logic production between two source (cause) predicates and one target (conclusion) predicate. The meaning is that if predicate no_dependencies is true (in other words if there aren't any dependencies between the list of the current state operations and those of the next – under examination – state), then predicate absorb is examined. If in turn absorb is true, which applies when it is executed to absorb the new list of operations into the current state, then process_state_ops is produced, which in turn means that the production rule is executed successfully. For the inference engine, this means that another fact of predicate process_state_ops is produced and the PARCS algorithm can continue with another state, if there are any more states left. The desirable side-effect of this production is that the list of operations of the newly examined state is absorbed into the current PARCS state, and PARCS processing can continue with the remaining states (of the initial schedule).

It is worthy to mention that although implemented with logic predicate rules, the PARCS optimizer is very efficient and fast. In most of benchmark cases that were run through the prototype hardware compiler flow, compilation did not exceed 1-10 minutes and the results of the compilation were very efficient as it will be explained bellow in the paragraph of experimental results. This cause and result/effect relation between these two predicates is characteristic of the way the backend compiler's inference engine rules work. By using this type of logic rules the backend compiler achieves provably-correct transformations on the source programs.

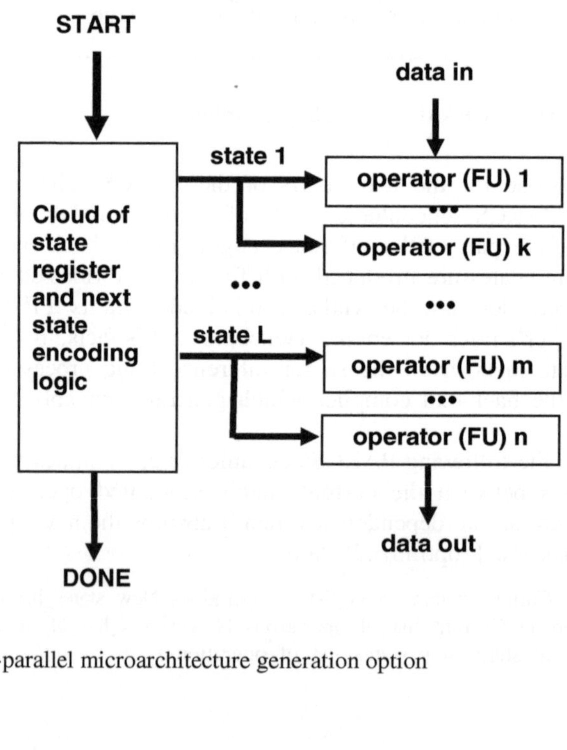

Fig. 8 Massively-parallel microarchitecture generation option

7.7 Generated Accelerator Architectures

As mentioned already the hardware architectures that are produced by the back-end synthesizer are platform and RTL tool – independent. This is because generalized, FSM-controlled datapath RTL VHDL descriptions are produced, without any macro details from particular technology libraries or any vendor specific tool directives. These descriptions (models) can be synthesized into hardware implementations by using any commercial or experimental RTL synthesizer, and without the slightest manual modifications of the generated RTL VHDL code. Moreover, the generated accelerator models do not depend on any hardware implementation technology. Thus, they can be implemented into any ASIC or FPGA technology.

The back-end stage of microarchitecture generation can be driven by command-line options. One of the options e.g. is to generate massively parallel architectures. The results of this option are shown in Figure 8. This option generates a single process – FSM VHDL description with all the data operations being enabled by different machine states. This means that every operator is enabled by single wire activation commands that are driven by different state register values. This in turn, means that there is a high degree of redundancy in the generated hardware, which results into a number of state-dedicated operators remaining idle, during parts of execution time. However, this redundancy is balanced by the fact that this option achieves the fastest clock cycle, since the state command encoder, as well as the data multiplexers are reduced to single wire commands without any additional circuit logic delay.

Another microarchitecture option is the generation of traditional FSM + datapath based VHDL models. The results of this option are shown in Figure 9. With this option activated, the generated VHDL models of the hardware accelerators include a next state process, as well as signal assignments with multiplexing which correspond to the input data multiplexers of the activated operators. Although this option produces more economic hardware structures in terms of the utilized functional units (FUs), it can result into longer minimum clock periods, due to larger delays through the data-steering multiplexers, which are used in the datapath and in order to route the data through the various allocated FUs.

Using the above microarchitecture options, the user of the inference-based CCC hardware compiler can select various solutions between the fastest and larger massively-parallel microarchitecture, which may be suitable for richer technologies in terms of resources such as large ASICs, and smaller and more economic (in terms of available resources) technologies such as the smaller FPGAs.

As it can be seen in Figure 8 and Figure 9, the generated accelerators (coprocessors) are initiated with the input command signal START. Upon receiving this command the accelerators respond to the controlling environment with the handshake output signal BUSY, right when they start processing the input data in order to execute the hardware's function(s). This process may take a number of clock cycles and it is controlled by a set of states (discrete control steps). When the accelerators complete their processing, they notify their environment with the output signal DONE. In order to conclude the handshake the controlling environment (e.g. a controlling central processing unit) responds with the handshake input RESULTS_READ, so as to notify the accelerator that the processed result data have been read by the environment. This handshake mechanism is followed also when one (higher-level) coprocessor calls the services of another (lower-level) coprocessor. The handshake is implemented between any number of accelerators (in pairs) using the START/BUSY and DONE/RESULTS_READ signals. Therefore, the set of running accelerators can be also hierarchical in this way.

Other environment options, which are passed to the back-end compiler, control the way that the data object resources are used, such as registers and memories. Using a memory port configuration file, the user can dictate that certain multi-dimensional data objects, such as arrays and array aggregates, are located and implemented inside external (e.g. central, shared) memories (e.g. system RAM). Otherwise, the default option is that all data objects are allocated onto hardware (e.g. on-chip) registers. Of course the external memory option is more economic in terms of accelerator compilation time and register and port use, but it causes a longer processing time, due to the communication protocols that need to be implemented, every time that a datum is accessed from/to the external shared memory. Nevertheless, all such memory communication protocols and hardware ports, are automatically generated by the backend synthesizer, and without the need for any manual editing of the RTL code by the user. Both synchronous and asynchronous memory communication protocol generation are supported in the memory port options file.

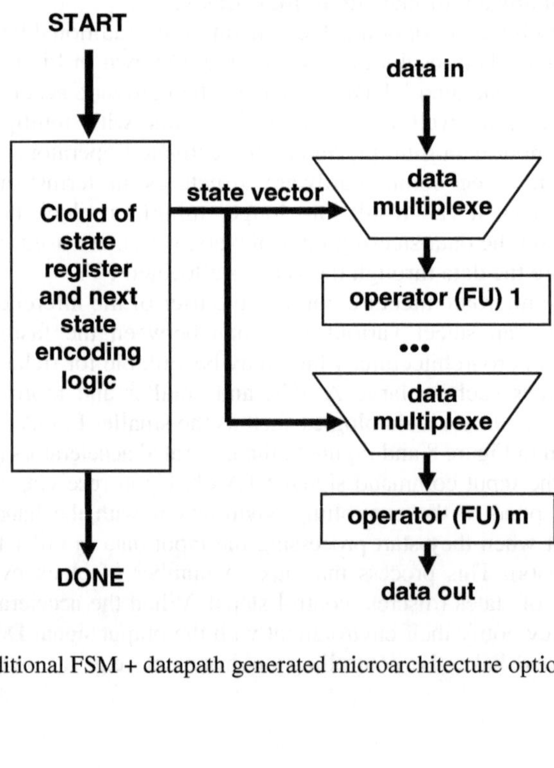

Fig. 9 The traditional FSM + datapath generated microarchitecture option

7.8 Accelerator Execution System

The generated accelerators can be placed inside the computing environment that they accelerate, or they can be executed standalone. This computing platform may be in the same ASIC SoC which contains the host processor core (e.g. an ARM or Microblaze core), an electronic board that contains a custom ASIC with the implemented coprocessors, or an electronic board that contains amongst other components and one or more FPGAs to accommodate the generated hardware. In the latter case, the same target resources (FPGAs) can be used to implement different logic, every time a product update or bug-fix is required, or when the specification (ADA code) is simply modified.

From every subprogram in the source specification code one accelerator is generated to speed up the particular system task. The whole system can be modeled in the ADA software code to implement both the software and hardware models of the system. The whole ADA code set can be compiled and executed with the host compiler and linker to run and thus verification can be executed in order to verify the operation of the whole system at the program code level. In this way, extremely fast verification can be achieved at the algorithmic level.

After the required accelerators are specified, coded in ADA, generated with the prototype hardware compiler and implemented with commercial back-end tools, they can be downloaded into the target computing system and executed to accelerate certain system tasks. This target may include ASICs or FPGAs to allow for downloading different coprocessors into the same hardware platform. This process is shown in Figure 10. Every accelerator can have both local (register file) and external (system RAM) memory resources. The accelerators can communicate with each other and with the host computing environment using synchronous handshake signals and connections with the system's handshake logic. In case that the host processor contains handshake capabilities, the communication can be done directly with the custom accelerator, which in this case operates as a coprocessor with the main processor. In a different case, some glue logic may be required to communicate the system's processor with the hardware shake interface of the CCC accelerators.

From Figure 10 it is obvious that the accelerators may contain shared memory ports to transfer data and results from and to the main memory. Also, there may be multiple external RAM memory blocks and each coprocessor may communicate with a separate memory block. Nevertheless, this increases the complexity of the target computing platform. Nevertheless, the followed design flow allows for a great flexibility on the type and number of processing components that can be part of the targeted computing platform. The latter may be a portable device, an embedded system or any other form of computing set.

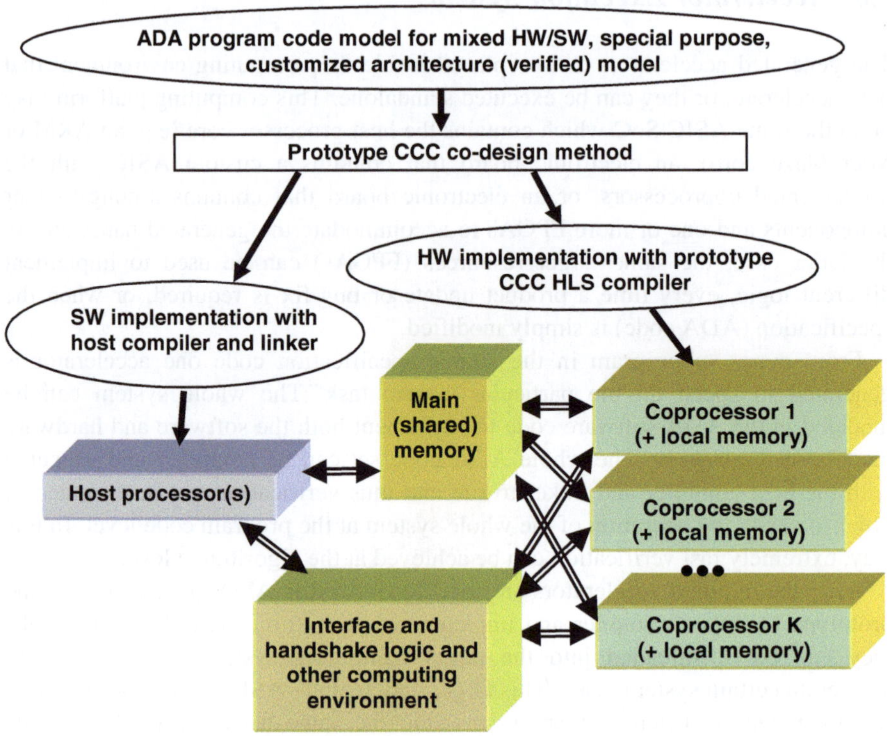

Fig. 10 Host computing environment and execution of generated accelerators

8 Experimental Results and Evaluation of the Formal Hardware Compilation Methodology

In order to evaluate the efficiency of the presented formal HLS method, many test designs from the area of hardware compilation and high-level synthesis were run through the front-end and the back-end compilers. Five such ADA programs are discussed here. The programs were executed on a Pentium-4 platform running the MS-Windows-XP-SP2 operating system. The five design benchmarks include a DSP FIR filter, a second order differential equation iterative solver, which has long been a well-known HLS benchmark, a RSA crypto-processor from cryptography applications, an application that uses two level nested for-loops, and a large MPEG video compression engine. The fourth benchmark includes subroutines with two-dimensional data arrays stored in external memories. These data arrays are processed within the bodies of 2-level nested loops.

All of the above generated accelerators were simulated and the RTL models confirmed the input source program's functionality, as it was expected, due to the formal nature of the hardware compilations. The state number reduction after applying the PARCS optimizer, on the various modules of the five benchmarks is shown in Table 1. State reduction rates of up to 35% were achieved with this HLS tool. Also, the number of lines of RTL code is orders of magnitude higher compared with the lines of the ADA source code model for each module (subprogram in the ADA code). This indicates the gain in engineering effort and performance when the prototype CCC HLS tool is used to automatically implement the computing products. This is due to the time gained by building and verifying (by program run) fast executable specifications in high-level programs, as opposed to run the system verification only when all the hardware details are fixed and then perform time-consuming detailed (RTL or structural) hardware design and simulations.

Table 1 State reduction statistics from the PARCS optimizer for the five benchmarks

Module name	Initial schedule states	PARCS parallelized states	State reduction rate
FIR filter main routine	17	10	41%
Differential equation solver	20	13	35%
RSA main routine	16	11	31%
nested loops 1st routine	28	20	29%
nested loops 2nd routine (with embedded mem)	36	26	28%
nested loops 2nd routine (with external mem)	96	79	18%
nested loops 3rd routine	15	10	33%
nested loops 4th routine	18	12	33%
nested loops 5th routine	17	13	24%
MPEG 1st routine	88	56	36%
MPEG 2nd routine	88	56	36%
MPEG 3rd routine	37	25	32%
MPEG top routine (with embed. mem)	326	223	32%
MPEG top routine (with external mem)	462	343	26%

It is widely accepted by the engineering community, that the verification time at the algorithmic program level is only a small fraction of the time required to verify designs at the RTL or the gate-netlist level. This gain is invaluable for developing complex computing systems. The formal nature of the CCC transformations can be utilized so as to eliminate the need for design cycle iterations, later on in the design flow. Such iterations, would be caused by functional errors and specification mismatches, which are very frequent in engineering practice, such as RTL coding or schematic design.

The relative reduction of the number of states in the largest design module of the application, before and after the PARCS parallelizing optimizer is shown graphically in Figure 11. The reduction of states reaches up to 30 to 40% at some cases which is a significant improvement. Such optimizations are usually very difficult to be done manually, even by experienced ASIC, HDL designers when the number of states exceed a couple of dozens in the designed application (there were more than 400 states in the initial schedule of the MPEG benchmark). In addition to this, manual coding (especially with so many FSM states) is extremely prone to errors which are very cumbersome and time-consuming to correct with (traditional) RTL simulations and debugging.

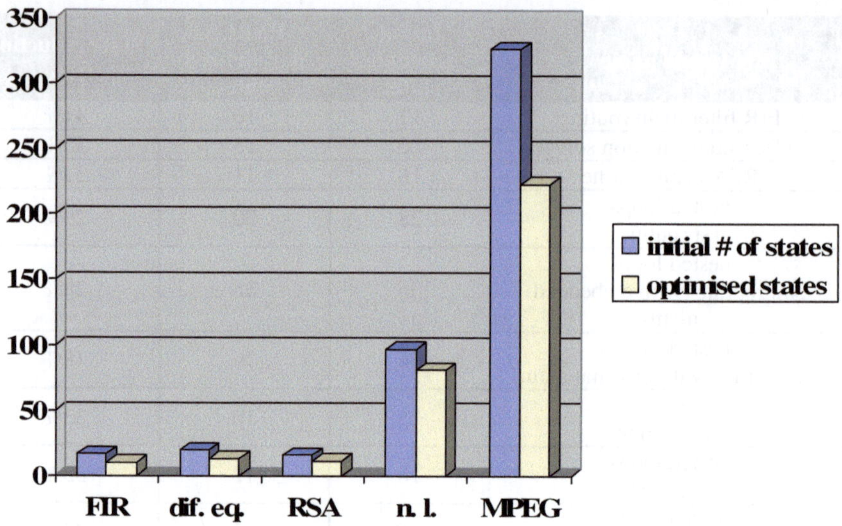

Fig. 11 Optimization (reduction) of number of states

The specification (source code) model of the various benchmarks it is an executable specification (since it is coded in ADA). All of the designs were tested using the CCC compilation flow. They include unaltered regular ADA program code, without additional semantics and compilation directives which are usual and common in other synthesis tools which compile code in SystemC, HandelC, or any other modified program code with additional object class and TLM primitive libraries. This advantage of the presented methodology eliminates the need for the system designers to learn a new language, a new set of program constructs or a new set of custom libraries.

Moreover, the programming constructs and semantics, that the prototype CCC compiler utilizes, belong to the subset, which is common to almost all of the imperative and procedural programming languages, such as ANSI C, Pascal,

Modula, Basic etc. Therefore, it is very easy for a user that is familiar with these other imperative languages, to also become familiar with the rich subset of ADA that the prototype CCC hardware compiler processes. From design experience it was found that this learning curve doesn't exceed a few days, if not hours for the very experienced software/system programmer/modeler.

9 Conclusions and Future Work

This work includes a survey and analysis of early and mature High-level and system synthesis techniques, with a special focus on formal techniques, which turn the compilation process into a formal one, and the produced hardware implementations, to correct-by-construction designs. Most of the existing HLS tools perform well in stream and data-flow oriented applications, but their performance on general problems, such as complex control-dominated and hierarchical designs, is still poor. Most of these tools translate a higher level of abstraction into RTL or structural models and they are tuned to perform well in a specific domain of designs and/or they target a specific type of microarchitecture templates. The input abstracted description can be sometimes proprietary (specialized languages or graphical/mathematical formats) or general and executable (e.g. ANSI-C or ADA). Moreover, work on specific problems of HLS such as scheduling, allocation and binding is analyzed here, as well as more specific problems such as testability in synthesis, synthesis of interconnections, synthesizing in order to reduce power consumption, synthesizing for low energy, and particular datapath vs controller trade-offs.

The main contribution of this work is a provably-correct, high-level hardware synthesis method and a unified prototype tool-chain developed by the author of this work. The CCC prototype tools are based on compiler-compiler and logic inference techniques. This makes the hardware compilation process formal. The hardware compiler accepts a rich subset of the ADA language, which is common to almost all of the imperative programming languages. The hardware compilation is enhanced with XML schema validation as well as RDF logic relations for the execution of the prototype hardware compiler via a web service.

The prototype tools transform a number of arbitrary and general input subprograms (at the moment coded in the ADA language) into a corresponding number of functionally-equivalent RTL VHDL hardware accelerator descriptions. In this way, application-specific hardware accelerators (coprocessors) are produced with the prototype CCC tools. These coprocessors can be used to accelerate specific functions of the host computing environment. Thus, custom hardware is automatically produced from algorithm-specific executable specification in the form of an ADA set of subprograms.

A large number of input applications were run through the hardware compiler, five of which were evaluated in this paper. In all cases, the functionality of the produced hardware accelerators matched that of the input subprograms. This was expected due to the formal definition/implementation of the various internal forms and transformations of the CCC compiler, including the intermediate IPF form and the inference rules of the back-end phase. Encouraging state-reduction rates of the

PARCS scheduler-optimizer were observed for five benchmarks in this paper, which exceed 30% in some cases. Using its formal flow, the CCC hardware compiler can be used to develop complex systems in orders of magnitude shorter time and with much less engineering effort, than that, which is usually required with conventional design approaches, such as RTL coding or IP encapsulation and schematic entry using custom libraries. Another advantage of the CCC method for design engineers is that the names of the ADA variables and other objects are maintained through the compilation process, which makes possible to back-track objects from the generated RTL into the source code ADA programs.

Future extensions of this work include upgrading the front-end phase to accommodate more input programming languages (e.g. ANSI-C, C++) and of the back-end HDL writer to include more back-end RTL languages (e.g. Verilog HDL), which are currently under development. Another extension could be the inclusion of more than 2 operand operations as well as multi-cycle arithmetic unit modules, such as multi-cycle operators, to be used for pipelining. Moreover, there is ongoing work to extend the IPF's semantics so that it can accommodate embedding of IP blocks (such as floating-point units) into the compilation flow, and enhance further the schedule optimizer algorithm for even more reduced schedules. Also, other compiler phase validation techniques based on formal semantic such as RDF and XML flows are used and they help to strengthen the formal features of the CCC design methodology. These techniques will be enhanced and be more generalized in their use in the framework of the CCC methodology. Furthermore, connection from the front-end phase to even more diagrammatic system modeling formats such as the UML design techniques are currently investigated.

References

[1] Pangrle, B.M., Gajski, D.D.: Design tools for intelligent silicon compilation. IEEE Trans. Comput.-Aided Des. Integ. Circuits Syst. 6(6), 1098–1112 (1987)

[2] Girczyc, E.F., Buhr, R.J.A., Knight, J.P.: Applicability of a subset of Ada as an algorithmic hardware description language for graph-based hardware compilation. IEEE Trans. Comput.-Aided Des. Integ. Circuits Syst. 4(2), 134–142 (1985)

[3] Paulin, P.G., Knight, J.P.: Force-directed scheduling for the behavioral synthesis of ASICs. IEEE Trans. Comput.-Aided Des. Integ. Circuits Syst. 8(6), 661–679 (1989)

[4] Park, I.C., Kyung, C.M.: Fast and near optimal scheduling in automatic data path synthesis. In: Proc. of the Des. Autom. Conf. (DAC), San Francisco, pp. 680–685 (1991)

[5] Park, N., Parker, A.: Sehwa: A software package for synthesis of pipelined data path from behavioral specification. IEEE Trans. Comput. Aided Des. Integrated Circuits Syst. 7(3), 356–370 (1988)

[6] Girczyc, E.F.: Loop winding—a data flow approach to functional pipelining. In: Proc. of the International Symp. on Circ. and Syst., pp. 382–385 (1987)

[7] Tseng, C.J., Siewiorek, D.P.: Automatic synthesis of data path on digital systems. IEEE Trans. Comput. Aided Des. Integ. Circuits Syst. 5(3), 379–395 (1986)

[8] Kurdahi, F.J., Parker, A.C.: REAL: A program for register allocation. In: Proc. of the Des. Autom. Conf. (DAC), Miami Beach, FL, pp. 210–215 (June 1987)
[9] Huang, C.Y., Chen, Y.S., Lin, Y.L., Hsu, Y.C.: Data path allocation based on bipartite weighted matching. In: Proc. of the Des. Autom. Conf. (DAC), Orlando, FL, pp. 499–504 (June 1990)
[10] Tsay, F.S., Hsu, Y.C.: Data path construction and refinement. In: Digest of Techn papers, Int. Conf. on Comp.-Aided Des. (ICCAD), Santa Clara, CA, pp. 308–311 (November 1990)
[11] Camposano, R., Rosenstiel, W.: Synthesizing circuits from behavioral descriptions. IEEE Trans. Comput.-Aided Des. Integr. Circuits Syst. 8(2), 171–180 (1989)
[12] Johnson, S.D.: Synthesis of Digital Designs from Recursion Equations. MIT Press, Cambridge (1984)
[13] Barbacci, M.R., Barnes, G.E., Cattell, R.G., Siewiorek, D.P.: The ISPS Computer Description Language. Rep. CMU-CS-79-137, dep. of Computer Science, Carnegie-Mellon Univ. (1979)
[14] Marwedel, P.: The MIMOLA design system: Tools for the design of digital processors. In: Proc. of the 21st Design Automation Conf., pp. 587–593 (1984)
[15] Casavant, A.E., d'Abreu, M.A., Dragomirecky, M., Duff, D.A., Jasica, J.R., Hartman, M.J., Hwang, K.S., Smith, W.D.: A synthesis environment for designing DSP systems. IEEE Des. & Test of Comput. 6(2), 35–44 (1989)
[16] Paulin, P.G., Knight, J.P.: Algorithms for high-level synthesis. IEEE Des. & Test of Comput. 6(6), 18–31 (1989)
[17] Gajski, D.D., Ramachandran, L.: Introduction to high-level synthesis. IEEE Des. & Test of Comput. 11(4), 44–54 (1994)
[18] Walker, R.A., Chaudhuri, S.: Introduction to the scheduling problem. IEEE Des. & Test of Comput. 12(2), 60–69 (1995)
[19] Berstis, V.: The V compiler: automatic hardware design. IEEE Des. & Test of Comput. 6(2), 8–17 (1989)
[20] Holub, A.: Compiler Design in C. Prentice-Hall Inc., Englewood Cliffs (1990)
[21] Fisher, J.: Trace Scheduling: A technique for global microcode compaction. IEEE Trans. on Comput. C-30(7), 478–490 (1981)
[22] Kuehlmann, A., Bergamaschi, R.A.: Timing analysis in high-level synthesis. In: Proc. of the 1992 IEEE/ACM International Conference on Computer-Aided Design (ICCAD 1992), pp. 349–354 (1992)
[23] Wilson, T.C., Mukherjee, N., Garg, M.K., Banerji, D.K.: An ILP Solution for Optimum Scheduling, Module and Register Allocation, and Operation Binding in Datapath Synthesis. VLSI Design 3(1), 21–36 (1995)
[24] Papachristou, C.A., Konuk, H.: A Linear program driven scheduling and allocation method followed by an interconnect optimization algorithm. In: Proc. of the 27th ACM/IEEE Design Automation Conf., pp. 77–83 (1990)
[25] Biesenack, J., Koster, M., Langmaier, A., et al.: The Siemens high-level synthesis system CALLAS. IEEE Trans. on Very Large Scale Integr (VLSI) Sys. 1(3), 244–253 (1993)
[26] The Electronic Design Interchange Format. Wikipedia, the free encyclopedia, `http://en.wikipedia.org/wiki/EDIF` (accessed February 4, 2011)
[27] Rubin, M.S.: Computer Aids for VLSI Design. Appendix D: Electronic Design Interchange Format (1994), `http://www.rulabinsky.com/cavd/text/chapd.html` (accessed February 4, 2011)

[28] Filkorn, T.: A method for symbolic verification of synchronous circuits. In: Comp. Hardware Descr. Lang. and their Application (CHDL 1991), Marseille, France, pp. 229–239 (1991)
[29] Kalavade, A., Lee, E.A.: A hardware-software codesign methodology for DSP applications. IEEE Des. & Test of Comput. 10(3), 16–28 (1993)
[30] Ernst, R., Henkel, J., Benner, T.: Hardware-software cosynthesis for microcontrollers. IEEE Des. & Test of Comput. 10(4), 64–75 (1993)
[31] De Micheli, G., et al.: The Olympus synthesis system. IEEE Des. & Test of Comput. 7(5), 37–53 (1990)
[32] Thomas, D.E., Adams, J.K., Schmit, H.: A model and methodology for hardware-software codesign. IEEE Des. & Test of Comput. 10(3), 6–15 (1993)
[33] Hoare, C.A.R.: Communicating sequential processes. Prentice-Hall, Englewood Cliffs (1985)
[34] Gupta, R.K., De Micheli, G.: Hardware-software cosynthesis for digital systems. IEEE Des. & Test of Comput. 10(3), 29–41 (1993)
[35] Bolsens, I., De Man, H.J., Lin, B., Van Rompaey, K., Vercauteren, S., Verkest, D.: Hardware/software co-design of digital telecommunication systems. Proc. of the IEEE 85(3), 391–418 (1997)
[36] Hilfinger, P.N., Rabaey, J., Genin, D., Scheers, C., De Man, H.: DSP specification using the SILAGE language. In: Proc. Int. Conf. on Acoust Speech Signal Process. Albuquerque, NM, pp. 1057–1060 (1990)
[37] Willekens, P., et al.: Algorithm specification in DSP station using data flow language. DSP Applicat. 3(1), 8–16 (1994)
[38] Halbwachs, N., Caspi, P., Raymond, P., Pilaud, D.: The synchronous dataflow programming language Lustre. Proc. IEEE 79(9), 1305–1320
[39] Van Canneyt, M.: Specification, simulation and implementation of a GSM speech codec with DSP station. DSP and Multimedia Technol. 3(5), 6–15 (1994)
[40] Buck, J.T., et al.: PTOLEMY: A framework for simulating and prototyping heterogeneous systems. Int. J. Computer Simulation, 1–34 (August 31, 1992/1994)
[41] Lauwereins, R., et al.: GRAPE-II: A system level prototyping environment for DSP applications. IEEE Computer 28(2), 35–43 (1995)
[42] Rafie, M.S., et al.: Rapid design and prototyping of a direct sequence spread-spectrum ASIC over a wireless link. DSP and Multimedia Technol. 3(6), 6–12 (1994)
[43] Semeria, L., Sato, K., De Micheli, G.: Synthesis of hardware models in C with pointers and complex data structures. IEEE Trans. VLSI Systems 9(6), 743–756 (2001)
[44] Wilson, R.P., et al.: Suif: An infrastructure for research on parallelizing and optimizing compilers. ACM SIPLAN Notices 28(9), 67–70 (1994)
[45] Kountouris, A., Wolinski, C.: Efficient Scheduling of Conditional Behaviors for High-Level Synthesis. ACM Trans. on Design Aut. of Electr. Sys. 7(3), 380–412 (2002)
[46] Gupta, S., Gupta, R.K., Dutt, N.D., Nikolau, A.: Coordinated Parallelizing Compiler Optimizations and High-Level Synthesis. ACM Trans. on Des. Aut. of Electr. Sys. 9(4), 441–470 (2004)
[47] Wang, W., Tan, T.K., et al.: A comprehensive high-level synthesis system for control-flow intensive behaviors. In: Proc. 13th ACM Great Lakes Symp. on VLSI GLSVLSI 2003, pp. 11–14 (2003)

[48] Gu, Z.P., Wang, J., Dick, R.P., Zhou, H.: Incremental exploration of the combined physical and behavioral design space. In: Proc. of the 42nd Annual Conf. on Des. Aut. DAC 2005, pp. 208–213 (2005)
[49] Zhong, L., Jha, N.K.: Interconnect-aware high-level synthesis for low power. In: Proc. IEEE/ACM Int. Conf. Comp.-Aided Des., pp. 110–117 (November 2002)
[50] Huang, C., Ravi, S., Raghunathan, A., Jha, N.K.: Generation of Heterogeneous Distributed Architectures for Memory-Intensive Applications Through High-Level Synthesis. IEEE Trans. on Very Large Scale Integr (VLSI) Sys. 15(11), 1191–1204 (2007)
[51] Wakabayashi, K.: C-based synthesis experiences with a behavior synthesizer, "Cyber". In: Proc. Des. Autom. and Test in Eur. Conf., pp. 390–393 (1999)
[52] Wang, W., Raghunathan, A., Jha, N.K., Dey, S.: High-level Synthesis of Multi-process Behavioral Descriptions. In: Proc. of the 16th IEEE International Conference on VLSI Design (VLSI 2003), pp. 467–473 (2003)
[53] Gal, B.L., Casseau, E., Huet, S.: Dynamic Memory Access Management for High-Performance DSP Applications Using High-Level Synthesis. IEEE Trans. Comput.-Aided Des. Integ. Circuits Syst. 16(11), 1454–1464 (2008)
[54] Wakabayashi, K., Tanaka, H.: Global scheduling independent of control dependencies based on condition vectors. In: Proc. 29th ACM/IEEE Conf. Des. Autom (DAC), Los Alamitos, CA, pp. 112–115 (1992)
[55] Gupta, S., Gupta, R., Dutt, N., Nicolau, A.: Dynamically increasing the scope of code motions during the high-level synthesis of digital circuits. In: Proc. IEEE Conf. Comput. Digit. Techn., vol. 150(5), pp. 330–337 (2003)
[56] Martin, E., Santieys, O., Philippe, J.: GAUT, an architecture synthesis tool for dedicated signal processors. In: Proc. IEEE Int. Eur. Des. Autom. Conf. (Euro-DAC), Hamburg, Germany, pp. 14–19 (1993)
[57] Molina, M.C., Ruiz-Sautua, R., Garcia-Repetto, P., Hermida, R.: Frequent-Pattern-Guided Multilevel Decomposition of Behavioral Specifications. IEEE Trans. Comput.-Aided Des. Integ. Circuits Syst. 28(1), 60–73 (2009)
[58] Avnit, K., D'silva, V., Sowmya, A., Ramesh, S., Parameswaran, S.: Provably correct on-chip communication: A formal approach to automatic protocol converter synthesis. ACM Trans. on Des. Autom. of Electr. Sys (TODAES) 14(2), article no: 19 (2009)
[59] Keinert, J., Streubuhr, M., Schlichter, T., Falk, J., Gladigau, J., Haubelt, C., Teich, J., Meredith, M.: System Co Designer—an automatic ESL synthesis approach by design space exploration and behavioral synthesis for streaming applications. ACM Trans. on Des. Autom. of Electr. Sys. (TODAES) 14(1), article no: 1 (2009)
[60] Kundu, S., Lerner, S., Gupta, R.K.: Translation Validation of High-Level Synthesis. IEEE Trans. Comput.-Aided Des. Integ. Circuits Syst. 29(4), 566–579 (2010)
[61] Falk, J., Haubelt, C., Teich, J.: Efficient representation and simulation of model-based designs in SystemC. In: Proc. of the Forum of Des. Lang., Darmstadt, Germany, September 19-22, pp. 129–134 (2006)
[62] Paik, S., Shin, I., Kim, T., Shin, Y.: HLS-l: A High-Level Synthesis framework for latch-based architectures. IEEE Trans. Comput.-Aided Des. Integ. Circuits Syst. 29(5), 657–670 (2010)
[63] Voros, N., Rosti, A., Hubner, M. (eds.): Dynamic System Reconfiguration in Heterogeneous Platforms: The MORPHEUS Approach. LNEE, vol. 40. Springer Science+Business Media B.V (2009)

[64] Rim, M., Jain, R.: Lower-bound performance estimation for the high level synthesis scheduling problem. IEEE Trans. Comput.-Aided Des. Integ. Circuits Syst. 13(4), 451–458 (1994)

[65] Chaudhuri, S., Walker, R.A.: Computing lower bounds on functional units before scheduling. IEEE Trans. Very Large Scale Integ (VLSI) Syst. 4(2), 273–279 (1996)

[66] Mecha, H., Fernandez, M., et al.: A method for area estimation of data path in high level synthesis. IEEE Trans. Comput. Aided Des. Integ. Circuits Syst. 15(2), 258–265 (1996)

[67] Fang, Y.M., Wong, D.F.: Simultaneous functional-unit binding and floorplanning. Digest of Technical Papers. In: Intern. Conf. on Comp.-Aided Des (ICCAD), Santa Clara, CA, pp. 317–321 (1994)

[68] Munch, M., Wehn, N., Glesner, M.: Optimum simultaneous placement and binding for bit-slice architectures. In: Proc. of the Asia and South-Pacific Des. Autom. Conf (ASP-DAC), Chiba, Japan, pp. 735–740 (1995)

[69] Ahmadi, A., Zwolinski, M.: Multiple-Width Bus Partitioning Approach to Datapath Synthesis. In: IEEE International Symposium on Circuits and Systems (ISCAS 2007), pp. 2994–2997 (2007)

[70] Gupta, R., Breuer, M.A.: Partial scan design of register-transfer level circuits. Journal of Electr. Test Theory and Appl. 7(1-2), 25–46 (1995)

[71] Flottes, M.L., Hammad, D., Rouzeyre, B.: High level synthesis for easy testability. In: Proc. European Des. and Test Conf., Paris, France, pp. 198–206 (1995)

[72] Dhodhi, M.K., Hielscher, F.H., Storer, R.H., Bhasker, J.: Datapath synthesis using a problem-space genetic algorithm. IEEE Trans. Comput-Aided Des. Integ. Circuits and Syst. 14(8), 934–944 (1995)

[73] Mujumdar, A., Jain, R., Saluja, K.: Incorporating testability considerations in high level synthesis. Journal of Electr. Test Theory and Appl. 5(1), 43–55 (1994)

[74] Potkonjak, M., Dey, S., Roy, R.K.: Synthesis-for-testability using transformations. In: Proc. of the Asia and South-Pacific Des., pp. 485–490 (1995)

[75] Hsu, F.F., Rudnick, E.M., Patel, J.H.: Enhancing high-level control-flow for improved testability. Dig. of Techn. Papers. In: Intern. Conf. on Comp.-Aided Des (ICCAD), San Jose, CA, pp. 322–328 (November 1996)

[76] Raghunathan, A., Dey, S., Jha, N.K.: Register-transfer level estimation techniques for switching activity and power consumption. Dig. of Techn. Papers. In: Intern. Conf. on Comp-Aided Des (ICCAD), San Jose, CA, pp. 158–165 (November 1996)

[77] Rabaey, J., Guerra, L., Mehra, R.: Design guidance in the power dimension. In: 1995 Intern. Conf. on Acoustics, Speech, and Signal Proc., pp. 2837–2840 (1995)

[78] Mehra, R., Rabaey, J.: Exploiting regularity for low-power design. Dig. of Techn. Papers. In: Intern. Conf. on Comp.-Aided Des (ICCAD), San Jose, CA, pp. 166–172 (November 1996)

[79] Raghunathan, A., Jha, N.K.: Behavioral synthesis for low power. In: Proc. of the Intern. Conf. on Comp. Des (ICCD), pp. 318–322 (1994)

[80] Goodby, L., Orailoglu, A., Chau, P.M.: Microarchitecture synthesis of performance-constrained low-power VLSI designs. In: Proc. of the Intern. Conf. on Comp. Des. (ICCD), pp. 323–326 (1994)

[81] Musoll, E., Cortadella, J.: Scheduling and resource binding for low power. In: Proc. the Eighth Symp. on Sys. Synth., pp. 104–109 (1995)

[82] Kumar, N., Katkoori, S., Rader, L., Vemuri, R.: Profile-driven behavioral synthesis for low-power VLSI systems. IEEE Des. Test Comput. 12(3), 70–84 (1995)

[83] Martin, R.S., Knight, J.P.: Power-profiler: Optimizing ASICs power consumption at the behavioral level. In: Proc. of the Des. Autom. Conf (DAC), San Francisco, CA, pp. 42–47 (1995)

[84] Issenin, I., Brockmeyer, E., Durinck, B., Dutt, N.D.: Data-Reuse-Driven Energy-Aware Cosynthesis of Scratch Pad Memory and Hierarchical Bus-Based Communication Architecture for Multiprocessor Streaming Applications. IEEE Trans. on Comp.-Aided Des. of Integr. Circ. and Sys. 27(8), 1439–1452 (2008)

[85] Rao, D.S., Kurdahi, F.J.: Controller and datapath trade-offs in hierarchical RT-level synthesis. In: Proc. of the Int. Symp. on High Level Synth., pp. 152–157 (1994)

[86] Huang, S.C.Y., Wolf, W.H.: How datapath allocation affects controller delay. In: Proc. of the Intern. Symp. on High Level Synth., Niagra Falls, Canada, pp. 158–163 (1994)

[87] Dossis, M.: Intermediate Predicate Format for design automation tools. Journal of Next Generation Information Technology (JNIT) 1(1), 100–117 (2010)

[88] Nilsson, U., Maluszynski, J.: Logic Programming and Prolog, 2nd edn. John Wiley & Sons Ltd. (1995)

Vision Based Semantic Analysis of Surveillance Videos

Virginia Fernandez Arguedas, Qianni Zhang, Krishna Chandramouli, and Ebroul Izquierdo

Multimedia and Vision Research Group,
School of Electronic Engineering and Computer Science,
Queen Mary, University of London,
Mile End Road, London, E1 4NS, UK
{virginia.fernandez,qianni.zhang,krishna.chandramouli, ebroul.izquierdo}@eecs.qmul.ac.uk

Abstract. As recent research in automatic surveillance systems has attracted many cross-domain researchers, a large-number of algorithms have been proposed for automating surveillance systems. The objective of this chapter is twofold: First, we present an extensive survey of different techniques that have been proposed for surveillance systems categorised into motion analysis, visual feature extraction and indexing. Second, an integrated surveillance framework for unsupervised object indexing is developed to study and evaluate the performance of visual features. The study focuses on two characteristics highly related with human visual perception, colour and texture. The set of visual features under analysis comprises two categories, new leading visual features versus state-of-the-art MPEG-7 visual features. The evaluation of the framework is carried out with AVSS 2007 and CamVid 2008 datasets.

1 Introduction

In a world of heightened vandalism and terrorist activity, video surveillance forms an integral part in ensuring safety and security from citizens. Due to the deployment of 24/7 video surveillance systems, trails of evidence prior to the incident are captured and recorded by various surveillance systems, such as traffic control systems, public transportation terminals systems (e.g. bus terminals, train stations, airports, etc) and private surveillance systems (e.g. installed in shops, banks ATMs, stadiums, etc). In all the developed countries of the world surveillance cameras are working 24/7. For instance, UK is considered one of the countries with the greatest amount of closed circuit television surveillance (CCTV). In 2003, Norris and McCahill [74] estimated UK accommodated 4.2 million surveillance cameras. As surveillance systems grow in scale, heterogeneity and utility, there is an increasingly critical need to provide automated and smart surveillance solutions. While combining archived (surveillance video) content from different compression formats, indexing systems, data storage format sources.

I.E. Anagnostopoulos et al. (Eds.): Semantic Hyper/Multimedia Adaptation, SCI 418, pp. 83–125.
springerlink.com

Dating back to the origins of video surveillance in 1950s, police started using such systems and, for instance, in 1960 the London Metropolitan police erected two pan-tilt-and-zoom cameras in Trafalgar Square to monitor the crowds [73]. At the beginning, officers were in charge of controlling surveillance cameras at all times. However, this method was a time consuming task as well as a high waste of resources. At present, CCTV has become ubiquitous and people are being watched by surveillance cameras almost everywhere. In fact, video surveillance systems are one of the main sources of information and security due to their wide-spread and increasing presence in all countries.

Due to the extensive usage of surveillance systems, a lot of efforts have been dedicated to develop algorithms and techniques for analysis, indexing and search in surveillance applications. However, it still remains a challenging problem due to many reasons. First of all, surveillance databases are usually of enormous sizes and contain a lot of redundancies. Thus, the ability to efficiently and effectively process large volumes of surveillance datasets is needed, without however requiring on too much human involvement. Secondly, surveillance data often have limited quality because of capturing conditions, compressions, etc. The semantic analysis of such videos requires intelligent algorithms for indexing and classification based on objects, events, or other types of semantic queries. Moreover, semantic events are usually complex and extremely difficult to define even in high-quality multimedia content. Nowadays, content based image retrieval (CBIR) in high-quality image databases is still an open research problem in the field. These challenges motivated our work on studying the state-of-the-art researches and developing new solutions for surveillance video analysis and management.

In this chapter, we first present a brief survey of state-of-the-art techniques that are related to surveillance video analysis. The aim is to make a non-exhaustive study on the popular algorithms and summarise their advantages and deficiencies. Then, as a foundation for our further research on surveillance data understanding, a content-based classifier based on *Support Vector Machines (SVMs)* is first developed for object detection in surveillance videos. The classification of objects is based on blobs extracted using motion based background subtraction. Furthermore, several established and novel visual descriptors are employed in order to study their characteristics and capabilities in the particular application scenario of surveillance video understanding. Performance evaluation and analysis of the object classifier and these visual features is presented.

The rest of the chapter is organised as follows. In Section 2, surveillance videos limitations and constraints are discussed as well as their repercussions. Section 3 presents a literature review of techniques involved in state-of-the-art classification algorithms and systems, as well as, a review of the popular techniques for object and event classification. In Section 4, the developed object classifier for surveillance scenarios is introduced and a brief description of the visual features used in this research is presented. Section 5 describes the experimental setup, database, and the evaluation and analysis of obtained results. Finally, Section 6 discusses the conclusions drawn from our studies and proposes ideas for future work.

2 Surveillance Videos Characteristics

CCTV video systems are limited due to their installation set-up and their high dependence on environmental factors. As a consequence, videos with low resolution and quality are provided, making CCTV data analysis a challenging research problem. Moreover, two different cases must be considered. First, the CCTV videos used by surveillance operators are generally good quality since they are directly connected to the control room and are therefore, live surveillance video footage. And second, surveillance videos that are stored onto a digital video recorder are excessively compressed to reserve disk space or to facilitate its transmission. There are a large number of installation variables that affect the end quality of the surveillance video, for instance, excessive digital video compression, incorrect configuration, low quality recording equipment, cameras' position or even their location. Due to such installation constraints, the general characteristics of surveillance videos include (i) low resolution of the videos, (ii) lack of contrast, (iii) different types of noise or disturbances, (iv) blurring caused by motion or lack of focus (v) object occlusions and (vi) geometric distortions [53].

Generally, two different types of surveillance videos can be distinguished, indoor and outdoor videos, each kind presents different characteristics and constraints. On one hand, indoor surveillance videos are affected by changes in the illumination of the room (switching the lights on and off), light perturbation introduced by noise in CCD cameras, reflections in windows or external changes in illumination are some examples. While on the other hand, outdoor videos are affected by many other external conditions such as the various levels of illumination at different times of the day, adverse weather condition (fog, rain, snow) or static objects that move due to the weather conditions, i.e. moving trees due to the wind.

3 Literature Review

Surveillance systems are gradually becoming more independent and the level of human intervention is steadily reducing. Surveillance video systems are categorised into three types which reflect the level of integration of the surveillance operator within the system:

- *Operator-controlled video surveillance systems* [92] are the most basic CCTV video surveillance systems. Typically, they consist of a collection of video cameras (mounted in fixed positions) that cover a circumscribed area defined by the field-of-view of the used video cameras. The video streams are transmitted to a central location where the video stream will be recorded and displayed on one or several video monitors. The main characteristic of this surveillance video system is the necessity of constant human supervision to determine whether the detected event requires a response or if it is a false alarm.

- *Basic automated video surveillance systems* aim to reduce the burden on the surveillance officer by employing video motion detectors. Consequently, some false alarms are detected and rejected by the system without manual/human intervention. However, a large amount of motion information is provided, relevant and non-relevant, in order to determine and extract only the relevant information smart video surveillance systems are necessary.
- *Smart video surveillance systems*, also called, *automated surveillance video systems* achieve more than motion detection. Object detection, classification, tracking, localisation and behaviour interpretation are aimed by automated surveillance video systems in order to classify objects, analyse object actions and interpret activities in real time.

Surveillance cameras have become pervasive in various spaces and record 24/7. The captured information is large, creating a critical need for isolating relevant information. A general scheme followed by surveillance object and event classifiers is shown in Figure 1. Surveillance cameras provide a large amount of information to process, however, information can be efficiently analysed if only the relevant information is isolated and extracted. Spatial segmentation is performed in the *motion detection stage* providing the extracted moving objects as the relevant information for its posterior analysis. However, moving objects include spatio-temporal information. In order to enhance the performance of any surveillance system, temporal information should be considered. *Object tracking* is performed to extract the temporal information of each individual moving object. Visual information is generally indexed for its storage and further use, in order to index moving objects, several features can be extracted, *visual analysis stage* consists of the moving object feature extraction as a useful compliment to represent each moving object.

In this section, a list of techniques commonly encountered in visual surveillance systems is presented. In general these techniques can be categorised into *object classification and event detection*. Moreover, a review of popular features and well-known machine learning algorithms for classification are presented. However, to facilitate the understanding of this review, several concepts must be clarified. First, all the relevant objects for surveillance analysis which include spatio-temporal information are defined as *moving objects*. Each moving object appears in several temporally consecutive frames, each representation of the moving object is denominated as *blob* and its geometrical representation is called *bounding box*. Finally, when the moving object has been indexed and classified, according to its temporal and visual information, it is defined as a *semantic object*.

The review is structured following the same flow as a general object or event detection framework in surveillance scenarios (refer to Figure 1). First, an analysis of the existing motion detection and spatial segmentation techniques is presented in Section 3.1. After, a review of the most important object tracking algorithms is presented in Section 3.2. In Section 3.3, a survey of some popular visual features is presented in conjunction with some of their most representative applications. An exhaustive analysis of the actual situation of object and event classification in surveillance scenarios is presented in Section 3.4.

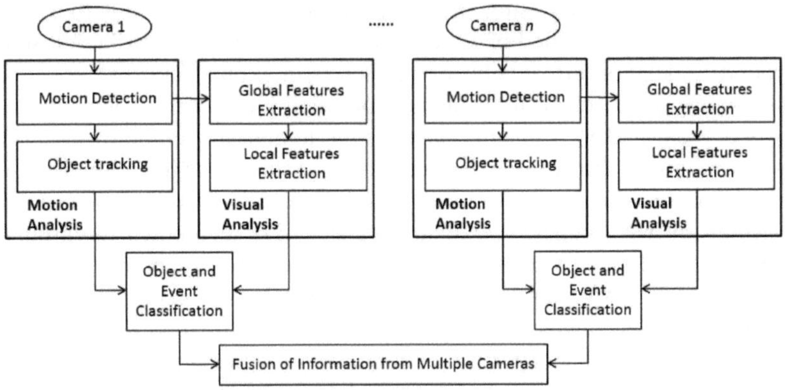

Fig. 1 Object and event detection framework for surveillance scenarios

3.1 Motion Analysis

Visual surveillance systems use fixed cameras providing videos with static background. Each surveillance video contains an enormous amount of data, requiring an exhaustive analysis. To reduce the analysis complexity, motion detection stage consists of spatially segmenting the moving objects which contain relevant information from the rest of a frame. To facilitate subsequent steps a background-related information removal is required, relieving difficulties in the posterior analysis stage. In the literature, motion detection algorithm is commonly categorised into environmental modelling and motion segmentation. Candamo et. al [16] proposed a motion detection classification which grouped algorithms in three categories, *background subtraction*, *temporal differencing* and *optical flow*.

Background subtraction is the most popular object spatial segmentation strategy. Background subtraction techniques generate a pixel-wise mathematical model to represent the background. Each frame of the video is compared with the simulated mathematical model so the foreground can be extracted. Consequently, the extracted foreground is composed of all the pixels or group of pixels with significant differences between the modelled background and the current image, considering them as moving blobs (moving information). Several challenges must be faced by background subtraction algorithms to protect the accuracy of the subsequent steps, for instance, changes in illumination, waving trees, camouflage, bootstrapping, sleep/walking person or shadows [104]. The most well-known background subtraction algorithms include methods by (i) Heikka and Silven [47] where the background model is updated using a recursive filter and monitoring the changes in activity during a few consecutive frames to eliminate pixels with random performance. The use of recursive filter enables the adaptation to slow changes in the lighting whilst monitoring each pixel performance provides robustness against waving trees effects. (ii) Stauffer and Grimson [98] proposed modelling each pixel as a mixture of Gaussians and using an on-line K-means approximation to update the model. Each pixel is classified based on whether its Gaussian distribution suits the background model most effectively or not. Such estimation is robust against changes in illumination, slight sensor movements, noise and long-term scene changes. (iii) Halevy and Weinshall [45] linearly subtract the temporal average of the previous frames from the new frame adjusting the model to slow changes in the image. (iv) Cutler and Davis [29] calculated a background model based on each pixel 1-D temporal median over N frames. (v) Toyama et al. [104] segmented foreground objects applying pixel-level Wiener filtering to make probabilistic prediction about the expected background, a frame-level component ensures its accurate performance against global changes and swaps in the image. (vi) Cheung and Kamath [25] presented a slow adapting Kalman filter in conjunction with statistics based on an elliptical moving object model to model the background over time. A detailed review of background subtraction techniques can be found in [80].

Temporal differencing is based on temporally local events. The procedure consists of establishing a sampling period to separate video frames and calculate pixel-wise differences between consecutive sampled frames to locate changed regions. Temporal differencing algorithms adapt to dynamic environments but cannot extract all the relevant pixels [48]. Due to such disadvantage, few approaches perform motion detection based on temporal differencing [60, 66], instead hybrid approaches based on a combination of temporal differencing and background subtraction have been presented providing more robust segmentation techniques [27].

Optical flow estimates motion in video by matching points on objects over time to detect moving regions in an image sequence. Optical flow based methods provide invariance to camera motion and robustness against simultaneous object and camera movement, enabling the analysis of crowds and dense motion situations [66]. However, most optical flow calculation methods are computationally complex and sensitive to noise. A detailed list of popular optical flow-based techniques is presented in [16].

3.2 Object Tracking

Once spatial segmentation has occurred and several moving blobs per frame have been detected, surveillance video systems require a further step, *object tracking*, to establish the segmented blobs temporal evolution and to focus the subsequent analysis on the essential information. Tracking blobs over time consists of matching such blobs in consecutive frames using representative features such as points, lines or blobs [115]. Object tracking algorithms can be classified depending on the features used to match the blobs within a video sequence. In the following paragraphs, a classification of the different existing tracking algorithms based on their object representation is presented. A detailed survey on this topic can be found in [114].

Tracking algorithms based on point detection focus the matching on information related to the blob position and movement. In general, point representation for tracking is suitable for objects that occupy small regions in an image. Occlusions, misdetections, entries and exits of objects affect point correspondence difficulting object tracking. Numerous approaches use point representation, categorised into two strategies, *deterministic* and *statistical*. The former, use qualitative motion heuristics to reduce the correspondence constraints [94]. While the latter, considers object measurements such as position, velocity or acceleration rate to create a statistical model. The most well-known model prediction techniques are based on (i) Kalman filter [83], (ii) particles filter [69], (iii) joint probabilistic data association filters [82] and (iv) multiple hypothesis tracking [50].

Kernel-based tracking techniques associate different blobs calculating the kernel movement over the frames of the video. Kernel-based techniques are typically classified in *model-based* and *multiview*. The former, model-based techniques, achieves blob matching performing an exhaustive search of a template over a whole frame [90] or modelling the blobs appearance within the frame [28]. Ease of implementation and low computational cost have enhance these techniques proliferation. The latter, multiview techniques, trains algorithms with different views of a blob to differentiate them. The most well-known approach applied *support vector machines* to distinguish and track an object [4].

Object trackers based on silhouette represent blobs using models of their shape and density of their appearance. The existing techniques are classified into two types depending on their matching algorithm, *silhouettes matching* and *outline evolution*. The former, *Silhouette matching techniques*, compares all existing shapes in a frame with the query's shape. Silhouettes are rigid features, however, any object silhouette evolves during time. In order to facilitate the matching, silhouettes are recalculated every certain time. Some examples of these techniques are the algorithms presented by Huttenlocher [51] and Haritaoglu [46]. The latter, *Outline evolution techniques*, represents each blob by their outline. Unlike *silhouettes matching*, outlines are dynamic and evolve according to the blob's energy to a range of possible states. Some examples of these techniques are the algorithms suggested by Chen [24], Bertalmio [9] and Yilmaz & Shah [115].

3.3 Visual Analysis

Following the extraction of moving objects, the next step involves recognising these objects for semantic indexing. In order to facilitate semantic indexing of objects, feature extraction is an important step. The extracted features should provide two advantages, compact and computationally not exhaustive. Two different applications are often considered in surveillance video systems, (i) real-time surveillance video processing and real-time detection of events and (ii) archiving surveillance moving objects and events for their posterior analysis or for future use. Each application have different crucial requirements, real time applications require a high computational efficiency and this advantage is not fundamental for archiving purposes whilst archiving surveillance applications require a greater compression and compactness of the data to store and the processing time is an optional advantage. Other desired characteristics include invariance to external factors to provide robustness to the surveillance system.

Over past several decades, many different approaches have been proposed to automatically represent moving objects in videos. However, surveillance videos characteristics set several limitations as previously discussed in Section 2. Considering such constraints several features have been popularly used in surveillance video applications, distinguishing two categories *global* and *local* features. The global features exploit temporal information and are a compact spatially segmented moving object representation whose performance is computationally less expensive to the local features when an object is well segmented [62]. However, global features require a good segmentation and are sensitive to occlusion and clutter. Unlike global features, local features do not require segmentation and are robust against occlusions, however, they are affected by object pose variety and require exhaustive object matching algorithms. These constraints show generally that no feature would describe perfectly every moving object in every scenario, instead the feature selection should be performed after a study of the scenario and its specific requirements. In the following paragraphs the most popular visual features in surveillance applications, both global and local, are described. Due to the extensive amount of developed global features, in the following paragraphs a descriptive study is presented. However, local features are more concrete, therefore, our study focuses on the description of the most well-known local features, *SIFT* and *SURF*.

3.3.1 Global Features

Visual features typically applied on surveillance video applications can be classified into three classes: colour, texture and shape. MPEG-7 standard [1] provides several global descriptors related to these features, enhancing the need of accurate compact representation [95]. The use of MPEG-7 to describe semantic objects produced in surveillance scenarios was first introduced in 2003 in the IEEE Symposium on Intelligent Distributed Surveillance Systems [8].

Before analysing some popular global features, two approaches on how to calculate global features must be distinguished. In literature, there are two levels where to calculate global features, from whole frames or from segmented moving objects. In this survey, we focus on global features calculated over segmented moving objects as visual analysis (see Figure 1) for object classifiers depend on the *motion analysis* component (see Section 3.1).

[1] MPEG-7, Multimedia Content Description Interface, is an ISO standard which facilitates image representation by using some image features such as the image dominant colour or its texture [95]. Moreover, MPEG-7 provides a simple and efficient numeric method to represent images through the usage of descriptors.

Among various types of visual descriptions, colour is one of the most researched. In the literature, many methods for moving object classification are based on colour similarity have been proposed [86, 89]. These methods are mostly derived from the basic idea of colour histogram, which shows the proportion of pixels in each segment of a colour spectrum within the image, and use histogram intersection for matching. Following this approach, several improved versions of it have been proposed. Cumulative colour histograms and colour moments were introduced with the motivation hat most of the information in a histogram could be summarised using low-order moments [99]. In [2], average colour vectors of some perceptually prominent colour areas extracted by a recursive HSV-space segmentation technique were used for image indexing. Finally, another pioneer approach was presented in [42], where the developed colour models were invariant to the viewpoint, geometry of the object and illumination. In surveillance scenarios, numerous approaches based on colour analysis have been presented. MPEG-7 colour descriptors are quite popular and several approaches perform object classification, indexing and retrieval extracting features like *Colour Layout Descriptor (CLD), Dominant Colour Descriptor (DCD) or Scalable Colour Descriptor (SCD)* [3, 40]. However, other colour descriptors have been used in the literature, for instance, in [14], authors presented a retrieval algorithm based on object cumulative colour histogram in bi-conic HSL space while in [112] objects are classified into one of eleven "culture" colours using a colour-drift table, which estimates the distance between individual colours, to train a support vector classifier operating on pixel, frame and sequence level.

Apart from colour features, texture is a key component of human visual perception and is one of the most popular characteristics in object classification, indexing and retrieval systems. Texture recognition is an easy task for human beings; however, it is difficult to define. Unlike colour features, texture features occur over a region rather than at a point. To make it orthogonal to colour, textures are usually defined purely by grey levels . In the literature, a variety of texture descriptors has been proposed. In [101], six texture features, named *coarseness, contrast, directionality, regularity, line-likeness* and *roughness*, were devised corresponding to human visual perception. In [63], images were decomposed into three mutually orthogonal components, *periodicity, directionality* and *randomness*, which were considered the most perceptually important texture properties. A popular alternative to the previously mentioned texture features, the Gabor filter-based multi-resolution representations were used to extract texture information [70, 71] considering the means and standard deviation of the Gabor transform coefficients as features. In surveillance scenarios, the variety of approaches is smaller compared with general videos, mainly due to the surveillance videos specific constraints. A numerous group of approaches use MPEG-7 texture features [3, 58, 40]. Another popular texture feature for surveillance scenarios is Gabor texture descriptor. In VISOR [44], authors used Gabor texture filters in conjunction with colour features to perform pedestrian recognition. In SUNAR [26], a set of global and local features were used to perform object recognition, its texture descriptor is based on extraction of energy from the frequency domain bands defined by a bank of Gabor filters. Other texture features have been proposed in surveillance applications, for instance, in [31], authors perform address human detection using a set of texture-based features such as histograms of oriented gradients, in fact, they compute high dimensional features based on edges and use *Support Vector Machines* to detect human regions.

Apart from colour and texture, the geometry of extracted moving objects has been investigated to classify objects in surveillance videos. Shape-based classification exploits features like size, compactness, aspect ratio and simple shape descriptors from the segmented objects [22]. In general scenarios, different shape descriptors have been presented. Zernike Moments [75] were presented as a suitable method to provide invariance to scale, translation and rotation and a good representation for more complex shape representations because of their independence to the boundary information. Grid descriptors [65] project a shape onto a grid of fixed size and depending on if the cells of the grid are partially or wholly covered by the shape or not, 1 or 0 is respectively assigned. In surveillance scenarios, shape features present two disadvantages, (i) high sensitivity to the presence of shadows and occlusions, as the object shape is altered; and (ii) low discriminant power between categories with similar shape such as vehicle and group of people. Lipton and Fujiyoshi [60] classify objects into *human, vehicles* and *clutter* describing moving objects by their dispersedness and area. To improve the results precision, temporal consistency constraints are applied. Collins and Lipton [27] extract several shape-based features, moving object dispersedness, area and bounding box apparent aspect ratio, to perform object classification using neural networks classifier and considering four classes, *human, vehicle, group of people* and *clutter*. In [10], authors classify semantic moving objects based on their calculated human height/width ratio based on data from the National Center for Statistics.

Several approaches based their classification on the idea that object motion characteristics and patterns are unique enough to distinguish between objects [16]. Humans have been shown to have distinct types of motion that can be used to recognise different human movements such as walking, running or fighting. In 1995, one of the earliest approaches in motion-based classification used trajectory representation for video retrieval [33]. Their approach involved constructing hierarchical motion descriptions from MPEG video data. Several approaches were proposed for compact motion representation exploiting wavelet or Fourier transforms, respectively [30, 85]. However, these approaches presented loss of valuable information. Bashir et al. [5] proposes a global trajectory representation consisting of PCA coefficients whose results provide superior precision-recall ratios than non-PCA based approaches. However, global trajectory representation is unable to handle partial trajectory matching, which is overcome by segmented trajectory representation [5]. Different algorithms have been presented for segmented trajectory representation, some based on MPEG-7 motion descriptors [23] and others based on curve fitting and string matching for segmented trajectory representation and retrieval [6, 57]. In [36], authors present a detailed review of motion-based object classifiers. Unlike in general videos, in surveillance scenarios, trajectory, velocity or, in general, motion features have not been deeply exploited for object classification [39].

3.3.2 Local Features

In recent years, local features have become popular due to the fact that they provide robustness against small viewpoint changes and partial occlusion. Moreover, local feature matching can recognise objects anywhere in an image with arbitrary size and rotated without need of a segmentation stage [89]. Two popular recent local features are *Scale Invariant Feature Transform (SIFT)* and *Speeded Up Robust Features (SURF)*.

Scale Invariant Feature Transform (SIFT) [64] appeared as a semantic object representation based on the object's appearance and the location of its keypoints. SIFT addresses object recognition detecting and extracting local feature descriptors which present some invariance to image noise, change of illumination, uniform scaling, rotation and minor changes in viewing direction. The local keypoints detection consists of a four-stage filtering approach:

1. *Scale-space extrema detection* pre-selects candidate keypoints using a cascade filtering approach. The identified candidates are relative locations independent to a change of scale and are detected searching for stable features across all possible scales, σ, using the Gaussian function. The scale space of an image, $L(x, y, \sigma)$, is defined as the convolution of a variable-scale Gaussian, $G(x, y, \sigma)$, with an input image, $I(x, y)$ (see Equation 1). In order to efficiently detect stable keypoint locations in the scale-space, Lowe proposed calculating scale-space extrema, $D(x, y, \sigma)$, as a difference of Gaussian functions convolved with the image (see Equation 2). The calculation of the maxima and minima of the scale space extrema consists of comparing each pixel of the pyramid with its neighbouring scales. The current pixel is compared with its 8 neightbours in the same scale, then with its 9 neighbours of the inferior scale and, finally, with its 9 neighbours of the superior scale. This process stops whenever the pixel is not detected as maximum or minimum, discarding the candidate pixel.
$$L(x, y, \sigma) = G(x, y, \sigma) * I(x, y) \quad (1)$$
$$D(x, y, \sigma) = (G(x, y, k\sigma) - G(x, y, \sigma)) * I(x, y) = L(x, y, k\sigma) - L(x, y, \sigma) \quad (2)$$
2. *Accurate keypoint localisation* provides the location of each candidate keypoint, $\hat{x}$ (see Equation 3). The location is calculated applying the Taylor expansion to the scale-space extrema, $D(x, y, \sigma)$, and setting its derivate to zero (considering $x = (x, y, \sigma)^T$). The extrema location can be interpolated adding the calculated offset, $\hat{x}$, and the location of its sample point. Besides the calculation of the keypoint location, in this stage an accurate candidate keypoint filtering is procured, rejecting unstable extrema with low contrast or located on a frame border.
$$\hat{x} = -\frac{\partial^2 D^{-1}}{\partial x^2}\frac{\partial D}{\partial x} \quad (3)$$

3. *Orientation assignment* provides invariance to image rotation by representing the candidate keypoints relative to their orientation. To calculate the orientation of each candidate keypoint, the gradient magnitude, $m(x,y)$, and gradient orientation, $\theta(x,y)$, are computed using pixel differences (refer to Equations 4 and 5 respectively). A 36-bin orientation histogram is formed from the gradient orientation. The histogram peaks, which correspond to the dominant direction of the local gradients, are detected. Thresholding is applied to the peak's magnitude, resulting in the storage of the keypoints coupled with their orientation. Finally, an interpolation of the gradient magnitude of the three closest peaks is calculated and assigned to the keypoint to achieve higher accuracy.

$$m(x,y) = ((L(x+1,y) - L(x-1,y))^2 + (L(x,y+1) - L(x,y-1))^2)^{\frac{1}{2}} \quad (4)$$

$$\theta(x,y) = tan^{-1}\left(\frac{(L(x,y+1) - L(x,y-1))}{((L(x+1,y) - L(x-1,y))}\right) \quad (5)$$

4. *Local image descriptor* calculation stage consists of computing a highly distinctive descriptor for the local image region, invariant to changes in illumination and 3D viewpoint. To calculate the keypoint descriptor, a 6-step process is applied to every candidate keypoint. First, the gradients' magnitude and orientation are sampled around the keypoint location. Second, a Gaussian weighted function is applied to all the grandiant magnitudes and orientations corresponding to the defined local area to emphasize the gradients close to the centre of the descriptor. Third, a 8-bin orientation histogram is calculated in the local area. Fourth, in order to avoid all boundary effects where the descriptor abruptly changes, trilinear interpolation is applied to distribute the value of each gradient sample into adjacent histogram bins. Fifth, the 128-descriptor is formed from a vector containing the values of all the orientation histogram entries. Finally, in order to reduce the effects of illumination changes and illumination changes over 3D surfaces, the vector is normalised to the unit length and all the gradient magnitudes overpassing a fixed threshold are removed.

Originally, SIFT appeared as a object recognition technique where no segmentation technique was required. However, in [38], authors demonstrated the benefits of a preprocessing stage consisting on spatial segmentation.

Speeded Up Robust Features (SURF) [7] is a robust point descriptor partially inspired in SIFT. Their main difference lie in the candidate keypoint detection, while SIFT uses a cascade Gaussian filtering approach, SURF is based on a Hessian matrix approximation. SURF is based on the same principles than SIFT using relative strengths and orientations of gradients to reduce the effect of illumination changes. However, in SURF its complexity is reduced.

SURF descriptor is extracted in a three-stage process. First, the keypoint candidates are located using a *Fast-Hessian Detector*. Second, the keypoint *orientation assignment* fixes a reproducible orientation based on information from a circular region around the candidate keypoint. Finally, a square region aligned to the selected orientation is built to extract the *SURF descriptor*.

1. *Fast-Hessian Detector:* SURF presents a detector based on the determinant of the Hessian matrix to calculate location and scale, providing a good performance in computation time and accuracy. Considering that SIFT descriptor substituted Gaussian filters with a Laplacian of Gaussians approximation, SURF proposes an approximation based on box filters. Such filters approximate the second order Gaussian derivatives, namely D_{xx}, D_{yy}, D_{xy}, achieving fast evaluation of images. The proposed approximation allows a fast approximate calculation of the determinant of the Hessian matrix for the estimation of the location and scale of the candidate keypoint (see Equation 6).
$$det(H_{approximation}) = D_{xx}D_{yy} - (0.9D_{xy})^2 \tag{6}$$
Scale spaces are typically implemented as image pyramids. However, the use of box filters and integral images, avoids the iterative application of the same filter to the output image. Instead, SURF applies box filters with different sizes in parallel accelerating the calculation process. Hence, the filter size is doubled between consecutive filter sizes, as the sampling intervals for the extraction of candidate keypoints. In order to localise interest keypoints in the image and over scales, a non-maximum suppression in a 3x3x3 neighbourhood is applied. Finally, the maxima of the Hessian matrix are interpolated in scale and image space, as an approximation to the difference between the scales computed in SIFT.
2. *Orientations Assignment:* invariance to rotation is provided by calculating the orientation of the candidate keypoints. Orientation assignment is addressed in a three-stage process. First, Haar-wavelet responses are computed in x and y directions and weighted with a Gaussian centred at the candidate keypoint. Second, the responses are represented as vectors in a 2D space, locating the horizontal responses on the abscissa and the vertical responses on the ordinate. These responses are used to calculate the dominant orientation as a sum of all the responses within one sliding orientation window covering an angle of $\pi/3$. Third, the horizontal and vertical responses obtained in the sliding window are summed yielding a new vector which defines the orientation of the keypoint candidate.
3. *SURF Descriptor* extraction is a process based on Haar wavelets. Once the keypoint orientation is computed, a square region centred in the keypoint and oriented according to its orientation is constructed. Such region is divided into 4x4 sub-regions. For each sub-region the Haar wavelet response in horizontal and vertical direction (d_x and d_y, respectively) is calculated in a subsampled space. Four measures are computed to form the feature vector, V, of each sub-region: (i) the sum of the Haar wavelet responses in the horizontal direction, $\sum d_x$; (ii) the sum of the Haar wavelet responses in the vertical direction, $\sum d_y$; (iii) the sum of the d_x absolute values, $\sum |d_x|$, and (iv) the sum of the d_y absolute values, $\sum |d_y|$. Finally, a four-dimensional descriptor vector is provided to define each sub-region, $V = \{\sum d_y, \sum d_x, \sum |d_x|, \sum |d_y|\}$, along with invariance to changes in scale, orientation and illumination.

An upright version of SURF descriptor, U-SURF, was also proposed in [7], providing a descriptor non-invariant for image rotation but computationally faster and more suitable for applications where the camera remains horizontal.

3.4 Object and Event Classification

Different segmented and tracked moving regions may correspond to different semantic objects in real scenarios. Typically two scenarios are distinguished in surveillance videos, *indoor* and *outdoor* (refer to Section 2 for their specific characteristics and constraints). Indoor surveillance videos detect a short list of object categories prevailing humans. Outdoor surveillance videos, unlike indoor videos, detect a wider range of objects, such as car, bus, bicyclist, pedestrian or animal. Object classification can be considered as a standard pattern recognition issue affected by problems coming from the motion detection stage, such as noise, occlusions, scene illumination variations or waving trees. Object classification aims to categorise each moving object segmented in the motion detection stage and eliminate false alarms[2]. Relevant work in this area includes numerous machine learning algorithms such as boosting, decision trees, *Support Vector Machines (SVMs)* or biologically inspired classifiers [18]. In surveillance scenarios, most commonly used machine learning algorithms include Support Vector Machines [77], Naive Bayes Classification [87] and AdaBoost [108].

In surveillance systems, event and activity recognition can be ambiguous and highly dependable on the scene context. The same behaviour might have different meanings depending on the video environment or the task under analysis. Several workshops have focused in event detection. For instance, PETS [3] provides annually a surveillance dataset to create an scenario where to compare different techniques from different researchers interested in the area of visual surveillance. Surveillance event detection is one of the task under evaluation in TRECVID since 2007 [4].

In the remainder of this section, an overview of some popular and innovative event classification techniques is presented in order to determine their specific application and the applied object classification algorithm. Events are grouped into three categories depending on the nature of the objects involved *human events*, *vehicle events* and *other events*. In Section 3.4.1, a review of some popular human event detectors and applications like human loitering, people counting or crowd detection, are presented. Section 3.4.2 describes the different applications and approaches for

[2] Eliminate false alarms or categorise such segmented objects as noise.

[3] International Workshop on Performance Evaluation of Tracking and Surveillance. `http://pets2010.net/`

[4] The TRECVID is an annual workshop co-sponsored by the National Institute of Standards and Technology (NIST) and the Intelligence Advanced Projects Activity (IARPA), focused on a set of content based video retrieval tasks such as *Semantic indexing, Known-item search, Content-based copy detection, Surveillance event detection* or *Instance search.* Moreover, TRECVID addresses various researches in information retrieval and provides an incomparable dataset for systems evaluation. `http://trecvid.nist.gov/`

vehicle event detection. Finally in Section 3.4.3, a summary of innovative applications of event detection is presented.

3.4.1 Human Events

At an early stage, event detection dealt with single pedestrians or a very limited number of pedestrians whose main activity consisted on entering, leaving and passing through, in a word, loitering. In [76], a single person event or two people interactions, such as altering one's path to meet another or following another person, are modelled by Hidden Markov Models (HMMs) and Coupled Hidden Markov Models (CHMMs). In this approach, for each moving object an appearance-based descriptor is generated and a Kalman filter tracks the object's location, coarse shape, colour pattern and velocity. From this temporally ordered stream of data, an object behavioural description is the input to a stochastic state-based behaviour models to detect interactions between objects. Both HMMs and CHMMs are used to classify the modelled behaviours. In [10], outdoor loitering detection is used as a cue to detect potential drug-dealing operations in bus stations. In this specific scenario, a suspicious activity is defined as individuals loitering for a longer time than the maximum waiting time for the bus. In this approach authors use a refined Gaussian mixture background subtraction algorithm to detect motion objects in a calibrated scene. A human vs non-human classification is made based on size and shape descriptors using an on-line classification technique that clusters incoming pedestrian images into classes comprised of images of a single pedestrian.

Outdoor videos are mainly recorded in crowded scenarios, like in public transportation or areas with high congestion. Accurate people detection has been exhaustively researched because of its many applications. One of the most important applications is crowd detection. Two main research challenges involved in crowd detection are crowd counting and crowd behaviour analysis.

Side-view surveillance cameras find difficulties in performing crowd segmentation, instead aerial cameras base their analysis in head detection and tracking. Since overhead views are prone to tracking errors across several cameras, counting crowds is usually performed in side-view multi-cameras systems. In the case of crowd segmentation, solutions based on several appearance-based features, face detection or skin colour have been proposed. As most of these techniques rely on appearance-based features, their accuracy depends on the image quality and frame rate. Shape indexing and skin colour are considered robust to poor video quality, whereas motion and face detection are most dependent on video quality [16]. As an example of crowd counting, in [35], authors address crowd detection and people counting. Frame differencing is used to spatially segment the moving objects. Crowd segmentation is considered a shape matching problem and addressed using an example-based algorithm which maps directly the global shape feature by Fourier descriptors to various configurations of humans stored in a look-up table. Moreover, authors use locally weighted regression (LWR) over the candidate parameter sets to quickly estimate the one that better explains the extracted shape. In this approach, crowd detection is converted into a shape matching problem and people counting depends on the shape assigned to the image under analysis.

Crowd behaviour analysis has drawn significant interest as a novel procedure to efficiently detect and deal with accidents or to control situations that potentially could lead to graver incidents. Recent crowd behaviour analysis methods include tracking of moving objects, motion models using optical flow and crowd density measurement using background reference images. A recent survey [116] focused on crowd analysis methods. However, the main application of crowd behaviour in surveillance scenarios is identifying individual events in crowded areas. In [55], authors use spatio-temporal shapes combined with Shechthtman and Irani's flow descriptor as moving objects indexing technique. In order to provide robustness against occlusions and actor variability, events are separated in parts. Authors divide their action templates into parts and extend the pictorical structures algorithm for recognition. For testing their approach, twenty minutes of video containing 110 events of interest was collected by the authors.

In surveillance applications, multiple-person interaction has gained importance due to the growing demand for recognising suspicious activities. In fact, projects like *Computer-assisted prescreening of video streams or unusual activities (BEHAVE)* [5] and *Context aware vision using image-based active recognition (CAVIAR)* [6] have produced several publications focusing on multiple-person interactions. Some outstanding approaches are presented in [41, 78]. In [41], the behaviour detection procedure consists of motion detection and object tracking followed by a semantic description of suspicious human behaviours based on sets of low-level object events, i.e., fights are defined as many moving objects moving together. In [78], multiple free-form moving objects and course models of the human body were used in a two-person interaction, which used a hierarchical Bayesian network to recognise human behaviours based on body part segmentation and motion. Authors extended this work to track multiple body parts of multiple people in order to detect and distinguish different human actions such as punching, handshaking, pushing and hugging [79].

3.4.2 Vehicle Events

In surveillance scenarios, events involving vehicles are conceived for specific applications such as control of traffic congestion or control and elimination of dangerous situations. Consequently, events such as accidents, illegal parking, congestion status or lane driving are considered. In literature, several approaches have been presented to detect the previous events. Most existing automated vehicle surveillance systems are based on trajectory analysis and are commonly learnt using expectation maximisation or modelled using semantic rules.

[5] BEHAVE project: `http://homepages.inf.ed.ac.uk/rbf/BEHAVE/`
[6] CAVIAR project: `http://homepages.inf.ed.ac.uk/rbf/CAVIAR/`

A first group of approaches targeted vehicle recognition, classification of different types of vehicles or discrimination between vehicles and other semantic objects. In [107], authors discriminate between pedestrians, bicycles, bus/truck, cars and motor using a naive Bayesian classifier based on a feature vector composed by (i) the semantic object-length, (ii) width, (iii) object's maximum speed (km/h), (iv) bounding box "filling degree" (distinguishing solid and "open" vehicles) and (v) the fraction of contour pixels. In [15], authors perform vehicle detection and classification using 3D wire frame models. Motion silhouettes are extracted and compared to a projected model silhouette to identify the ground plane position and class of the vehicle. Other existing approaches are [61, 102].

Vehicle detection and recognition is a fundamental task and the basis for vehicle event detection. Several approaches analyse the vehicles' trajectory in order to detect events. In [59], a traffic event detection based on Gaussian Mixture Hidden Markov Model (GMHMM) is presented. Authors build a feature vector based on Discrete Cosine Transform (DCT) coefficients and macro-block motion vectors. Six traffic patterns are defined, heavy congestion, high density with low speed, high density with high speed, low density with high speed, low density with low speed and vacancy. Each traffic pattern is modelled by a separate GMHMM that is trained using the EM algorithm. In [106], a multiple cue-based approach combined with a switching Kalman filter for robust estimation of targets is presented. The results of tracking are analysed in an event detection stage, where domain rule-based algorithm detects several predefined events, turning, lane changes, slow moving, stopped and speeding vehicles. In [56], authors perform semantic object classification using Bayesian Networks. Based on the semantic object classification results, an event detector extracts some features, such as shape, motion, position and track, and includes *a priori* knowledge of the spatial context in order to detect a list of traffic events, i.e., moving towards the checkpost, stopped in front of the checkpost and crossing the checkpost.

Significant research has been done in anomaly behaviour detection, some examples can be found in [61, 113]. Moreover, a complete review of on-road vehicle detection systems can be found in [100].

Exhaustive research has been devoted to develop *License Plate Recognition* (LPR) algorithms as core modules for intelligent infrastructure systems and freeway managing systems for traffic surveillance. Generally, LPR algorithms consist on three steps named: (i) license plate detection, (ii) segmentation of the characters and (iii) character recognition. Many methods to locate license plates have been proposed in recent years, such as edge detection method, line sensitive filters to extract the plate areas, the window method and morphology methods. Despite their ability to locate license plates, these algorithms are affected by brightness, processing time or variable environment. Once, the license plate is located, character segmentation algorithms are applied. There have been a large number of character segmentation techniques reported based on projection, morphology and connected components. Finally, numerous character recognition techniques have been reported using model match, Bayes classifier, artificial neural network or Support Vector Machines. Most of the proposed character segmentation techniques are limited by two constraints, only single-line characters can be processed and only English and numeric characters can be recognised. Considering these challenges, some recent LPR algorithms proposed: (i) in [17], licence plate localisation is based on Gabor filters followed by thresholding and connected components, while for character recognition Self Organising Maps neural networks were applied, (ii) in [54], authors recognise characters using edge analysis followed by feed forward neural networks and (iii) in [109], authors propose an improved Bernsen algorithm for license plate location while for the character recognition a feature comparison based on SVM is presented. A complete review of license plate recognition algorithms can be found in [1].

A summary of the presented systems developed for human and vehicle events can be found in Table 1.

Table 1 Summary of event and object detection systems over surveillance videos

Task	Classifier	Dataset
Human interaction [11]	HMMs and CHMMs	20h outdoor videos
Pedestrian loitering [10]	On-line cluster classification	Outdoor videos
Crowd detection & people counting [35]	Locally weighted regression	outdoor videos
Event recognition [55]	Pictorical structures algorithm	Outdoor/indoor videos
Event detection [41]	Rule-based classifier	CAVIAR project dataset
Human interactions[78, 79]	Hierarchical Bayesian network	Indoor videos
Semantic object discrimination [107]	Naive Bayesian classifier	Outdoor videos
Vehicle classification [15]	3D wire frame models	i-LIDS dataset
Traffic event detection [59]	GMHMM	12h outdoor videos
Traffic event detection [106]	Kalman filters	Outdoor videos
Traffic event detection [56]	Bayesian networks	Outdoor videos
License Plate Recognition [17]	Self Organising Maps neural networks	Outdoor videos (parking lots/highways)

3.4.3 Other Events

Due to the wide range of surveillance scenarios, from airports to underground video surveillance, a broad spectrum of specific situations is analysed and targeted in dedicated surveillance systems. Among several projects focused on solving specific surveillance tasks, PETS 2004 and 2006 targeted the *detection of dangerous objects*. In table 2, a summary of some surveillance oriented specific tasks is presented.

4 Vision Based Semantic Analysis of Surveillance Videos

The pervasive presence of surveillance video cameras in crowded or strategic places requires a dedicated analysis. Typically these places are wide areas covered by numerous surveillance video cameras generating a large amount of information. In

Table 2 Summary of surveillance oriented specific tasks

Task	References		
	First Author	Year	Reference
Intrusion/Trespassing	Black	2005	[11]
	Schwerdt	2005	[91]
	Seyve	2005	[93]
Vandalism	Sacchi	2001	[84]
	Fuentes	2004	[41]
	Ghazal	2007	[43]
Suspicious stationary objects	Black	2005	[11]
	Spagnolo	2006	[96]
	Lu	2006	[67]
Smoke detection	Toreyin	2005	[103]
	Maruta	2010	[72]
	Ma	2010	[68]

order to extract useful information, the surveillance system must be robust enough to eliminate *false alarms* [7]. In the previous Section 3, a review of popular object and event detection systems for surveillance videos is presented. In this section, a content-based classifier based on *Support Vector Machines* is presented for object detection in surveillance videos (refer to Figure 2). The classification of objects is based on blobs extracted using motion-based background subtraction. Furthermore, several established and novel visual descriptors are employed in order to study their capabilities in the particular application scenario of surveillance video understanding.

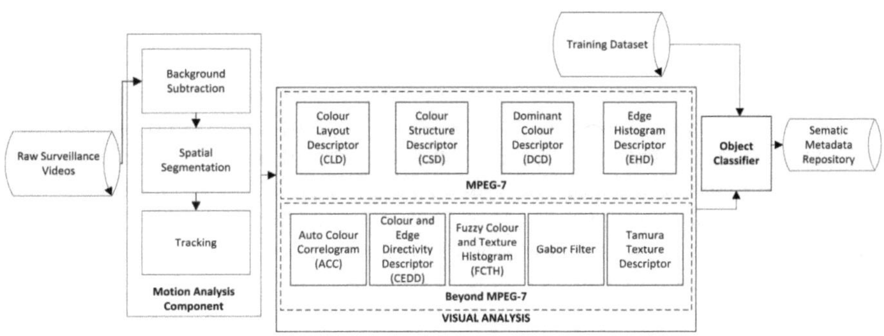

Fig. 2 Content-based object classifier based on Support Vector Machines

[7] *False alarms* are events that do not imply any danger. Moreover, false alarms should not be considered in posterior stages since they do not represent any moving object and do not provide any additional information. Some false alarm examples are moving trees, rain, small camera motion or variation in the illumination conditions.

The remaining of this Section is structured as follows. In Section 4.1, spatial object segmentation based on Gaussian Mixture Models (GMMs) is performed in the Motion Analysis Component as well as object tracking based on Kalman filters. In Section 4.2, the proposed semantic object classifier based on *Support Vector Machines (SMVs)* is further explained. Finally, the selection of visual features used for evaluation is introduced in Section 4.3.

4.1 Motion Analysis

Due to the nature of surveillance videos, a huge proportion of the information is redundant, but requires a very time-consuming process to be analysed. *Motion analysis component* aims to improve the computational efficiency of the system and to extract objects exhibiting motion characteristics. A three-step real-time *Motion Analysis Component* is presented to procure individual blobs for further processing through *Visual analysis component*.

The first processing stage in the motion analysis component is an adaptive background subtraction technique based on Stauffer and Grimson algorithm [97]. The objective of this stage is to remove all the redundant information of the surveillance videos, allowing a faster analysis and providing robustness against external factors, such as changes in illumination or camouflage. Adaptive background subtraction algorithm involves (i) modelling the background as a mixture of Gaussians and (ii) modelling each pixel of an image as a weighted mixture of Gaussians and classifying it into foreground or background according to the persistence and variance of each of the Gaussians of the mixture. Thus, pixels are classified as foreground if their values do not fit in the background distributions formerly calculated. Second, spatial segmentation of moving objects is performed grouping the resulting Gaussian mixtures. Consequently, a two-pass connected component algorithm (as in [98]) assuming an 8-connection is applied. As a result, foreground moving objects are isolated. Third, temporal segmentation is performed establishing the correspondence of the spatially segmented objects between frames using a linearly predictive multiple hypothesis tracking algorithm based on a set of Kalman filters. Moreover, Kalman filters are used to predict the tracks related to each frame as well as the assignment between the available tracks and the detected blobs in each frame. Despite many advantages of the use of motion analysis component, as highlighted in Figure 3, object detection from surveillance video is affected by a lot of noise generated from (i) the low quality of the image, (ii) lack of contrast or the image blurring due to the camera motion and (iii) external factors due to an open environment. As a result, some of the detected and segmented moving objects are *false alarms*.

4.2 Object Classification

Subsequent to the extraction of the foreground moving objects, the next step involves the indexing process. In general such indexing schemes involve the extraction

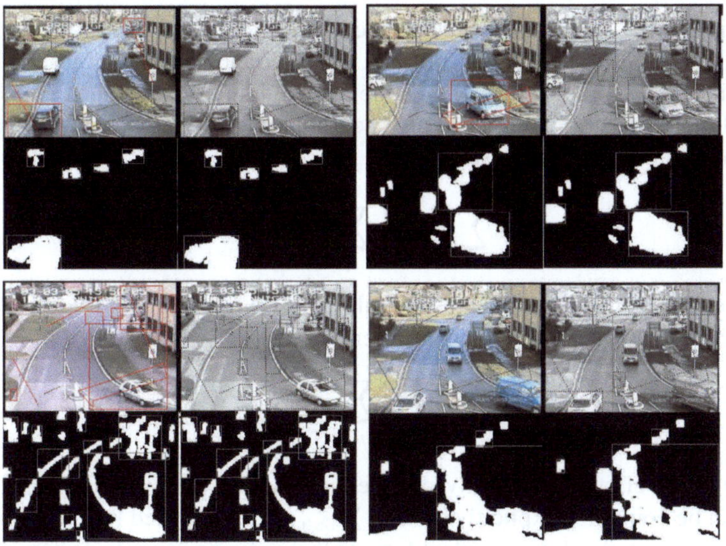

Fig. 3 Motion analysis component results. Background subtraction and spatial segmentation techniques results can be observed for four different problematic situations as low quality image (top-left),videos with inaccurate background subtraction (top-right), videos with camera movement (bottom-left) and objects merged due to noise and shadows (bottom-right)

of visual features and classifying these features into a set of predefined object categories. To this end, *Support Vector Machine (SVM)* algorithm has been popularly chosen among many researchers as they are based on the *Structural Risk Minimisation principle* [34]. In the simplest form, SVMs provide a binary classification method which generates a hyperplane or decision surface for class-separation, maximising the distance between features from different classes. The goal is to discover the inherent most representative characteristics of each class, which at the same time maximises the inter-class dissimilarity, by analysing the training dataset. As a result, a set of labelled patterns representing each class is generated. SVM aims to discover the mapping able to predict the label of any unseen pattern based on the analysis and comparison of each new query to the previously calculated model.

Given the labelled binary dataset $(X, Y) = \{(x_i, y_i)|_{i=1,\dots,N}, y_i \in \{-1, +1\}\}$, the *Linear SVM (LSVM)* classifier recovers an optimal separating hyperplane $w^T x + b = 0$, which maximises the margin of the classifier. This can be formulated as the following constrained optimisation problem [88]:

$$\min_{w} \frac{||w||^2}{2} + C \sum_{i} l(w; x_i, y_i) \tag{7}$$

The first term on the right-hand side is the regularisation term on the classifier weights, which is related to the classifier margin through the inverse distance between the marginal hyperplanes $w^T x = 1$ and $w^T x = -1$. The second term on the right-hand side is related to the classification error, where $l(w; x_i, y_i) = max(1 - y_i w^T x_i, 0)$ is the Hinge loss that upper bounds on the empirical error of the classifier. The parameter C controls the relative importance of the regularisation term and the error term. SVMs can also be trained by solving the Lagrangian dual shown in Equation 7, which results in

$$\max_{\alpha} \sum_{i} \alpha_i - \frac{1}{2} \sum_{i,j} y_i \alpha_i x_i^T x_i y_i \alpha_i \tag{8}$$

where the following conditions must be fulfilled, $0 \leq \alpha \leq C$ and $\sum_i y_i \alpha_i = 0$.

The classifier for LSVM is then represented by

$$f(x) = w^T x + b = \sum_{\alpha_i > 0} \alpha_i y_i K(x_i, x) + b \tag{9}$$

where w is the classifier weight vector defined by the dual variables. The weight vector can be computed explicitly by $w = \sum_{\alpha_i > 0} \alpha_i y_i x_i$ and used for prediction.

The advantage of solving the dual form is that only inner products between data points are needed. Consequently, we can derive the nonlinear SVM by implicitly mapping the input data x into the feature space and training the SVM for the mapped features $\phi(x)$. This is achieved by the *Kernel trick*, where the implicit feature space is induced by the kernel function governed by the inner product for feature space maps $K(x_i, x_j) = \phi(x_i)^T \phi(x_j)$. Nonlinear SVMs can then be trained by replacing the inner products in Equation 8 with the corresponding kernel $K(x_i, x_j)$. The resulting classifier for the nonlinear SVM is then represented by

$$f(x) = \sum_{\alpha_i > 0} \alpha_i y_i K(x_i, x) + b \tag{10}$$

where the α's are the Lagrangian multipliers. Conceptually, the only difference between the nonlinear SVM and their linear counterparts is in the use of kernel function instead of the inner product in Equation 9. Computationally, LSVMs can be directly evaluated by using Equation 9, which is much more efficient than nonlinear SVMs for purposes of prediction. Note that only those instances with positive value of α_i's, called support vectors, will contribute to classification. Geometrically, these correspond to the points lying on or outside the marginal hyperplanes which incur nonzero hinge losses, i.e., $y_i f(x_i) < 1$. Different kernels can be used for nonlinear SVMs. The most common ones include the radial basis function, polynomial and the sigmoidal kernels.

SVMs can operate either explicitly in the input space leading to the linear SVM (LSVM), or implicitly in the feature space via the kernel mapping giving rise to the kernel SVM. LSVMs are computationally simple to train and use, as they involve only inner product operations with the input data. However, they can be quite restricted in discriminative power and cannot be applied to nonlinear data. This limits their application to many real-world problems where the data distributions are inherently nonlinear. The nonlinear SVM, on the other hand, can handle linearly inseparable data but is not as efficient as the linear classifier. Its complexity is multiplied with the number of support vectors. This is unfavourable for prediction tasks on large-scale datasets.

Among the different algorithms proposed in the literature review (refer to Section 3.4), such as boosting, decision trees or biologically inspired classifiers. SVMs have exhibited great performance against different feature spaces [32, 81, 52, 111]. These characteristics could be largely attributed to the risk minimisation principle and the kernel trick, as previously mentioned.

4.3 Visual Analysis

As often noted in the literature, the performance of the indexing framework is not only dependent on the machine learning algorithm but also on the visual features used to represent objects. A good descriptor should fulfil several criteria. For example, it should be equally discriminant for any type of media (equal variance for different media parameters), the descriptor extraction process should be robust against different levels of quality and detail, it should produce an equally distributed "net" of measurements over a well-defined media collection, etc [37].

In this section, a set of global appearance-based features are exhaustively evaluated. The selected features focus in the representation of two characteristics highly related with human visual perception, *colour* and *texture*. The descriptors used to extract the colour characteristics are *Colour Layout Descriptor (CLD)*, *Colour Structure Descriptor (CSD)*, *Dominant Colour Descriptor (DCD)* and *Auto Colour Correlogram (ACC)*. Whilst texture is represented by *Edge Histogram Descriptor (EHD)*, *Gabor texture filters* and *Tamura descriptors*. Other two features have been selected for this study, *Colour and Edge Directivity Descriptor (CEDD)* and *Fuzzy Colour and Texture Histogram (FCTH)*, because they provide a combined representation of colour and texture characteristics. Among these features, *CLD*, *CSD*, *DCD* and *EHD*, belong to the family of MPEG-7, which has been extensively studied for other applications and hence we briefly discussed. As the remaining features are quite recent, a more extended analysis is presented for completeness.

4.3.1 MPEG-7 Descriptors

Colour Layout Descriptor (CLD) is a very compact and resolution-invariant representation of the spatial distribution of colour in an arbitrarily-shaped region [95]. Its representation is based on coefficients of the Discrete Cosine Transform (DCT) assuring its compactness.

Colour Structure Descriptor (CSD) describes spatial distribution of colour in an image. Unlike colour histograms, CSD encodes not only information about the frequency of occurrence of colours in an image but also their spatial layout, providing *CSD* with a sensitivity to features that cannot be captured by a colour histogram.

Dominant Colour Descriptor (DCD) describes global as well as local spatial colour distribution in images for fast search and retrieval. For DCD extraction, the representative colours (dominant colours) are computed from each image instead of being fixed in the colour space, allowing the feature representation to be accurate as well as compact.

Edge Histogram Descriptor (EHD) provides a description for non-homogeneous texture images and captures spatial distribution of edges whilst providing ease of extraction, scale invariance and support for rotation-sensitive and rotation-invariant matching.

4.3.2 Auto Colour Correlogram (ACC)

The ACC feature extracts the spatial correlation of pairs of colour changes with distance, while the colour histogram captures only the colour distribution but does not contribute with any spatial correlation information [49]. As the autocorrelogram, $\alpha_c^{(k)}$ (see Equation 11), shows a subset of the information provided by a correlogram, $\gamma_{c,c}^{(k)}$,it is computationally efficient for large databases.

$$\alpha_c^{(k)}(I) = \gamma_{c,c}^{k}(I) \tag{11}$$

ACC similarity measure is L_1. Such similarity measure was chosen for its simplicity and robustness [49].

4.3.3 Colour and Edge Directivity Descriptor (CEDD)

CEDD extraction [20] is a block-based three stage procedure. Firstly, colour information is extracted using a fuzzy-linking histogram [19] . Secondly, texture information is extracted using 5 digital filters. Thirdly, a quantisation functionality to reduce the resulting 144 byte vector into a 432 bits descriptor is performed.

CEDD uses Tanimoto coefficient (see Eq. 12) as similarity measure

$$T_{ij} = \frac{x_i^T x_j}{x_i^T x_i + x_j^T x_j - x_i^T x_j} \tag{12}$$

where x represents each descriptor and x^T is the transpose vector of x.

CEDD procures a compact low level feature which combines colour and texture information in one histogram. Its limited size (maximum of 54 bytes) makes it suitable for large image databases. However, its main advantage is the low computational power needed for its extraction versus MPEG-7 features requirements [110].

4.3.4 Fuzzy Colour and Texture Histogram (FCTH)

FCTH extraction process is the combination of three fuzzy units [21]. Firstly, a set of 20 fuzzy rules are applied over each of the channels of HSV colour space to generate a 10-bin histogram. Secondly, an expansion fuzzy system is applied to convert the 10-bin histogram into a 24-bin in order to include information related to the hue of each colour. Finally, a third fuzzy system converts the 24-bin histogram into a 192-bin to insert the Haar Wavelet of each block of the image and a set of texture elements.

FCTH is a 72 bytes descriptor which includes in one quantised histogram colour and texture information and results from the combination of three fuzzy systems. FCTH is an accurate descriptor valid even in appearance of distortion and deformations such as noise and smoothing [21]. Similar to CEDD, Tanimoto coefficient is applied as a similarity measure (refer to Eq. 12) .

4.3.5 Gabor Filter

Gabor filter (or Gabor wavelet) has been used to extract texture features from images in indexing [117]. Gabor texture features extraction procedure consists of pyramid structured wavelet transform which captures the energy at a specific frequency and direction [70]. Discrete Gabor wavelet transform, $G_{m,n}(x, y)$ (refer to Equation 13), is calculated by a convolution between the image and the complex conjugate, $\psi_{m,n}(s, t)$ (refer to Equation 14). The texture representation using Gabor filters consists of applying Gabor filters on an image with different orientation and scales, m and n respectively. Afterwards, the mean and standard deviation of the magnitude of the transformed coefficients are computed in the search for regions with homogeneous texture.

$$G_{mn} = \sum_s \sum_t I(x-s, y-t)\psi^*_{mn}(s,t) \tag{13}$$

$$\psi(x,y) = \frac{1}{2\pi\sigma_x\sigma_y} exp[-\frac{1}{2}(\frac{x^2}{\sigma_x^2} + \frac{y^2}{\sigma_y^2} +)]exp(j2\pi Wx) \tag{14}$$

Considering the extracted texture features using Gabor filter, similarity is calculated as follows:

$$D(Q,T) = \sum_m \sum_n d_{mn}(Q,T) = \sum_m \sum_n \sqrt{(\mu^Q_{mn} - \mu^T_{mn})^2 + (\sigma^Q_{mn} - \sigma^T_{mn})^2} \tag{15}$$

4.3.6 Tamura Texture Descriptor

In [101], authors proposed the extraction of six textural features relevant to human visual perception, namely *coarseness, contrast, directionality, line-likeness, regularity* and *roughness*.

Coarseness (F_{crs}) is calculated from a large size blob (unlike microtexture, coarseness cannot be determined in small blobs). Coarseness extraction procedure is a four-stage procedure. First, an average over the gray levels in the neighbourhood of size $2^k x 2^k$. Second, for each point differences between pairs of averages corresponding to pairs of non-overlapping neighbourhoods just on opposite sides of the point in both horizontal and vertical orientations is calculated. Third, at each point, the best size, which gives the highest output value, is selected using $S_{best}(x, y) = 2^k$, where k maximises the differences calculated previously in either direction. Fourth, the average of S_{best} over the picture is calculated to obtain the coarseness measure F_{crs}:

$$F_{crs} = \frac{1}{mxn} \sum_{i}^{m} \sum_{j}^{n} S_{best}(i, j) \tag{16}$$

Contrast (F_{con}) considers two different characteristics of the images, (i) its dynamic range of grey-levels and (ii) the polarisation of the distribution of black and white areas. Tamura calculates the variance, σ^2, or the standard deviation, σ, of the image to take into consideration the dynamic range of grey-levels of an image. After, the polarisation of the distribution of black and white areas in the image is calculated using the kurtosis α_4,

$$\alpha_4 = \frac{\mu_4}{\sigma^4} \tag{17}$$

where μ_4 is the fourth moment about the mean and σ^2 is the variance. Finally, the measure is normalised. For the contrast measure calculation, $F_{contrast}$, both the kurtosis and the variance are considered:

$$F_{con} = \frac{\sigma}{(\alpha_4)^n} \tag{18}$$

where n is a positive number.

Directionality (F_{dir}) is calculated using a histogram of local edge probabilities against their directional angle, which detects long lines and simple curves. The desired histogram is based on the gradient calculation:

$$|\Delta G| = (|\Delta_H| + |\Delta_V|)/2\theta \tag{19}$$

$$= tan^{-1}(\Delta_V/\Delta_H) + \frac{\pi}{2} H_D(k) \tag{20}$$

$$= N_\Delta(k) / \sum_{i=0}^{n-1} N_\Delta(i); k = 0, 1, ..., n-1 \tag{21}$$

where $N_\Delta(k)$ is the number of points at which $(2k-1)\pi/2n \leq \theta < (2k+1)/2n$ and the magnitude of the gradient is greater than 1, $|\Delta G| > t$.

A way of measuring the directionality quantitatively from H_D is to compute the sharpness of the peaks. The approach adopted en Tamura [101] is to sum the second moments around each peak from valley to valley, if multiple peaks are determined to exist. The directionality measure, F_{dir} is calculated as follows:

$$F_{dir} = 1 - r \cdot n_p \cdot \sum_{p}^{n_p} \sum_{\phi \in w_p} (\phi - \phi_p)^2 \cdot H_D(\phi) \tag{22}$$

where n_p is the number of peaks, ϕ_p is the pth peak position of H_D and w_p is the range of pth peak between valleys.

Line-likeness (F_{lin}) is measured counting the co-occurrences in the same direction weighted by +1 and those in the perpendicular direction by −1:

$$F_{lin} = \sum_{i}^{n} \sum_{j}^{n} P_{Dd}(i,j) cos \left| (i-j)\frac{2\pi}{n} \right| / \sum_{i}^{n} \sum_{j}^{n} P_{Dd}(i,j) \tag{23}$$

where P_{Dd} is the nxn local direction co-occurrence matrix of points at distance.

Regularity (F_{reg}) is considered when any feature of a texture varies over the whole image. Hence taking partitioned subimages, the variation of each feature is considered. As a regularity measure, the sum of the variation for each of the features previously explained:

$$F_{reg} = 1 - r(\sigma_{crs} + \sigma_{con} + \sigma_{dir} + \sigma_{lin}) \tag{24}$$

Roughness (F_{rgh}) is a measure to emphasise the effects of coarseness and contrast:

$$F_{rgh} = F_{crs} + F_{con} \tag{25}$$

Tamura Similarity Measure consists of an Euclidean distance function.

5 Experimental Results

A study of the performance of some appearance-based global features over raw surveillance videos is presented. An evaluation framework for studying the performance of multiple visual descriptor spaces is presented (refer to Figure 2). In addition to studying the classical features, the framework also includes an extensive analysis of MPEG-7 and beyond MPEG-7 features [8]. Among the different visual characteristics of an object, this study is focused on colour and texture as these two characteristics are highly related with human visual perception.

A binary classification-based indexing framework is presented to study the performance of visual features. Therefore for the object class model generation, semantic objects are only labelled as $\{+1, -1\}$. The object classification of the surveillance

[8] These features are published subsequent to the publication of MPEG-7 and are shown to be robust.

indexing framework is performed with non-linear Support Vector Machines (SVMs) with a radial kernel (refer to Section 4.2). An implementation of Vapnik's SVM [105] for the problem of pattern recognition is applied [9].

In the remainder of the section, an extensive evaluation of different features on outdoor surveillance videos with static and dynamic background is presented. Two datasets were used for this evaluation, *AVSS 2007* and *CamVid* datasets. Both datasets contain videos from outdoor surveillance systems but deployed for different applications, i.e. *AVSS 2007* has static background while *CamVid* has dynamic background.

AVSS 2007 Dataset

AVSS 2007 dataset [10] consists of both indoor and outdoor surveillance videos. The dataset has been widely used in the research community for the purposes of object tracking, detection, etc. In total the dataset contains more than 35000 frames (equivalent to 23.3 minutes of surveillance video footage). Further to a careful observation, a subset of the dataset has been chosen with more than 13000 frames (equivalent to 8.67 minutes of video footage) under varying lighting conditions. Also, the chosen dataset show videos with different levels of difficulty depending on the number of events to detect and the environmental conditions (refer to Figure 5). As previously illustrated in Section 2, external noisy environmental conditions severely affect the analysis of the chosen dataset. To investigate the performance of our surveillance video indexing approach, the ground truth annotations were manually assigned to all the moving objects extracted from the dataset, two were the predefined concepts, *Car* and *Person* (see Figure 6). A total of 1377 objects were included and annotated in the ground truth. Besides, the ground truth was partially selected to form

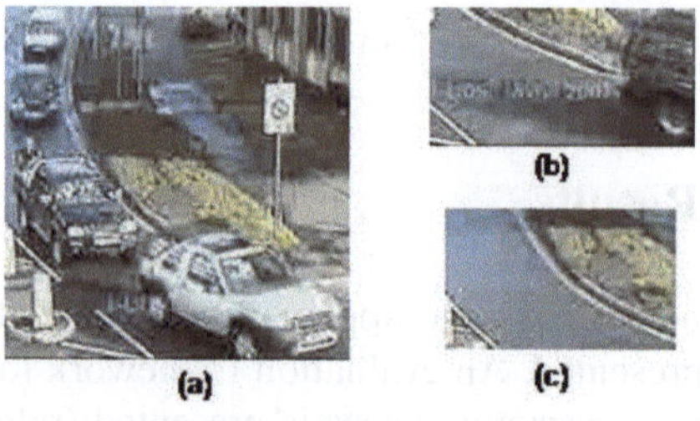

Fig. 4 Several were the reasons included in the interannotator agreement to label a blob as *Unknown*: (a) coexistence of several semantic objects within a blob, (b) appearance of less than a 50% of a semantic object and (c) appearance of blobs without relevant information. The Motion Analysis Component presented several challenges (refer to Section 4.1) generating some of the blobs later labelled as unknown

[9] The module used to compute the *SVM* is based on Cornell University's module, www.cs.cornell.edu/People/tj/svm_light

[10] http://www.eecs.qmul.ac.uk/ãndrea/avss2007_d.html

(a) Easy (b) Medium (c) Hard

Fig. 5 Examples of the surveillance videos provided in AVSS 2007 dataset. Each image shows a frame of a video with different levels of difficulty depending on the environmental factors that affect them. (a) Easy level shows shadows and a low quality video. (b) Medium level presents low quality video, a higher number of events and objects and shadows. (c) Hard level shows a higher number of events, shadows and small camera shakes. These limitations affect to the spatial segmentation procedure and therefore to the subsequent steps

Fig. 6 Examples of the segmented moving objects provided by AVSS 2007 dataset

the training dataset to construct the training models used for semantic object classification afterwards. Less than a 6% of the ground truth was selected for the training dataset, where 50% of the objects were annotated as *Car* against the 30% as *Person*. The remaining 20% of the objects were annotated as *Unknown*. Largely, these unknown blobs are composed of noise and those blobs which could not be assigned to either one of the above concepts (refer to Figure 4). Instead of ignoring the blobs labelled as "Unknown", our dataset included these blobs to explicitly study the effect of noise on the performance of the retrieval models. An overview of the dataset used for the evaluation of the proposed framework is presented in Fig. 6. Each of these blobs are represented along with the resolution as extracted from the motion analysis component. As it can be noted, the blobs vary in the resolution and often contains visual disturbance which are often found to be difficult for human annotators to assign concept labels.

The results obtained from the object classifier over the AVSS 2007 Dataset are presented in Figure 7. AVSS dataset was composed of two semantic concepts, *Car* and *Person*. The performance of the concept *Car* show a high performance of each individual feature. However, among the nine features used in the evaluation, two features namely *ACC* and *EHD*, has shown to outperform rest of the features by achieving more than 80% precision until 30% recall. In contrast, *Gabor filters* and *CSD* features have resulted 50% precision for 10% recall and further decreasing subsequently.

On the other hand, the feature performance for the concept *Person*, which has in general been quite positive with five features out of nine, has resulted with 100% precision until 45% recall. However, the performance of *FCTH, Gabor filters* and

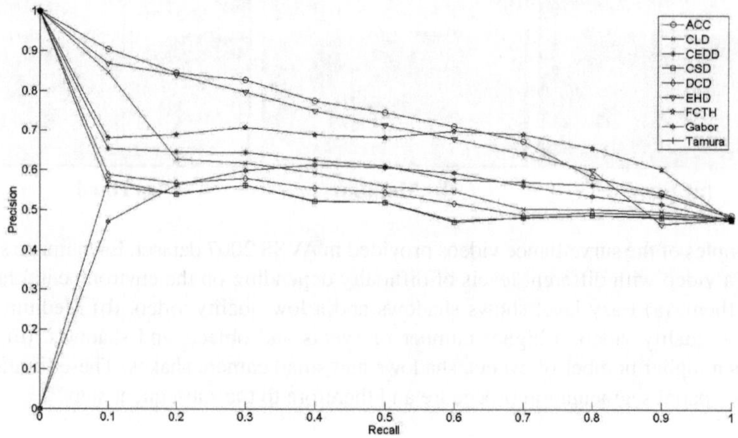

(a) Car

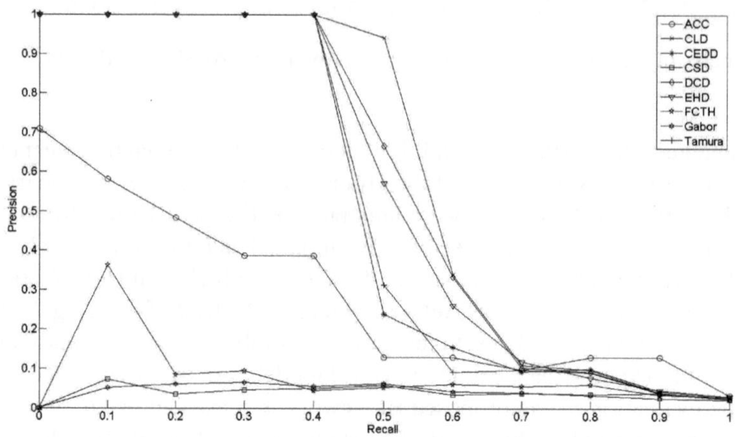

(b) Person

Fig. 7 Analysis of the presented features performance over the individual semantic concepts considered in AVSS 2007 dataset

CSD have stagnated approximately 10% precision throughout all recall. In general from the analysis of the features from *AVSS 2007 dataset*, among the nine features studied except for *CSD* and *Gabor filters*, other features has shown positive results when classified with SVM. Therefore, it is worth noting that the low-performance of *CSD* and *Gabor filters* could be largely attributed to the low ability to discriminate of the spatial distribution of colour and the energy extracted when the size of the image (blob) under analysis is too small, as the examples provided for this dataset in Figure 6.

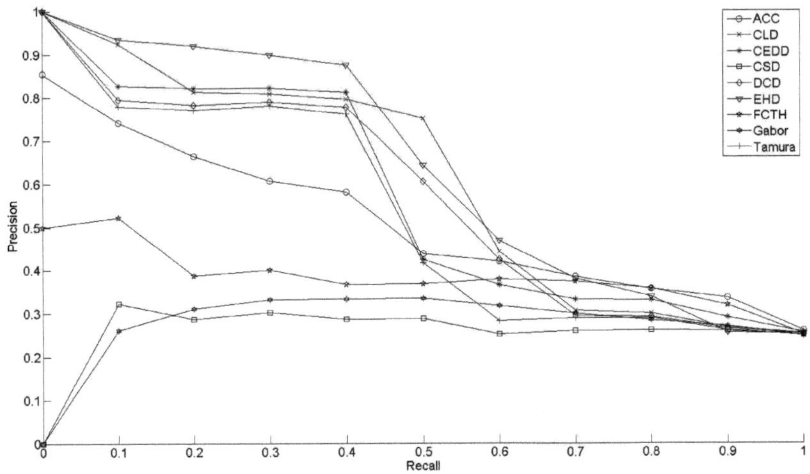

Fig. 8 Average precision-recall across all concepts for AVSS 2007 dataset

Cambridge-Driving Labeled Video Database (CamVid)

The second dataset used in the evaluation framework is CamVid[11] [12, 13]. As opposed to AVSS 2007 dataset which contains static background, CamVid dataset has been filmed with fixed-position CCTV-style cameras, captured from the perspective of a driving automobile. The driving scenario increases the number and heterogeneity of the observed objects. The dataset is composed of five sequences with over 18000 frames in total.

CamVid dataset provides a manual pixel-level segmentation of over 700 images. Moreover, the high quality and colour resolution of videos provide a valuable footage for analysis and is immune to external factors such as camera movement. A set of 32 semantic concepts grouped into four different categories namely *moving objects, road, ceiling* and *fixed objects*, are available in the dataset. Of the 32 concepts, eleven semantic classes correspond to the moving objects, namely *animal, pedestrian, child, rolling cart, bicyclist, motorcycle, car, SUV/pickup truck, truck/bus, train* and *miscellaneous*. For our study, we selected only a representative subset of moving objects from the dataset extracting a total of 3702 blobs. In Figure 9, an example of a frame and its corresponding ground truth is presented. The concept distribution in the dataset is represented with a pie chart in Figure 10.

The performance evaluation of seven semantic concepts over nine features is presented individually in Figures 11, 12 and 13. Due to the unbalanced nature of the semantic concept distribution (ranging from 30% to 0.78%), five different indexing runs have been performed. The performance of the presented features over each individual semantic concept shown in Figures 11, 12 and 13, represents an average precision at different recalls, 10%, 20%, 30%, etc.

[11] `http://mi.eng.cam.ac.uk/research/projects/VideoRec/CamVid/`

Fig. 9 Example of the surveillance scenarios studied in CamVid dataset and the manual pixel-based annotation provided for evaluation purposes

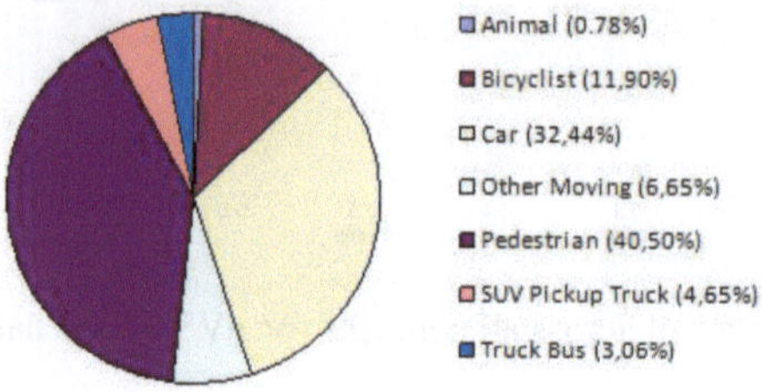

Fig. 10 Concept distribution of the moving objects included in the CamVid dataset

CamVid dataset provides an ideal scenario for evaluating the performance of different datasets, due to its completeness presenting seven semantic concepts related to moving objects in outdoor surveillance videos. Generally, features beyond MPEG-7 standard provide a better performance. Moreover, three features, namely *CEDD*, *FCTH* and *ACC*, outperformed the rest in almost all semantic concepts. For instance, for the concept *SUV Pickup Truck*, these three features exceeded the rest achieving above a 90% precision under 20% recall and showing a smooth evolution while the rest of the features performance drops drastically as the recall increases. However, one MPEG-7 standard descriptor provides similar performance or even beat the previous features, *CLD*. For instance, *CLD* feature presents a high performance in highly representative concepts such as *Car*, where it achieves over 90% precision under 60% recall. Moreover, in another highly documented concept, *Person*, *CLD* obtains above a 90% precision under 50% recall.

Even though, before it was stated that non-MPEG-7 descriptors outperformed MPEG-7 descriptors, such generalisation has two exceptions, *Gabor filters* and *Tamura texture features*. The former, shows a low performance in almost all semantic concepts, not overcoming 35% precision at any recall, for instance, in concepts like *Animal*, *bicyclist*, *SUV pickup truck* and *truck/bus*. However, such concepts are not the most documented, in concepts like *Car* or *Pedestrian*, *Gabor filters* presents fair performance, achieving over 60% precision under 70% recall. While the latter, *Tamura*, obtains a precision over 80% for the concept *bicyclist*, however, its performance is lower in concepts such as *car, other moving, pedestrian* or *SUV pickup truck*.

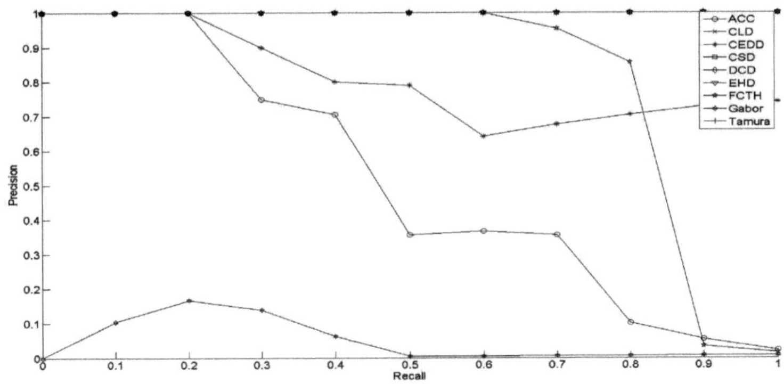

(a) Animal

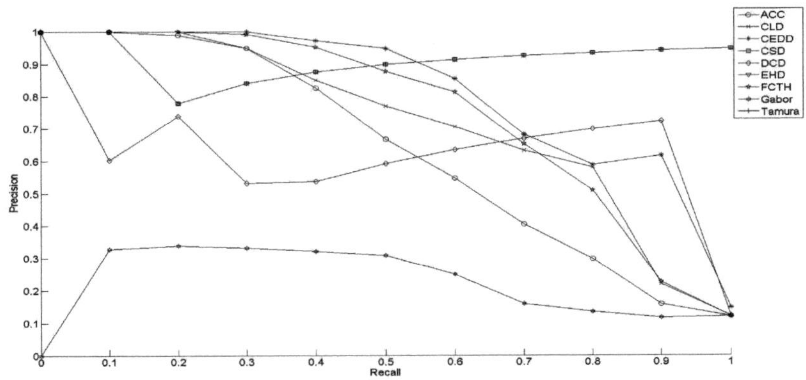

(b) Bicyclist

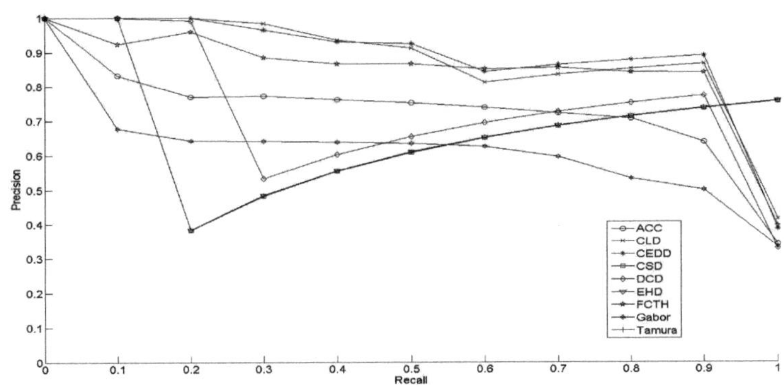

(c) Car

Fig. 11 Analysis of the presented features performance over CamVid dataset for several semantic concepts: *animal, bicyclist* and *car*

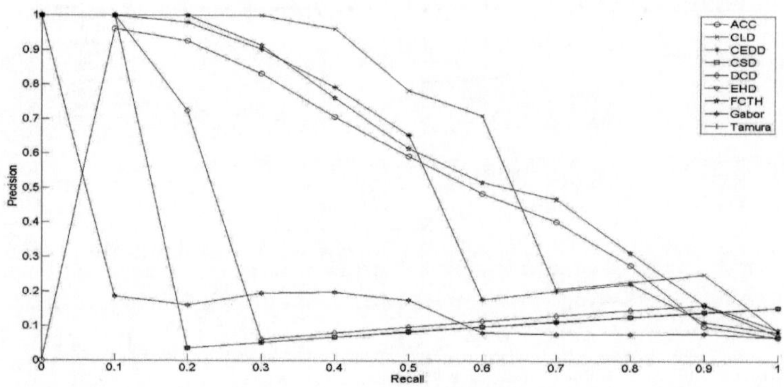

(d) Other

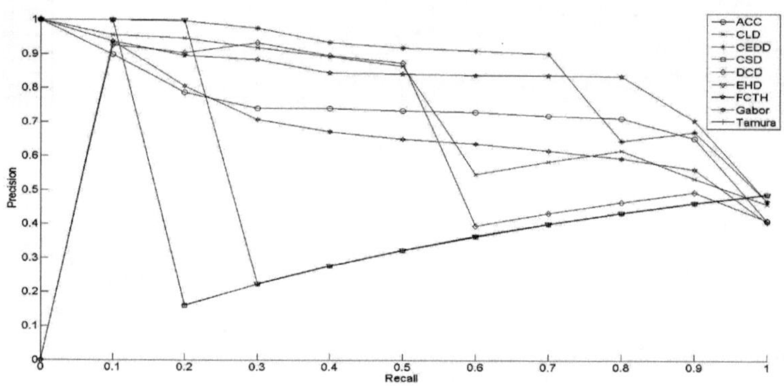

(e) Pedestrian

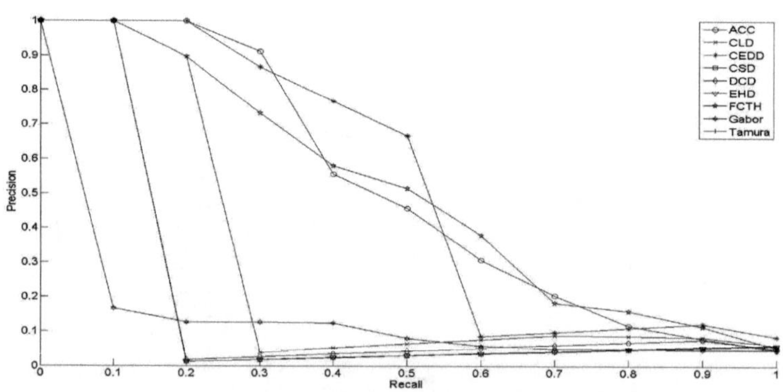

(f) SUV

Fig. 12 Analysis of the presented features performance over CamVid dataset for several semantic concepts: *other moving, pedestrian* and *SUV pickup truck*

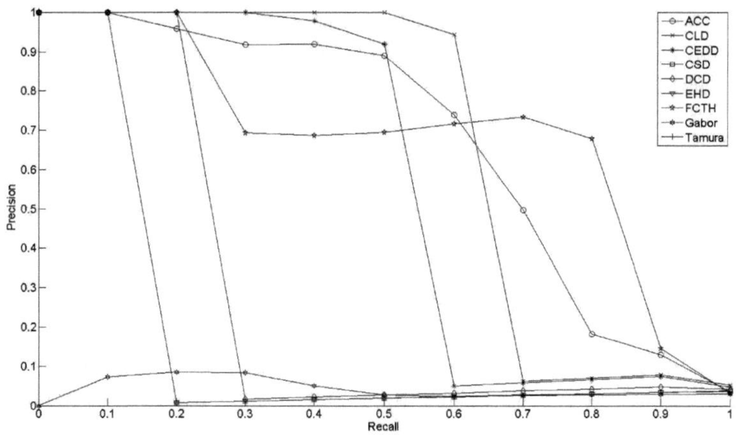

(g) Truck

Fig. 13 Analysis of the presented features performance over CamVid dataset for the semantic concept *truck*

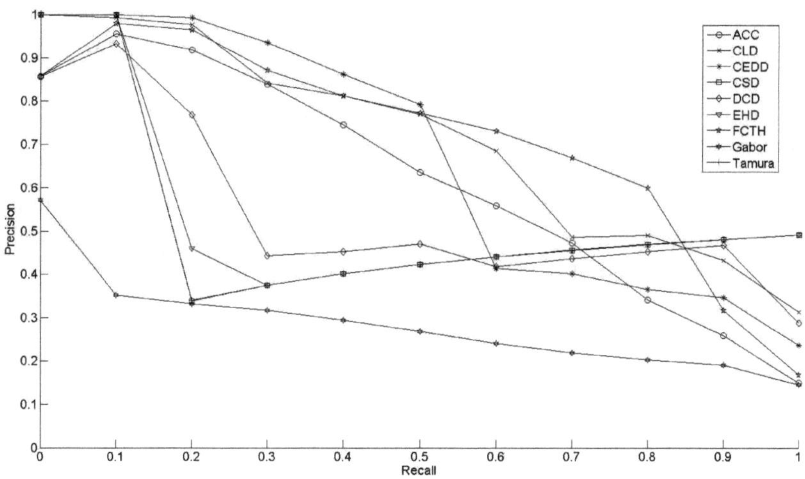

Fig. 14 Average precision-recall across all concepts for CamVid dataset

The concept *animal* has not been considered in the previous analysis due to their exceptional results. For this concept, MPEG-7 descriptors achieve a 100% precision across all recall as well as *Tamura texture features*. The remaining non-MPEG-7 features, except *Gabor filters*, obtain also a high performance, (i) *FCTH* achieves 100% precision under 60% recall, (ii) *CEDD* presents over 90% precision under 30% recall and (iii) *ACC* obtains over 70% precision under 30% recall. However, an exhaustive study of the CamVid dataset reveals a plausible reason for such an outstanding performance, the small set presents repeated examples for this concept in the dataset which could have partially biased the classification results. Finally, the average performance of all features on seven concepts is presented in Figure 14.

6 Conclusions and Future Work

In recent times, the number of research methodologies developed towards automated surveillance systems has been exponential. Such increment could be largely attributed to the ever-increasing need for safety and security of citizens. Thus in this chapter, we present a survey of most commonly occurring research components in automated surveillance systems. Also, we present a practical implementation of an automatic surveillance system framework which closely resembles the generic architectures. As many studies have focused on the importance of the machine-learning algorithms for achieving accurate performances of the system, we presented an extensive study of a set of selected visual features over surveillance datasets. The studied visual features were classified into two categories, MPEG-7 and beyond MPEG-7 descriptors. The feature selection depended on their performance and benefits, such as their compactness and perceptual representation. The selection went beyond MPEG-7 standard to substantiate the existence of more suitable visual features for object representation even in particular scenarios with special constraints like surveillance scenarios, and more specifically, outdoor surveillance. From our analysis, features derived from combining colour and texture have shown to outperform the remaining features.

In this chapter, visual features were studied individually, however, in the future, we plan to tackle object and event classification using techniques for optimal combination of features from multiple feature spaces. Another considered research direction for object and event classification includes the study of semantic object behaviour as a suitable feature in conjunction with the visual information.

Acknowledgement. The research was partially supported by the European Commission under contract FP7-SEC 261743 VideoSense.

References

1. Anagnostopoulos, C.N.E., Anagnostopoulos, I.E., Psoroulas, I.D., Loumos, V., Kayafas, E.: License plate recognition from still images and video sequences: A survey. IEEE Transactions on Intelligent Transportation Systems 9(3), 377–391 (2008)

2. Androutsos, D., Plataniotis, K.N., Venetsanopoulos, A.N.: A novel vector-based approach to color image retrieval using a vector angular-based distance measure. Computer Vision and Image Understanding 75(1-2), 46–58 (1999)
3. Annesley, J., Orwell, J., Renno, J.P.: Evaluation of mpeg7 color descriptors for visual surveillance retrieval. In: IEEE International Workshop on Visual Surveillance and Performance Evaluation of Tracking and Surveillance, pp. 105–112. IEEE (2005)
4. Avidan, S., LTD, M.E.V.T., Jerusalem, I.: Support vector tracking. IEEE Transactions on Pattern Analysis and Machine Intelligence 26(8), 1064–1072 (2004)
5. Bashir, F.I., Khokhar, A.A., Schonfeld, D.: Segmented trajectory based indexing and retrieval of video data. In: IEEE International Conference on Image Processing, vol. 2. IEEE (2003)
6. Bashir, F.I., Khokhar, A.A., Schonfeld, D.: Real-time motion trajectory-based indexing and retrieval of video sequences. IEEE Transactions on Multimedia 9(1), 58–65 (2007)
7. Bay, H., Ess, A., Tuytelaars, T., Van Gool, L.: Speeded-up robust features (surf). Computer Vision and Image Understanding 110(3), 346–359 (2008)
8. Berriss, W.P., Price, W.G., Bober, M.Z., et al: The use of mpeg-7 for intelligent analysis and retrieval in video surveillance. In: IEEE Symposium on Intelligent Distributed Surveillance Systems (2003)
9. Bertalmio, M., Sapiro, G., Randall, G.: Morphing active contours. IEEE Transactions on Pattern Analysis and Machine Intelligence 22(7), 733–737 (2000)
10. Bird, N.D., Masoud, O., Papanikolopoulos, N.P., Isaacs, A.: Detection of loitering individuals in public transportation areas. IEEE Transactions on Intelligent Transportation Systems 6(2), 167–177 (2005)
11. Black, J., Velastin, S., Boghossian, B.: A real time surveillance system for metropolitan railways. In: IEEE Conference on Advanced Video and Signal Based Surveillance, pp. 189–194. IEEE (2005)
12. Brostow, G.J., Fauqueur, J., Cipolla, R.: Semantic object classes in video: A high-definition ground truth database. Pattern Recognition Letters (2008)
13. Brostow, G.J., Shotton, J., Fauqueur, J., Cipolla, R.: Segmentation and recognition using structure from motion point clouds. In: Forsyth, D., Torr, P., Zisserman, A. (eds.) ECCV 2008, Part I. LNCS, vol. 5302, pp. 44–57. Springer, Heidelberg (2008)
14. Brown, L.M.: Color retrieval for video surveillance. In: IEEE Conference on Advanced Video and Signal Based Surveillance, pp. 283–290. IEEE (2008)
15. Buch, N., Orwell, J., Velastin, S.A.: Detection and classification of vehicles for urban traffic scenes. In: International Conference on Visual Information Engineering, pp. 182–187. IET (2008)
16. Candamo, J., Shreve, M., Goldgof, D.B., Sapper, D.B., Kasturi, R.: Understanding transit scenes: a survey on human behavior-recognition algorithms. IEEE Transactions on Intelligent Transportation Systems 11(1), 206–224 (2010)
17. Caner, H., Gecim, H.S., Alkar, A.Z.: Efficient embedded neural-network-based license plate recognition system. IEEE Transactions on Vehicular Technology 57(5), 2675–2683 (2008)
18. Chandramouli, K., Ebroul, I.: Image retrieval using particle swarm optimization, pp. 297–320. Auerbach Publications (2009)
19. Chatzichristofis, S., Boutalis, Y.: A hybrid scheme for fast and accurate image retrieval based on color descriptors. In: International Conference on Artificial Intelligence and Soft Computing, pp. 280–285. ACTA Press (2007)

20. Chatzichristofis, S.A., Boutalis, Y.S.: CEDD: Color and Edge Directivity Descriptor: A Compact Descriptor for Image Indexing and Retrieval. In: Gasteratos, A., Vincze, M., Tsotsos, J.K. (eds.) ICVS 2008. LNCS, vol. 5008, pp. 312–322. Springer, Heidelberg (2008)
21. Chatzichristofis, S.A., Boutalis, Y.S.: Fcth: Fuzzy color and texture histogram-a low level feature for accurate image retrieval. In: International Workshop on Image Analysis for Multimedia Interactive Services, pp. 191–196. IEEE (2008)
22. Chen, L., Feris, R., Zhai, Y., Brown, L., Hampapur, A.: An integrated system for moving object classification in surveillance videos. In: IEEE International Conference on Advanced Video and Signal Based Surveillance, pp. 52–59. IEEE (2008)
23. Chen, P.Y., Chen, A.L.P.: Video retrieval based on video motion tracks of moving objects. In: Proceedings of SPIE, vol. 5307, pp. 550–558 (2003)
24. Chen, Y., Rui, Y., Huang, T.S.: Jpdaf based hmm or real-time contour tracking (2001)
25. Cheung, S.C.S., Kamath, C.: Robust background subtraction with foreground validation for urban traffic video. Eurasip Journal on Applied Signal Processing 2005, 2330–2340 (2005)
26. Chmelar, P., Lanik, A., Mlich, J.: SUNAR Surveillance Network Augmented by Retrieval. In: Blanc-Talon, J., Bone, D., Philips, W., Popescu, D., Scheunders, P. (eds.) ACIVS 2010, Part II. LNCS, vol. 6475, pp. 155–166. Springer, Heidelberg (2010)
27. Collins, R., Lipton, A., Kanade, T., Fujiyoshi, H., Duggins, D., Tsin, Y., Tolliver, D., Enomoto, N., Hasegawa, O., Burt, P., et al: A system for video surveillance and monitoring: Vsam final report. Technical report, Carnegie Mellon University, Pittsburgh, PA, CMU-RI-TR-00-12 (August 2000)
28. Comaniciu, D.: Bayesian kernel tracking. Pattern Recognition, 438–445 (2002)
29. Cutler, R., Davis, L.: View-based detection and analysis of periodic motion. In: International Conference on Pattern Recognition, vol. 1, pp. 495–500. IEEE (1998)
30. Dagtas, S., Al-Khatib, W., Ghafoor, A., Kashyap, R.L.: Models for motion-based video indexing and retrieval. IEEE Transactions on Image Processing 9(1), 88–101 (2000)
31. Dalal, N., Triggs, B.: Histograms of oriented gradients for human detection. In: IEEE Computer Society Conference on Computer Vision and Pattern Recognition, vol. 1, pp. 886–893 (2005)
32. Danafar, S., Gheissari, N.: Action Recognition for Surveillance Applications Using Optic Flow and SVM. In: Yagi, Y., Kang, S.B., Kweon, I.S., Zha, H. (eds.) ACCV 2007, Part II. LNCS, vol. 4844, pp. 457–466. Springer, Heidelberg (2007)
33. Dimitrova, N., Golshani, F.: Motion recovery for video content classification. ACM Transactions on Information Systems 13(4), 408–439 (1995)
34. Djordjevic, D., Izquierdo, E.: An object- and user- driven system for semantic-based image annotation and retrieval. IEEE Transactions on Circuits and Systems for Video Technology 17(3), 313–323 (2007)
35. Dong, L., Parameswaran, V., Ramesh, V., Zoghlami, I.: Fast crowd segmentation using shape indexing. In: International Conference on Computer Vision, pp. 1–8. IEEE (2007)
36. Dyana, A., Das, S.: Trajectory representation using gabor features for motion-based video retrieval. Pattern Recognition Letters 30(10), 877–892 (2009)
37. Eidenberger, H.: How good are the visual mpeg-7 features? In: SPIE Proceedings, pp. 476–488. Society of Photo-Optical Instrumentation Engineers (2003)
38. Fernandez Arguedas, V., Chandramouli, K., Izquierdo, E.: Semantic object based retrieval from surveillance videos. In: International Workshop on Semantic Media Adaptation and Personalization, pp. 79–83. IEEE (2009)
39. Fernandez Arguedas, V., Izquierdo, E.: Object classification based on behaviour patterns. In: International Conference on Imaging for Crime Detection and Prevention (2011)

40. Fernandez Arguedas, V., Zhang, Q., Chandramouli, K., Izquierdo, E.: Multi-feature fusion for surveillance video indexing. In: International Workshop on Image Analysis for Multimedia Interactive Services. IEEE (2011)
41. Fuentes, L.M., Velastin, S.A.: Tracking-based event detection for cctv systems. Pattern Analysis & Applications 7(4), 356–364 (2004)
42. Gevers, T., Smeulders, A.W.M.: Pictoseek: Combining color and shape invariant features for image retrieval. IEEE Transactions on Image Processing 9(1), 102–119 (2000)
43. Ghazal, M., Vázquez, C., Amer, A.: Real-time automatic detection of vandalism behavior in video sequences. In: International Conference on Systems, Man and Cybernetics, pp. 1056–1060. IEEE (2007)
44. Gray, D., Tao, H.: Viewpoint Invariant Pedestrian Recognition with an Ensemble of Localized Features. In: Forsyth, D., Torr, P., Zisserman, A. (eds.) ECCV 2008, Part I. LNCS, vol. 5302, pp. 262–275. Springer, Heidelberg (2008)
45. Halevy, G., Weinshall, D.: Motion of disturbances: detection and tracking of multi-body non-rigid motion. Machine Vision and Applications 11(3), 122–137 (1999)
46. Haritaoglu, I., Harwood, D., Davis, L.S.: W: Real-time surveillance of people and their activities. IEEE Transactions on Pattern Analysis and Machine Intelligence 22(8), 809–830 (2000)
47. Heikkila, J., Silvén, O.: A real-time system for monitoring of cyclists and pedestrians. In: IEEE Workshop on Visual Surveillance, pp. 74–81. IEEE (1999)
48. Hu, W., Tan, T., Wang, L., Maybank, S.: A survey on visual surveillance of object motion and behaviors. IEEE Transactions on Systems, Man, and Cybernetics 34(3), 334–352 (2004)
49. Huang, J., Ravi Kumar, S., Mitra, M., Zhu, W.J., Zabih, R.: Image indexing using color correlograms. In: IEEE Computer Society Conference on Visual and Pattern Recognition. IEEE (1997)
50. Hue, C., Le Cadre, J.P., Pérez, P.: Sequential monte carlo methods for multiple target tracking and data fusion. IEEE Transactions on Signal Processing 50(2), 309–325 (2002)
51. Huttenlocher, D.P., Noh, J.J., Rucklidge, W.J.: Tracking non-rigid objects in complex scenes. In: International Conference on Computer Vision, pp. 93–101 (1993)
52. Ivanov, I., Dufaux, F., Ha, T.H., Ebrahimi, T.: Towards generic detection of unusual events in video surveillance. In: IEEE Conference on Advanced Video and Signal Based Surveillance, Los Alamitos, CA, USA, pp. 61–66. IEEE Computer Society (2009)
53. Jerian, M., Paolino, S., Cervelli, F., Carrato, S., Mattei, A., Garofano, L.: A forensic image processing environment for investigation of surveillance video. Forensic Science International 167(2-3), 207–212 (2007)
54. Jiao, J., Ye, Q., Huang, Q.: A configurable method for multi-style license plate recognition. Pattern Recognition 42(3), 358–369 (2009)
55. Ke, Y., Sukthankar, R., Hebert, M.: Event detection in crowded videos. In: International Conference on Computer Vision, pp. 1–8. IEEE (2007)
56. Kumar, P., Ranganath, S., Weimin, H., Sengupta, K.: Framework for real-time behavior interpretation from traffic video. IEEE Transactions on Intelligent Transportation Systems 6(1), 43–53 (2005)
57. Le, T.-L., Boucher, A., Thonnat, M.: Subtrajectory-Based Video Indexing and Retrieval. In: Cham, T.-J., Cai, J., Dorai, C., Rajan, D., Chua, T.-S., Chia, L.-T. (eds.) MMM 2007. LNCS, vol. 4351, pp. 418–427. Springer, Heidelberg (2007)
58. Le, T.L., Boucher, A., Thonnat, M., Bremond, F.: Surveillance video retrieval: what we have already done? In: International Conference on Communications and Electronics (2010)

59. Li, X., Porikli, F.M.: A hidden markov model framework for traffic event detection using video features. In: International Conference on Image Processing, vol. 5, pp. 2901–2904. IEEE (2004)
60. Lipton, A.J., Fujiyoshi, H., Patil, R.S.: Moving target classification and tracking from real-time video. In: IEEE Workshop on Applications of Computer Vision, pp. 8–14 (October 1998)
61. Lipton, A.J., Haering, N.: Commode: An algorithm for video background modeling and object segmentation. In: International Conference on Control, Automation, Robotics and Vision, vol. 3, pp. 1603–1608. IEEE (2002)
62. Lisin, D.A., Mattar, M.A., Blaschko, M.B., Learned-Miller, E.G., Benfield, M.C.: Combining local and global image features for object class recognition. In: IEEE Computer Society Conference on Computer Vision and Pattern Recognition. IEEE Computer Society (2005)
63. Liu, F., Picard, R.W.: Periodicity, directionality, and randomness: Wold features for image modeling and retrieval. IEEE Transactions on Pattern Analysis and Machine Intelligence 18(7), 722–733 (1996)
64. Lowe, D.G.: Distinctive image features from scale-invariant keypoints. International Journal of Computer Vision 60(2), 91–110 (2004)
65. Lu, G., Sajjanhar, A.: Region-based shape representation and similarity measure suitable for content-based image retrieval. Multimedia Systems 7(2), 165–174 (1999)
66. Lu, N., Wang, J., Wu, Q.H., Yang, L.: An improved motion detection method for real-time surveillance. International Journal of Computer Science 35(1), 119–128 (2008)
67. Lu, S., Zhang, J., Feng, D.: A knowledge-based approach for detecting unattended packages in surveillance video. In: IEEE International Conference on Video and Signal Based Surveillance, p. 110. IEEE (2006)
68. Ma, L., Wu, K., Zhu, L.: Fire smoke detection in video images using kalman filter and gaussian mixture color model. In: International Conference on Artificial Intelligence and Computational Intelligence, pp. 484–487. IEEE (2010)
69. MacKay, D.J.C.: Introduction to monte carlo methods. Learning in graphical Models, 175–204 (1998)
70. Manjunath, B.S., Ma, W.Y.: Texture features for browsing and retrieval of image data. IEEE Transactions on Pattern Analysis and Machine Intelligence 18(8), 837–842 (2002)
71. Manjunath, B.S., Ohm, J.R., Vasudevan, V.V., Yamada, A.: Color and texture descriptors. IEEE Transactions on Circuits and Systems for Video Technology 11(6), 703–715 (2001)
72. Maruta, H., Nakamura, A., Kurokawa, F.: A new approach for smoke detection with texture analysis and support vector machine. In: IEEE International Symposium on Industrial Electronics, pp. 1550–1555 (2010)
73. McCahill, M., Norris, C.: Cctv in britain, pp. 1–70. Center for Criminology and Criminal Justice-University of Hull, UK (2002)
74. McCahill, M., Norris, C.: Estimating the extent, sophistication and legality of cctv in london. In: CCTV, pp. 51–66 (2003)
75. Mokhtarian, F., Abbasi, S., Kittler, J.: Robust and efficient shape indexing through curvature scale space. In: British Machine Vision Conference, vol. 62 (1996)
76. Oliver, N.M., Rosario, B., Pentland, A.P.: A bayesian computer vision system for modeling human interactions. IEEE Transactions on Pattern Analysis and Machine Intelligence 22(8), 831–843 (2000)
77. Papageorgiou, C., Poggio, T.: Trainable pedestrian detection. In: International Conference on Image Processing, vol. 4, pp. 35–39. IEEE (1999)

78. Park, S.: JK Aggarwal. Recognition of two-person interactions using a hierarchical bayesian network. In: ACM SIGMM International Workshop on Video Surveillance, pp. 65–76. ACM (2003)
79. Park, S., Aggarwal, J.K.: Simultaneous tracking of multiple body parts of interacting persons. Computer Vision and Image Understanding 102(1), 1–21 (2006)
80. Piccardi, M.: Background subtraction techniques: a review. In: International Conference on Systems, Man and Cybernetics, vol. 4, pp. 3099–3104. IEEE (2004)
81. Qu, L., Tian, X., Chen, Y.: Svm-based interactive retrieval for intelligent visual surveillance system. In: International Conference on Natural Computation, vol. 5, pp. 619–623 (2008)
82. Rasmussen, C., Hager, G.D.: Probabilistic data association methods for tracking complex visual objects. IEEE Transactions on Pattern Analysis and Machine Intelligence 23(6), 560–576 (2002)
83. Rosales, R., Sclaroff, S.: 3d trajectory recovery for tracking multiple objects and trajectory guided recognition of actions. In: IEEE Computer Society Conference on Computer Vision and Pattern Recognition, vol. 2. IEEE (2002)
84. C. Sacchi, C. Regazzoni, and G. Vernazza. A neural network-based image processing system for detection of vandal acts in unmanned railway environments. In *International Conference on Image Analysis and Processing*, page 0529. Published by the IEEE Computer Society, 2001.
85. Sahouria, E., Zakhor, A.: Motion indexing of video. In: International Conference on Image Processing, vol. 2, pp. 526–529. IEEE (1997)
86. Schettini, R., Ciocca, G., Zuffi, S., et al.: A survey on methods for colour image indexing and retrieval in image databases. In: Color Imaging Science: Exploiting Digital Media, pp. 183–211 (2001)
87. Schneiderman, H., Kanade, T.: A statistical method for 3d object detection applied to faces and cars. In: IEEE Conference on Computer Vision and Pattern Recognition, p. 1746. IEEE Computer Society Press (2000)
88. Scholkopf, B., Smola, A.J.: Learning with kernels: support vector machines, regularization, optimization, and beyond. The MIT Press (2002)
89. Schügerl, P., Sorschag, R., Bailer, W., Thallinger, G.: Object Re-detection Using SIFT and MPEG-7 Color Descriptors. In: Sebe, N., Liu, Y., Zhuang, Y.-t., Huang, T.S. (eds.) MCAM 2007. LNCS, vol. 4577, pp. 305–314. Springer, Heidelberg (2007)
90. Schweitzer, H., Bell, J.W., Wu, F.: Very Fast Template Matching. In: Heyden, A., Sparr, G., Nielsen, M., Johansen, P. (eds.) ECCV 2002. LNCS, vol. 2353, pp. 358–372. Springer, Heidelberg (2002)
91. Schwerdt, K., Maman, D., Bernas, P., Paul, E.: Target segmentation and event detection at video-rate: the eagle project. In: IEEE Conference on Advanced Video and Signal Based Surveillance, pp. 183–188. IEEE (2005)
92. Sedky, M.H., Moniri, M., Chibelushi, C.C.: Classification of smart video surveillance systems for commercial applications. In: IEEE Conference on Advanced Video and Signal Based Surveillance, vol. 1, pp. 638–643 (2005)
93. Seyve, C.: Metro railway security algorithms with real world experience adapted to the ratp dataset. In: IEEE Conference on Advanced Video and Signal Based Surveillance, pp. 177–182. IEEE (2005)
94. Shafique, K., Shah, M.: A non-iterative greedy algorithm for multi-frame point correspondence. IEEE Transactions on Pattern Analysis and Machine Intelligence, 51–65 (2005)
95. Sikora, T.: The mpeg-7 visual standard for content descriptor - an overview. IEEE Transactions on Circuits and Systems for Video Technology 11, 696–702 (2001)

96. Spagnolo, P., Caroppo, A., Leo, M., Martiriggiano, T., D'Orazio, T.: An abandoned/removed objects detection algorithm and its evaluation on pets datasets. In: International Conference on Video and Signal Based Surveillance, p. 17. IEEE (2006)
97. Stauffer, C., Eric, W., Grimson, W.E.L.: Learning patterns of activity using real-time tracking. IEEE Transactions on Pattern Analysis and Machine Intelligence 22, 747–757 (2000)
98. Stauffer, C., Grimson, W.E.L.: Adaptive background mixture models for real-time tracking. In: IEEE Computer Society Conference on Computer Vision and Pattern Recognition, vol. 2. IEEE (1999)
99. Strieker, M., Orengo, M.: Similarity of color images. In: Storage and Retrieval for Image and Video Databases, p. 381. Society of Photo Optical (1995)
100. Sun, Z., Bebis, G., Miller, R.: On-road vehicle detection: A review. IEEE Transactions on Pattern Analysis and Machine Intelligence, 694–711 (2006)
101. Tamura, H., Mori, S., Yamawaki, T.: Textural features corresponding to visual perception. IEEE Transactions on Systems, Man and Cybernetics 8(6), 460–473 (1978)
102. Tao, H., Sawhney, H.S., Kumar, R.: Object tracking with bayesian estimation of dynamic layer representations. IEEE Transactions on Pattern Analysis and Machine Intelligence, 75–89 (2002)
103. Toreyin, B.U., Dedeoglu, Y., Cetin, A.E.: Wavelet based real-time smoke detection in video. In: European Signal Processing Conference (2005)
104. Toyama, K., Krumm, J., Brumitt, B., Meyers, B.: Wallflower: Principles and practice of background maintenance. In: International Conference on Computer Vision, p. 255. IEEE Computer Society (1999)
105. Vapnik, V.N.: The nature of statistical learning theory. Springer-Verlag New York, Inc., New York (1995)
106. Veeraraghavan, H., Schrater, P., Papanikolopoulos, N.: Switching kalman filter-based approach for tracking and event detection at traffic intersections. In: IEEE International Symposium on Mediterranean Conference on Control and Automation Intelligent Control, pp. 1167–1172. IEEE (2005)
107. Vijverberg, J.A., de Koning, N.A.H.M., Han, J., Cornelissen, D.: High-level traffic-violation detection for embedded traffic analysis. In: IEEE International Conference on Acoustics, Speech and Signal Processing, vol. 2, pp. II-793–II-796. IEEE (2007)
108. Viola, P., Jones, M.J., Snow, D.: Detecting pedestrians using patterns of motion and appearance. International Journal of Computer Vision 63(2), 153–161 (2005)
109. Wen, Y., Lu, Y., Yan, J., Zhou, Z., von Deneen, K.M., Shi, P.: An algorithm for license plate recognition applied to intelligent transportation system. IEEE Transactions on Intelligent Transportation Systems 12(99), 1–16 (2011)
110. Won, C.S., Park, D.K., Park, S.J.: Efficient use of mpeg-7 edge histogram descriptor. ETRI Journal 24(1), 23–30 (2002)
111. Woo, J.W., Lee, W., Lee, M.: A traffic surveillance system using dynamic saliency map and svm boosting. International Journal of Control, Automation and Systems 8(5), 948–956 (2010)
112. Wu, G., Rahimi, A., Chang, E.Y., Goh, K., Tsai, T., Jain, A., Wang, Y.F.: Identifying color in motion in video sensors. In: IEEE Computer Society Conference on Computer Vision and Pattern Recognition, vol. 1, pp. 561–569. IEEE Computer Society (2006)
113. Xiang, T., Gong, S.: Video behavior profiling for anomaly detection. IEEE Transactions on Pattern Analysis and Machine Intelligence, 893–908 (2007)
114. Yilmaz, A., Javed, O., Shah, M.: Object tracking: A survey. ACM Computing Surveys 38(4), 1–45 (2006)

115. Yilmaz, A., Li, X., Shah, M.: Contour-based object tracking with occlusion handling in video acquired using mobile cameras. IEEE Transactions on Pattern Analysis and Machine Intelligence 26(11), 1531–1536 (2004)
116. Zhan, B., Monekosso, D.N., Remagnino, P., Velastin, S.A., Xu, L.Q.: Crowd analysis: a survey. Machine Vision and Applications 19(5), 345–357 (2008)
117. Zhang, D., Wong, A., Indrawan, M., Lu, G.: Content-based image retrieval using gabor texture features. IEEE Transactions on Pattern Analysis and Machine Intelligence, 13–15 (2000)

On the Use of Semantic Technologies to Support Education, Mobility and Employability

Valentina Gatteschi, Fabrizio Lamberti, and Claudio Demartini

Politecnico di Torino – Dipartimento di Automatica e Informatica,
Corso Duca degli Abruzzi 24, 10129 Torino, Italy
{valentina.gatteschi,fabrizio.lamberti,claudio.demartini}
@polito.it

Abstract. The technological development that has taken place over recent years has affected the way that the Web is exploited by users. In fact, a user is no longer seen as a mere observer; instead, he or she actively takes part in the production of new knowledge, which is then made available to other users. In this scenario, which is characterized by ever-growing information, new approaches (e.g., based on semantic technologies) for processing large amounts of content must be developed. Other sectors, including education and learning, as well as job seeking and hiring activities, could benefit from the exploitation of tools that can be developed within the Semantic Web initiative. As a consequence of the use of semantic technologies, learners could find more efficient training paths that provide them with their missing areas of competence and training institutes could analyze and modify existing qualifications according to a market's requirements, while companies could effectively identify the best candidates for a given job, not only within national boundaries but also worldwide. This study aims to present an overview of recent research in the field of semantic technologies applied to education and job-seeking contexts. Issues that are related to students' and workers' mobility, job seeking and hiring and the improvement of qualification offers will be analyzed and compared, by distinguishing three fields of research: knowledge base creation, the development of strategies for the integration of heterogeneous systems and the definition of inference rules, and the identification of methodologies for the visualization of qualification outcomes and curricula.

1 Introduction

Recent years have been characterized by a large change in the type of data that is stored and accessed on the Web. In fact, technological developments have transformed the Web from a pure information repository to a world where the contents are increasingly based on users' preferences, and needs are generated and disseminated. From this perspective, the user is no longer seen as a mere observer;

I.E. Anagnostopoulos et al. (Eds.): Semantic Hyper/Multimedia Adaptation, SCI 418, pp. 127–150.
springerlink.com

instead, he or she also plays an actively part in the production of new knowledge, which is then made available to others users.

It is evident that this type of context requires new approaches to be developed, to effectively process a large amount of content. In fact, although for a long time the possibility for users to find information of interest strictly depended on their ability to properly use the available search engines, today the ever-growing amount of information gathered on the Web makes experience-based data processing extremely time-consuming; hence, new information mapping systems, e.g., systems that are based on semantic technologies, must be exploited.

From this viewpoint, the education and learning field could benefit from the exploitation of instruments that are created within the Semantic Web initiative. Solutions developed until now range from the use of semantic tools to support learning, such as the so-called *Computer-Supported Collaborative Learning (CSCL)* systems, to instruments that are devised to improve interactions among learners, to allow them to reduce the effort that is needed to complete a topic individually [Devedžić, 2005, Devedžić, 2006a], and to the exploitation of metadata to improve the e-learning experience, such as *Learning management systems (LMSs)*, software packages developed to help trainers to foster quality in online courses and manage learning outcomes [Aydin and Tirkes, 2010].

However, despite the main focus of recent research on the use of instruments such as ontologies and taxonomies to improve the way teachers and learners approach education [Devedžić, 2006b, Bittencourt et al. 2008 and Anderson et al. 2004], issues that are related to the definition of occupational profiles, the construction of professional qualifications and the improvement of job seekers employability, from a lifelong learning viewpoint, should also be considered to be of primary importance.

In fact, even though in Europe, during recent years, several instruments to support mobility and employability were defined (e.g., EQF "European Qualification Framework" [EQF, 2008] and ECVET "European Credit System for Vocational Education and Training" [ECVET, 2009]), students and workers who decide to spend a working period abroad still encounter several difficulties in the recognition/validation of their qualifications, which can be mainly associated with information asymmetries among students, job seekers, employers and training centers. This asymmetry results from the fact that, usually for a student who decides to continue his/her study career abroad, it is difficult to find (within course descriptions) areas of competence that he/she needs to achieve or classes that he/she needs to attend to obtain a given qualification.

This difficulty arises from the heterogeneity of the qualifications structure and from the lack of well-established definitions. For example, two countries can show qualifications that are articulated in different ways, in which the contents could be described heterogeneously; moreover, qualification pillars (such as knowledge, areas of competence and skills) could assume different meanings in a specific national domain, with serious consequences to mutual understanding.

Similar considerations apply to the working dimension: frequently, human resources staff (who have a substantial psychology background but are possibly lacking in technical knowledge), when examining a curriculum, e.g., in digital form, often perform a type of "keyword-based" analysis on the basis of requirements that are indicated by the business department that is looking for new workers to hire; however, when specialized areas of competence that are required to fully understand the curriculum are not available, a possibly adequate match between a demand and an offer could be lost, e.g., because of the use of non-aligned vocabularies. An example could be a job seeker who declared his or her knowledge of "Java" and "C++" but would not be selected as a possible candidate if the requested knowledge was related to "object-oriented programming languages".

This study aims at presenting an overview of recent research in the field of semantic technologies applied to education and job seeking contexts. Specifically, we show approaches for improving educational offers as well as for the comparison of curricula and for fostering students' and workers' mobility; these approaches are described and analyzed. Section 2 presents an overview of existing studies that investigate whether Semantic Web instruments could support education and job recruitment processes and analyzes the mobility issue from different perspectives. In addition, three fields of research in which Semantic Web technologies could be exploited are identified. In Section 3, the first field of research, knowledge base creation, is presented, by describing it in conjunction with recent studies. Section 4 reports a list of research activities that have been performed in the second field, the integration of heterogeneous systems and the definition of inference rules, whereas studies within the third field, visualization, are depicted in Section 5. Finally, conclusions are reported in Section 6.

2 The Relevance of Semantic Web Technologies to Education

In recent decades, the way that people approach learning has changed. Currently, increasing numbers of students participate in forums or surf the internet to gain knowledge or to find resources for their homework; they have learned how to search, reuse and (often) publish information.

A recent survey [Netcraft, 2011] identified, in May 2011, 324.697.205 websites that are accessible on the Internet. Last year, during the same period, the number of active websites was only 206.026.787. Even though some people argue that "The more information the better!", identifying searched information is sometimes difficult; hence, available data should somehow be processed, to simplify teachers' and students' searches.

A first study [Ohler, 2008] investigated how Semantic Web technologies could be exploited to support learners and identified three areas of impact: knowledge construction, personal learning network maintenance and personal educational administration. According to the author, Semantic Web instruments could be used to accomplish the following: a) produce, in reply to a search, a multimedia report that is drawn from many sources such as websites, chapters in textbooks, and speeches posted on YouTube, rather than a list of hits (knowledge construction area); b) identify relevant information from blogs, podcasts and other semantically accessible sources and provide an information synthesis tailored to the student's personal learning objectives, thus creating personal learning networks that are mainly built around subjects rather than around services (personal learning network maintenance); and c) compare courses that are provided by different institutions to improve student mobility (personal educational administration).

Other authors [Tiropanis et al. 2009] conducted a survey of semantic tools and services that were relevant to UK higher education and identified learning and teaching challenges to which they could be relevant. Specifically, the working group involved in this research activity recognized that the main areas that could benefit from Semantic Web technologies are the following: a) course creation and revision, b) the recommendation of resources that match the topics of students' assignments, c) the creation of groups for collaborative work on the basis of a learner's background, personal preferences and successful prior collaboration, d) the creation of links between discussions for enhancing critical thinking, e) the match of people and resources across schools in the same or different institutions, as support for cross-curricular activities, f) the creation of personalized knowledge, and g) the support of group knowledge construction.

Starting from the above challenges, the authors tested the available tools and services and identified best practices for four main categories. Specifically, for the first category, collaborative authoring and annotation tools, 7 services were identified. For the second category, searching and matching tools using semantic technologies, 2 tools were found. For the third category, repositories for import/export of data using semantic technologies, 5 tools were identified, whereas for the fourth category, infrastructural tools and services for the integration of data sources across organizations in interoperable semantic formats, 4 services were found. Hence, whereas several tools for collaborative authoring and annotation already exist, only a few instruments for performing searches within institutions have been developed.

The results of the above studies denote to what extent Semantic Web technologies could be used to support education. In addition, job seekers or human resources staff could benefit from these instruments, because an adequate exploitation of semantic tools could increase market transparency, lower the transaction costs for employers, and change the business models of the intermediaries that are involved, as reported in [Bizer et al. 2005].

The focus of this chapter will be mobility and job recruitment: specifically, an investigation on how existing tools could improve activities such as the identification of training modules that a learner/worker should attend, to increase his areas of competence, so that the choice of the right candidate for a given job position will be accomplished.

The mobility issue is multi-faceted. In fact, it affects both students and workers.

Usually, people decide to spend a studying or working period abroad to acquire missing areas of competence or to find better working opportunities. In this context, students must identify the set of competences that could be gained abroad, whereas workers, to have more chances to be employed, must communicate as best they can what are their qualities. Unfortunately, in addition to language barriers, other difficulties arise because countries usually present different training systems, and it is extremely difficult to identify competences that are provided by a qualification or are held by workers.

The first instrument that supports European mobility is the European Qualification Framework, which is a framework for classifying qualifications according to 8 levels that precisely identify the semantic of *qualification*, *learning outcome*, and *knowledge*, *skill*, and *competence* concepts, thus enhancing the creation of an expected shared understanding in a life-long learning domain. Thanks to EQF, training systems (or workers) defining their courses (or curricula) according to the above guidelines have more possibilities for being understood by other actors. Nevertheless, because of the very large amount of data, a mere application of EQF principles conducted on a manual basis, in most cases, is not sufficient. Hence, ad hoc instruments addressing the following issues should be created:

- concerning *students' mobility*, EQF-based matchmaking systems that allow students to analyze qualifications that are provided by other training institutions and to identify classes that they need to attend, to fill their gaps, must be developed;
- with regard to the *quality of the qualification offer*, instruments that enable training authorities to develop qualifications that answer to the needs of the labor market and that could combine training modules by taking into account their level of difficulty, as well as relations among them, should be created;
- for what concerns *workers' mobility*, tools supporting the owner of a qualification in communicating to employers his/her qualities and in identifying the best jobs to apply for, as well as human resources staff in identifying the more suitable job applicant for a job, must be provided.

These instruments should be capable of comparing heterogeneous training systems, curricula and job offers and should allow the user to perform not only keyword-based searches but also more complex queries that are based on relations among concepts; hence, they should exploit Semantic Web instruments, such as taxonomies (classifications arranged in a hierarchical structure) or ontologies (*explicit specifications of a conceptualization*, as defined in [Gruber, 1993]).

Moreover, the above tools should a) rely on a knowledge base that collects relevant information or on different merged knowledge bases, because of the heterogeneity of the information to be processed; b) foresee the definition of inference rules and perform matchmaking, to provide end-users with a ranked list of potential qualifications and job applicants answering their needs; and c) present to end-users the above results in an easy and understandable way.

In the following Sections, research activities that answer the needs of fostering students' and workers' mobility and that improve the quality of qualification offers will be investigated according to the three requisites identified above: a) *knowledge base creation*, b) *integration of heterogeneous systems and definition of inference rules* and c) *visualization*.

3 Knowledge Base Creation

To allow the automatic processing of information, collected data should be expressed according to a shared formalism [Staab and Studer, 2004]. This task could be accomplished at two different levels of detail: the first level of detail concerns the formalization of the way that the information is expressed (which allows a machine to identify where to retrieve given data), and a second level of detail allows a software tool, once it identifies the searched information, to analyze its contents (by describing concepts that express its meaning).

The steps to be conducted to formalize existing knowledge are expressed well in [Tao et al. 2005]: in this study, the authors divide the knowledge lifecycle into 4 phases: *knowledge acquisition (KA)*, which requires interviews with experts or desk analysis to develop a domain vocabulary of the most important concepts, *knowledge management (KM)*, in which, starting from concepts identified in the previous phases, an ontology is built, *knowledge annotation*, in which resources from the domain are annotated with the ontological metadata, and, finally, *knowledge reuse*, which is achieved when new applications reuse the resources. In this study, for the KA phase, learning domain experts and teaching and learning materials were interviewed and investigated, to identify key concepts; then, in the KM phase [Protégé, 2000], an ontology-building tool, was used for the definition of concepts and hierarchical relationships among them. For the knowledge annotation phase, instances were generated and exported in OWL [W3C, 2004] files, which were exploited in the knowledge reuse phase for recommending learning materials or for combining simple services to realize more complex customized functionalities.

While the above work provides an overview of the whole knowledge lifecycle process and focuses mainly on the reuse of learning materials, other authors investigated whether semantic tools could be used to support the personalization of training paths, with the aim of improving students' curricula. Specifically, two papers [Poyry et al. 2002, Poyry and Puustjarvi, 2003] showed how metadata could be exploited to support learners who are looking for higher education courses that match their needs: with this aim, the authors first investigated whether the Learning Objects Metadata (LOM) [LOM, 2002] could be used to describe courses provided by European universities and then suggested an extension of it, by identifying four aggregation levels (study material, study course, study package and study program).

In a more recent study, [Sampson and Fytros, 2008], realized that the LOM specification did not directly support the description of learning resources in terms of their relevance to competence-based learning programs and suggested an IEEE Competence Application Profile for tagging educational resources: the authors proposed to use the Purpose (Nr 9.1) element to specify that acquiring a competence is the outcome of a learning object and the Difficulty (Nr 5.8) element to communicate the degree of difficulty of a learning object; they created a competence category that consisted of three main elements: title, description and context.

While the above approaches focus on methodologies for expressing the structure of information, other research activities aim at identifying the best ways of expressing the content of a qualification in a machine-understandable format. This aspect is of primary importance in all of the cases in which phrases that are syntactically heterogeneous but that have similar meaning must be compared, such as the identification of the best job applicant for a given job (because curricula are not necessarily expressed through the same terms that appear in a job offer) and the comparison of different training offers.

A first solution, mostly applied in the training domain, is presented by Bloom's taxonomy [Bloom, 1956]. Bloom's taxonomy was initially developed for educational assessments, and was subsequently exploited to evaluate courses according to the type of subjects that were taught. Bloom identified 6 levels of learning mastery (from less to more complex); namely, (1) *knowledge*, involving the mere recall of methods and processes; (2) *comprehension*, representing the lowest level of understanding; (3) *application*, requiring abstraction in specific and concrete situations; (4) *analysis*, the identification of constituent elements of communication and the understanding of relations between them; (5) *synthesis*, the combining of elements to form a whole; and (6) *evaluation*, the judgement about the value of the methods for a given purpose. A revised version [Anderson et al. 2001], slightly modified the original taxonomy by inverting the synthesis and evaluation concepts (renamed *create* and *evaluate*, respectively), thus better reflecting engineering disciplines, and by expressing other concepts with a verb, instead of a noun. Moreover, the revised version defined also a new dimension, the knowledge dimension, which describes factual knowledge, conceptual knowledge, procedural knowledge and metacognitive knowledge. It is worth remarking that this new representation allows training authorities to describe each element of a course as a paired verb-noun, thus defining 24 possible combinations. An application of Bloom's revised taxonomy is presented [Spivey, 2007]: in this study, the author presented how Digital Logic Design courses could be modeled according to [Anderson et al. 2001]. A matrix reporting the knowledge dimensions (nouns) and the cognitive process dimension (verbs) was drafted, and course objectives, activities, homework and quizzes were written in cells according to their degree of difficulty and the type of knowledge that was required. This representation allowed teachers to not only produce homework that was tailored to tough subjects but to also make it possible to compare homework of common Digital Logic lessons, thus enabling the reuse of existing material.

Another example of how Bloom's taxonomy could be exploited for the comparison of qualifications is reported in [Bourque et al. 2003]: in this study, the authors compared areas of competence that were acquired by new graduates, graduates with four years of experience, and experienced software engineers, by assigning them to one of the categories defined in [Bloom, 1956].

[Starr et al. 2003] presented a different nuance of Bloom's taxonomy, because they introduced a meta-level structure: according to the authors, three meta levels could be distinguished: *memorization and basic understanding* (beginner level), *use or competent application* (intermediate level), and *design or creation and critique* (expert level). Each level is represented by two phases, a production of learning and an explanation phase. Moreover, the authors identified an interesting phenomenon, namely concept shifting, that occurs when concepts describing a competence are inadvertently switched with a related, less abstract concept: in this case, the associated level of Bloom's taxonomy may denote a different degree of difficulty when applied to higher or lower-level concepts (e.g., the "iteration" concept at the Synthesis level implies the ability to design new ways to perform a loop, whereas the "for loop" at the same level entails only the ability to write loops to perform given tasks).

While the authors of the above studies presented strategies that were based on the exploitation of Bloom's taxonomy for a qualitative comparison of study courses, [Hoffman, 2003] defined a quantitative approach to computing the level of skills that belong to a qualification. In his study, the author started from the matrix, reporting the knowledge dimension and the cognitive process dimension and, for each intersection, he defined a skill level that explains how much a given skill is acquired by students who attend the course. In addition, he suggested calculating the center of gravity of both the knowledge and cognitive process dimensions. This representation was extremely useful, because it quickly communicated the characteristics of a given course as well as the level of the skills that were acquired by the students.

This approach, expressing skills as a combination of a coupled noun-verb, could be exploited also from the EQF perspective. However, while the knowledge and skills dimensions could be represented with the proposed methodology, the competence element was still missing. [Pernici et al. 2006] suggested a way to overcome this limitation and to cope with EQF guidelines by introducing *knowledge objects* (*KO*), *action verbs* (*AV*) and *context* (*CX*) concepts. According to the authors, knowledge could be seen as a set of KOs, a skill could be defined as a KO that was "put into action" through an AV, hence by one or more pairs KO - AV, and a competence could be represented as a triple KO - AV - CX, thus describing the ability of putting into action a given KO in a specific context.

In [Gatteschi et al. 2011a], the above approach was exploited for strengthening the relation between the labor and the educational worlds, by defining a methodology for comparing occupational and educational profiles and for identifying their common elements, to create a common profile that accounts for the requirements of both fields. Specifically, this study is aimed at lowering not only the transnational barriers but also the differences between labor market needs and educational system outputs by creating a common profile that is shared at the European level within the commerce field. The methodology proposed by the authors consisted of linking the elements that belong to occupational and educational descriptions to a set of concepts that were organized into a taxonomy, and it could be divided into four stages: *information collection*, *taxonomy and ontology construction*, *definition of inference rules and approaches for semantic comparison* and, finally, *common profile creation and examination of results*. In the first phase, labor and educational markets of a number of European countries were investigated by means of interviews, with the aim of collecting business requirements (i.e., a list of knowledge, skills and areas of competence that a worker must possess for fulfilling a task) and training outputs (i.e., learning outcomes that are achieved by the students at the end of a specific training route), both of which are characterized by the corresponding EQF level, depicting the degree of complexity. Collected information was then inspected, to identify the core elements (knowledge objects, action verbs and context elements) to be structured into a taxonomy; then, for each instance of knowledge, skill and competence elements, one or more concepts were selected. These elements were then linked to each other by exploiting subsumption relations in a taxonomic representation, identifying three families of concepts (KO, AV and CX), which were organized into a hierarchical tree. For the creation of the AV tree, Bloom's taxonomy was exploited and adapted, thus identifying six families of verbs, *arrange*, *act*, *prepare*, *check*, *assess*, and *react*, which were ordered according to their level of difficulty, whereas the KO and the CX families were created from scratch, with the support of experts. After the creation of the taxonomy, qualifications, tasks and subtasks that were collected from interviews were described by linking their composing elements (knowledge, skills and areas of competence) to the corresponding concepts (knowledge objects, action verbs and context elements). The newly created ontology was exploited in the third phase, where different strategies for matchmaking were defined. Results provided by the different algorithms were analyzed in the fourth stage, to evaluate the improvement that was associated with the use of the taxonomy.

[Harzallah et al. 2002] present another way to strengthen the links between the working and the training/job seeking contexts: in fact, instead of focusing on strategies to create a shared profile, which would be understood by both worlds, the authors present the CommOn framework, a Competency-Based System targeted to recruiters, job seekers or training providers. By the exploitation of such instruments, users of the Healthcare and Information and Telecommunication domains can formally represent competency profiles and compute the matching of them. The basic idea behind this study is that a curriculum vitæ, a job offer or a training program usually represents a synthetic view of a richer network of areas of competence; hence, the underlying areas of competence should be made explicit and should be formally represented, to allow an e-recruitment service to perform automatic analysis and matching. For this purpose, two ontologies were developed: the first ontology was dedicated to the definition of a competency reference system for each domain (e.g., it defined composing elements of a Job Position, such as Objectives and Resources), while the second ontology was devoted to the definition of competency profiles (hence, to the identification of the quintuplet *knowledge*, *behavioral aptitudes*, *basic competencies*, *aspects of context* and *objective* characterizing a competency). Users of the CommOn platform could then express their curriculum vitæ or job requirements by combining the above elements.

Another interesting study that is devoted to the improvement of students' and workers' mobility is [De Meo et al. 2007]. Here, the authors propose an XML-based multiagent recommender system to support online recruitment services. Specifically, ontologies for two agents (the User Agent, storing users' profiles, preferences and constraints and past queries and the Recruitment Agent, consisting of a set of ranked job proposals) are developed and stored as XML documents. The advantage of the proposed approach consists of the fact that the system integrates different e-recruitment services and learns users' preferences by storing their reactions to past proposals.

[Hexin and Bin, 2006] propose a different approach to recruitment activities: in their study, the authors present an intelligent recruitment platform with a skill ontology-based semantic model and its matching algorithm. By exploiting such a platform, recruitment websites can automatically find the most suitable job seeker for a given job position and vice versa. According to the authors, the proposed ontology is composed of a set of sub-ontologies that are expressed in the form of a weighted graph, denoting concepts (skills, such as C++, C, and Java) and weighted relationships, expressing the degree of relatedness between two nodes (e.g., an expert of C++ is quite familiar with C while someone who knows C also knows something of C++; hence, the weights assigned to relations between the two concepts will be 0.2 and 0.7). As a result, the platform can compare job offers and job seekers' curricula even in those situations in which they are heterogeneously expressed.

[Gatteschi et al. 2011b] propose another way to tackle the heterogeneity that characterizes job seekers' curricula and companies' requirements, based on the WordNet database [WordNet, 1985]. More specifically, the authors present the LO-MATCH platform[1], a web-based tool that allows learners to effectively find qualifications that fulfill their education and training needs, while job seekers can find the job offers that better match their abilities, and companies can identify the best candidates for a given job position. In addition, in this study, a tag cloud-based visualization technique, allowing the system to quickly depict aspects to be considered in the mobility and job-seeking phases, is presented. The functioning of such a platform is the following: when a user specifies a new learning outcome (e.g., as part of a qualification/curriculum vitæ or an occupational profile), the system automatically detects and suggests to him/her the relevant concepts that could be linked to each word in the newly introduced element. The user can then decide whether to use the suggested concepts to annotate his/her learning outcomes. Each time that a concept is specified, the platform also records concepts that are semantically linked to it, according to the WordNet lexical database. Moreover, the user can also specify new terms that could be used for the annotation and eventually links them to other concepts that belong to the WordNet ontology, in case they are not contained in it. It is worth remarking that the advantage of this strategy, which is to rely on a database containing a very large number of semantically linked terms, such as WordNet, instead of on an ontology built from scratch, allows the platform to be adopted by workers/companies acting in different sectors.

The different approaches devoted to support learners, job seekers and job recruiters presented in this Section show how the exploitation of a semantic tool could improve mobility and employability as well as existing training offers. Devised solutions include a) the use of metadata for describing training courses [Poyry et al. 2002, Poyry and Puustjarvi, 2003 and Sampson and Fytros, 2008]; b) the exploitation of Bloom's taxonomy for the description/comparison of learning paths, to allow students to improve their employability by personalizing their curriculum [Spivey, 2007, Bourque et al. 2003, Starr et al. 2003 and Hoffman, 2003];

[1] http://www.lo-match.polito.it

c) the creation of ad hoc ontologies to express the relations among terms of a given domain [Gatteschi et al. 2011a and Hexin and Bin, 2006] or among the composing elements of curricula/job positions [Harzallah et al. 2002]; and d) the use of existing semantic thesauri to annotate and compare job offers and demands [Gatteschi et al. 2011b].

The utilization of metadata is particularly useful in those cases in which information is unstructured (such as courses from different providers or curricula expressed with a variety of structures): in this view, they could improve search and comparison phases by providing more correct results. However, if only metadata are used, users can only perform keyword-based searches, because the system would not be able to compare sentences that are expressed through different terms. To perform such a study, ontologies defining relations among terms should be exploited. In this view, Bloom's taxonomy is certainly useful for the description of skills because it identifies families of verbs that are grouped according to the degree of complexity. However, it lacks relations among terms that belong to the same group, and it is still limited to a small set of concepts. A possible way to overcome this limitation could be the extension of it in a given context, by the creation of an ad hoc ontology for the domain of interest: this solution could allow a software system to automatically address job offers and qualifications, which are expressed in a heterogeneous way. However, when a broad context such as job recruitment or qualification comparison is considered, this solution is no longer sustainable because an ad hoc ontology should be developed for each working/learning domain, thus resulting in an extremely time-consuming activity. Research and future studies in this field should then focus on the adaptation of existing ontologies/semantic thesauri, such as WordNet and DBpedia [Auer and Lehmann, 2007] and on their exploitation for automatically annotating job offers, curricula and training courses. The only drawback of this solution is that, in some contexts, the above ontologies are not sufficient, because they lack concepts or relations. Hence, applications that allow users to eventually extend them should be developed. Moreover, because a word can assume one or more meanings, according to the context in which it is used, the possibility of automatically annotating the inserted data according to information that is already provided by users (e.g., the domain of interest) should be investigated.

4 Integration of Heterogeneous Systems and Definition of Inference Rules

While Section 3 presents different strategies for the creation of a knowledge base for collecting curricula, different approaches could be devoted to the integration of existing academic management systems, for the definition of rules for matchmaking (a process that queries a knowledge base and returns all of the elements that potentially match the requirements that are expressed by the user), to support students/workers who are looking for courses to attend, to improve curricula, to help human resources staff identify more suitable candidates for a given job, or to help training institutes improve their training offerings. As presented in [Ronchetti and Sant, 2007], ontologies could be used for managing, inspecting and monitoring a full study curriculum of courses, because they could allow a system to verify overlaps between courses, to find out areas that are not covered and to analyze possible synergies with courses offered in other schools. The authors started from the ontology extracted by [Saini and Ronchetti, 2003], which was based on the ACM Computing Curricula 2001 for CS (CC2001), a comprehensive work defining Computer Science curricula for undergraduate students and specifying prerequisites and syllabi for courses. In the CC2001 view, each course could be defined by topics (the smallest-grain elements, divided into core and elective), units (collections of topics) and areas (collections of units). The objective of [Ronchetti and Sant, 2007] was to analyze syllabi of the courses provided for the Computer Science Bachelor degree at the University of Trento, Italy. The pursued approach consisted of matching all of the syllabi against the ontology, asking teachers to identify the topics that were taught, and showing the results of the mapping. Results of these activities allowed trainers to identify the non-covered areas, thus improving the training that was offered.

A different methodology, which was developed to enable learners to integrate classes from other institutions into their curriculum, in the Bologna Process view [Bologna Declaration, 2001], is shown in [Hackelbusch, 2006]. In this study, the author presented a system that provided students with a ranked list of classes offered by other academic institutes, by including only classes that were identified as interchangeable, from an organizational and semantic point of view. With this aim, a curriculum mapping ontology, which was used as a common basis for formalizing academic programs, was developed. However, to find similar classes, a mere comparison of the title of the modules was not sufficient, because modules with the same title could contain completely different subject matters. Hence, the author proposed to exploit methods for indexing texts, to summarize module content.

While the above study starts from the definition of an ontology and requires a formalization of academic programs, the strategy proposed in [Cubillos et al. 2006, Cubillos et al. 2007a, Cubillos et al. 2007b] starts from the assumption that, due to political reasons, in some cases, the creation of a common ontology could not be possible; hence, local meta-ontologies, allowing an automatic tool to compare heterogeneous qualification systems, should be developed. The aim of these studies was to define a methodology exploiting meta-ontologies to tackle the problem of integrating qualification systems in the Vocational Education and Training (VET) context, to improve the transparency and mobility of students across Europe. The steps for the formalization and construction of the local meta-ontologies are shown in Fig. 1: specifically, an initial document-gathering phase was conducted, to collect and analyze all of the relevant information; then, a UML [UML, 1996] model was drafted. Finally, a third step, specifically the construction of templates and the compilation of case studies, was performed. Based on the above documents, an ontology was built with the OWL language.

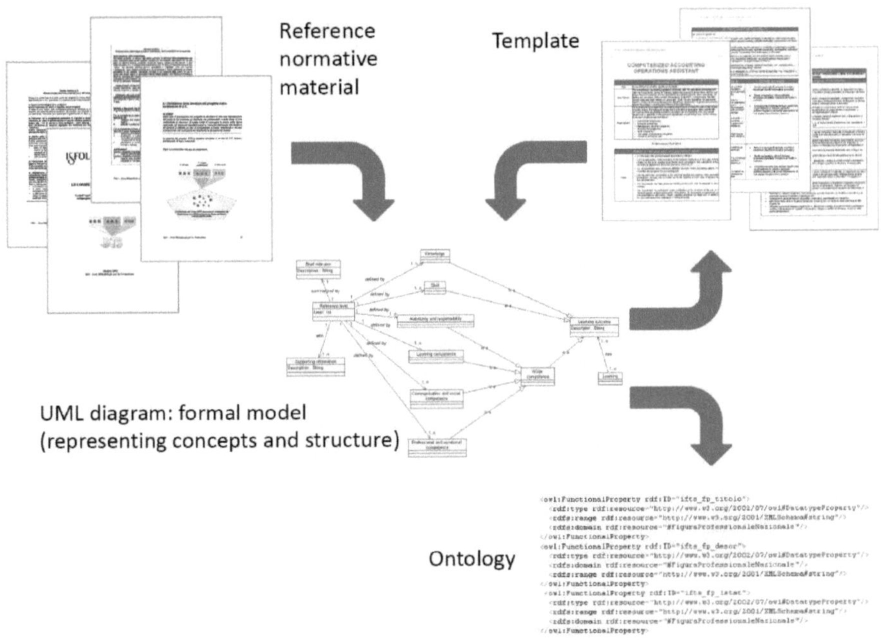

Fig. 1 The ontology construction steps [Cubillos et al. 2007a].

For the document-gathering phase, we collected information about the national educational system (reporting an overview of the general educational system in the country), the post-secondary non-academic education (detailing different types of schools providing this type of qualification), the post-secondary non-academic education system (a deeper analysis for selected systems) and the profiles comprising the parts and description of the concepts (a detailed study on how a profile was articulated in terms of competences and modules). The UML formalization phase was performed to provide a formal and easy-to-read representation of the gathered documents; in this phase, classes and attributes for each qualification model were detected, and relations among them and foreign qualifications were drafted. At the end of this step, one UML diagram was provided for each couple of countries to be compared. However, because this representation was not familiar to all of the partners who were involved in the development of the tool, formalized information was also structured in MS-Word templates, to be checked also by a non-skilled person. These templates were also used for the compilation of study cases. Fig. 2 reports an excerpt of the meta-ontology for the French and Dutch systems; this representation was able to immediately indicate where (in a "target", e.g., foreign, qualification) a semantic engine should look to find information that was comparable with data belonging to the "source" (e.g., national) profile.

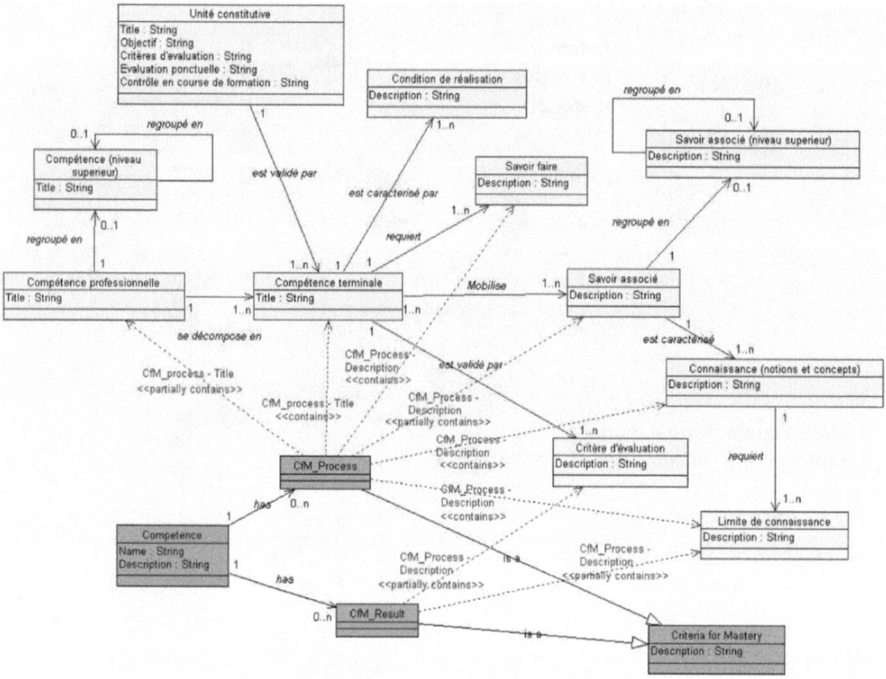

Fig. 2 UML-based representation of the rules for referencing the Dutch and French models.

Once the meta-ontologies were defined, qualification profiles were annotated by referring them to a common glossary of concepts, and for each concept, a weight expressing its relevance was inserted [Gatteschi et al. 2009] that was aimed at exploiting the above models to help students to identify, starting from their missing areas of competence, courses to attend at a foreign institution; two different strategies were proposed for matchmaking. A first approach, exploiting country-to-country relations, was pursued. However, it was pointed out that this strategy is not suitable when the number of qualification systems to be compared is high, because, for each comparison, a meta-ontology such as the one shown in Fig. 2 must be developed. Hence, a country-to-EQF strategy was suggested. According to this approach, classes composing a qualification must be referenced to classes that are identified by the EQF (such as learning outcome, knowledge, skills, and areas of competence), thus exploiting the EQF as a translator device.

A different strategy for matchmaking, which was applied in the context of job offers and demands, is presented in [Hexin and Bin, 2006]. Here, the authors start from an ontology that depicts skills and their weighted relations and computes the match between a job offer and job candidates' characteristics by finding the shortest path between the required and the offered skills. The proposed system allows human resources staff to perform a first analysis of candidates' curricula and accounts for the priority (i.e., the importance of a skill for a given job), to provide results that could better satisfy a company's requirements. It is worth remarking that, in this study, even if the authors refer to workers' characteristics as "skills", only the "knowledge" dimension (according the EQF definition of learning outcomes) is analyzed, while elements regarding the ability to use the nodes of the ontology or to apply them in a given context are missing.

[Gatteschi et al. 2011c] attempts to overcome this limitation by presenting an ontology-based matchmaking algorithm, to make a comparison of the qualifications that are expressed, according to the EQF guidelines. The proposed approach exploits subsumption relations among three sets of concepts, KO, AV and CX, similar to the definition in [Pernici et al. 2006], and addresses also the relations among the above sets, which are specified by the users (e.g., couples KO-AV or triples KO-AV-CX). As a result, the matchmaking tool can identify "similar" skills such as *to be able to program operating systems* and *to be able to develop Linux OS* (because one of them is more specific than the other) and to infer that they have a different level of difficulty, with respect to the skill *to be able to use operating systems*.

As shown by the study results presented in this Section, tools developed within the Semantic Web initiative could be used for performing the following activities: a) analyzing and verifying training offers [Ronchetti and Sant, 2007 and Saini and Ronchetti, 2003]; b) comparing training modules to allow learners to personalize their curricula [Hackelbusch, 2006]; c) expressing heterogeneous qualifications through a common formalism [Cubillos et al. 2006, Cubillos et al. 2007a, Cubillos et al. 2007b and Gatteschi et al. 2009]; and d) computing matchmaking between qualifications/curricula and job offers [Gatteschi et al. 2009, Hexin and Bin, 2006 and Gatteschi et al. 2011c].

Concerning the usage of ontologies for verifying missing competences in a training module, the study presented in [Ronchetti and Sant, 2007] provides interesting results. However, an approach such as the one shown by those authors requires an ad hoc ontology for each training domain; hence, the feasibility of this strategy for different sectors should be investigated.

The system presented by [Hackelbusch, 2006] can compare module organizational requisites and attempts to investigate whether a comparison could also been performed on the content of the modules themselves. However, this approach appears to be able to work only on courses owning the same structure.

Learners could benefit from the exploitation of a meta-ontology that links heterogeneous training systems, through the identification of a common framework [Gatteschi et al. 2009]. In fact, even if this solution requires the definition of a shared framework, as well as a considerable amount of work from experts of the training domain of different countries to create ontologies for each coupled country-common framework, it could be used as a translator device when learners or companies want to compare foreign qualifications for the purpose of finding interesting training paths or the characteristics of a job applicant having a specific qualification.

The definition of the rules for matchmaking could simplify the above activities, which could be performed automatically instead of on a manual basis. However, to provide correct results, they should rely on a semantic representation of relations among terms. Studies presented in [Hexin and Bin, 2006 and Gatteschi et al.

2011c] allow us to perform this task; however, they exploit a small and ad hoc ontology of terms. From this viewpoint, future research activities should be devoted to the application of matchmaking algorithms in those situations in which learning outcomes are annotated through terms that belong to shared and existing ontologies. Moreover, an interesting field of research could be the identification of rules for automatically assigning a numerical value to relations among concepts (as already proposed in [Hexin and Bin, 2006]), to improve the matchmaking results.

5 Visualization

While Section 3 and Section 4 focus on the representation of knowledge in a machine-understandable way and on strategies to perform automatic reasoning on it, another aspect to be considered is how very large amounts of data, such as the datasets that characterize a context such as the one analyzed in this chapter, could be effectively presented to end-users.

In fact, even though a system for the comparison of qualifications or for job matchmaking could provide users with a list of ranked results, a user could be interested in analyzing why a given qualification/module or candidate obtained the specific ranking position (i.e., is the job candidate suitable for the job because he/she has a low level of knowledge of all of the aspects of interest or because he/she possess a deep knowledge of only one aspect that is required for the job). Consequently, in the following, different approaches for the representation of qualifications and curricula will be investigated.

Several research activities have been performed to provide a graphical representation of the qualifications, outcomes or relationships between courses. A first approach is presented in [Gestwicki, 2008]; in this study, the author suggests a curriculum visualization application in which a curriculum was modeled as a directed graph, courses were represented by nodes, and relationships between them were depicted by edges. By looking at the graph, it was then possible for students and program administrators to identify the flow of a curriculum in terms of course prerequisites or the amount of time that is needed to satisfy course requirements. According to this representation, elective courses were represented by shaded nodes, whereas required courses were depicted as unfilled nodes. Nodes were then linked by different types of arrows, representing, for example, prerequisites. By selecting a shaded node, and by analyzing arrows pointing to it, it was possible to identify prerequisites, and by finding the longest path between nodes, the user could determine the number of semesters it would take to fulfill the course requirements.

Another interesting strategy that was exploited to manage and maintain Medicine curricula is reported in [Dexter and Davies, 2009]: this research aimed at easing the continual process of review, characterizing curricula that require frequent revisions and adopting the Problem Based Learning approach. The authors first built a visual model of the curriculum by exploiting the UML formalism, thus identifying, among others, the competency class; they then tagged each competence by assigning it values for different dimensions and built the model as an ontology in OWL. Finally, to visualize the created knowledge base, they suggested a visual metaphor that was derived from the London Underground map. On this map, elements of the ontology were represented as stations, whereas object properties (i.e., relations) were depicted by lines. Similar to the London Underground map, not all stations are linked to all others, and there could be different routes that may be taken between stations; hence, the user can browse the knowledge base by selecting a station and by following trails that connect related items.

While the above studies adopt a two-dimensional representation of the qualifications, other authors suggested a three-dimensional depiction. In fact, [Hoffman, 2003] represented skills composing a learning objective as blocks defined by a triple of values characterized by the knowledge dimension (x axis), the cognitive process dimension (y axis) and the skills level (z axis). According to this description, the height of a block communicates the degree of acquisition of a given skill, whereas the spatial position provides information on its level of complexity.

[Sommaruga and Catenazzi, 2007] took a step forward in this direction, by representing university undergraduate education programs in a 3D environment, based on a city metaphor. In this view, each department is depicted as a district, whereas each curriculum offered by the department is represented by a block, divided into 6 areas, one for each semester (because bachelor courses are three years long). Each block contains one or more buildings (modules of a curriculum), which have heights and widths that are proportional to the number of credits and the duration of the module, respectively. An ad hoc exploitation of colors (different colors for each department and different nuances for each semester) completed this representation, by making it easier to distinguish the main characteristics of a training path and by allowing users to quickly compare different educational programs.

While authors of the above studies suggested strategies to visualize relations among courses and prerequisites and to compare training paths by analyzing graphical depiction of skills or modules, an interesting field of research is represented by the visualization of qualification content.

In fact, to compare qualifications, concepts that are taught should be accounted for. A quick way to produce a graphical representation of the main subjects that are provided by a training institute is to create a tag cloud (a visual overview of textual data, often corresponding to a set of tags in which the font size used for drawing the tag is generally linked to its frequency) based on course specifications. However, [Gatteschi et al. 2011a] remarked that this representation, if produced by submitting qualification specifications to an online tool, such as Wordle [Wordle, 2009], does not account for relations among terms; hence, it is not suitable when the courses are described with a variety of terms (because synonyms are not grouped together). They suggest a representation of concepts composing qualifications or working activities exploiting UML notation (for this purpose, the open-source tool UMLGraph [UMLGraph, 2003] was used). Fig. 3 depicts an excerpt of the working subtask *To welcome the customer and understand the customer's needs and requests*, which belongs to a Shop Assistant profile, defined as presented in Section 3; more specifically, the diagram displays the knowledge *Communication techniques knowledge*, the two skills *To be able to apply selling techniques* and *To be able to communicate in English*, and the competence *Full responsibility in identifying the customer and his needs.* To better distinguish knowledge, skill and competence elements, the corresponding classes are painted black, whereas the concepts of the taxonomy they are linked to are colored light gray. Subsumption relationships are characterized by a solid line with a hollow arrowhead pointing from the class that is subsumed to the class that subsumes, whereas relationships that show that a knowledge, skill or competence is expressed by one or more concepts belonging to the taxonomy are expressed by a dashed line. It is worth remarking that dashed lines have been exploited to make the diagram more readable, thus making it immediately understandable which relationships define the subsumption of terms and which ones link knowledge, skills and competences to the taxonomy (in this case, a dashed line is drawn for showing the link between an element of a subtask and a concept in the taxonomy, or among knowledge elements, action verbs and context, and not for indicating dependency relationships). Information expressed by the diagram is the following: the knowledge *Communication techniques knowledge* is expressed by the knowledge object *communication techniques,* which is a type of *selling techniques*, i.e., another knowledge object. The subsumption relationship between *selling techniques* and *communication techniques* indicates that if someone has knowledge of a *communication techniques* component, then he or she also has knowledge of a *selling techniques component.* Furthermore, the skill *To be able to apply selling techniques* is characterized by the pair of concepts *apply*, an action verb that specifies the action verb *act*, and *selling techniques*, a knowledge object, while the skill *To be able to communicate in English* is expressed by the action verb *communicate*, a specification of the action verb *react*, which is applied to the *English* concept, and a specification of a *foreign language* knowledge object. Finally, the competence *Full responsibility in identifying the customer and his needs* is defined by a *full responsibility* context, applied to the *identify* action verb, which is linked to the *customer* and *customer needs* knowledge objects.

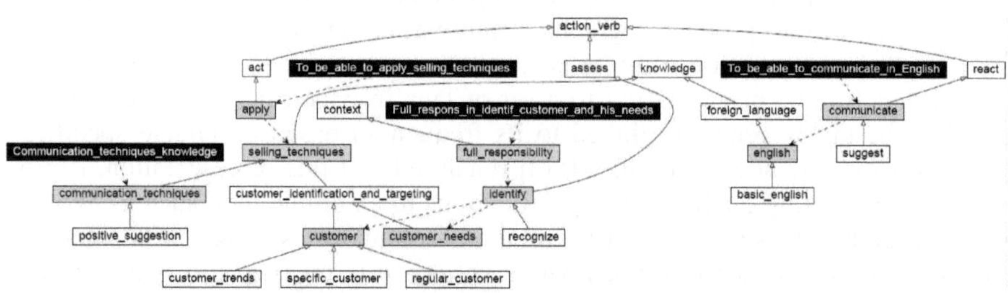

Fig. 3 Ontology related to the subtask: To welcome the customer and understand the customer's needs and requests [Gatteschi et al. 2011a].

Even though tag clouds that exploit online tools present some limitations, because they are not designed to account for relations among concepts, ad hoc tag cloud generators could ease the semantic annotation of curricula. In this view, [Mirizzi et al. 2009] presented a system to automatically produce a semantically annotated résumé exploiting tag clouds and an ontology, to better suggest to the user which concepts could be added to his curriculum to better characterize it. The developed system first collects a list of skills and competences that are expressed as a set of tags and then recommends tags that are semantically related to those concepts. The size of each tag is computed according to the popularity of the suggested tag in other profiles that are stored in the knowledge base and the distance of the selected class and the suggested class in the ontology. Another application of tag clouds is reported in [Gatteschi et al. 2011b]. In this study, the authors propose a tag cloud-based representation of qualifications, curricula and job profiles that can show the two dimensions of a matchmaking problem (i.e., how much a learning outcome is needed/possessed/required by a learner/job applicant/company and how much a concept is taught/recognized/possessed by a given qualification/company/job applicant). More specifically, the authors first start from a knowledge base that stores concepts and relationships among them and also a degree of importance/mastery of a learning outcome; then, they represent the above information in a tag cloud that is structured on a circular form. The importance of a concept (i.e., how much the given concept is relevant for the one who wants to find the best match with respect to his requirements) is represented by means of the font size (i.e., larger fonts indicate more relevant concepts), whereas the degree of mastery (i.e., how much the concept is possessed by actors who compete in the market) is linked to the distance from the center of the cloud (e.g., for applicants with an exhaustive knowledge of the requested subjects, a compact tag cloud would be generated).

It has been shown that there are several strategies that quickly depict the main aspects of a qualification/curriculum vitae or the reasons behind the result of a matchmaking process, and these strategies could be categorized into the following groups: a) approaches that exploit a graph-based representation of information through nodes and relations among them [Gestwicki, 2008 and Dexter and Davies, 2009]; b) three-dimensional representations of skills provided by a qualification [Hoffman, 2003] or of the composition (e.g., in terms of modules, credits) of education programs [Sommaruga and Catenazzi, 2007]; c) tag cloud-based representations depicted by exploiting online tools [Gatteschi et al. 2011a], or ad hoc tools [Mirizzi et al. 2009 and Gatteschi et al. 2011b]; and d) UML graphs [Gatteschi et al. 2011a].

Because the visualization of information strictly depends on the type of data to be displayed and on the characteristics of end-users, it is difficult to identify a priori the best approach to be pursued. In general, graph-based representations turned out to be able to provide an overview of a qualification, but there is the possibility that some end-users would not be able to fully understand them. However, the approach suggested by [Dexter and Davies, 2009], consisting of the creation of a visual metaphor of the London Underground map, could allow even non-skilled users to comprehend them. Similarly, three-dimensional representations have the advantage of communicating to the user different information about a given qualification/curriculum, but they should be simplified to be fully understandable by each actor involved, such as in the study of [Sommaruga and Catenazzi, 2007].

In contrast, a tag cloud is quite easy to comprehend, because the most important concepts are written with larger fonts; however, they can communicate only one aspect of interest. In this view, ad hoc tag cloud-based representations could represent more dimensions; however, their understanding might not be immediate.

Finally, UML graphs have the advantage of displaying concepts together with relations among them, but they risk being difficult to comprehend for a non-skilled user. Moreover, the UML representation of a qualification and its composing elements could risk being too wide and, hence, not very efficient.

6 Conclusions

This study presents recent research activities that support the mobility of students and workers, conducted in the Semantic Web field. Specifically, an analysis of learning and job recruitment challenges and an evaluation of benefits linked to the exploitation of Semantic Web technologies applied in these contexts has been performed.

Issues related to students' and workers' mobility and to the improvement of qualifications to offer, as well as of curricula/job offer comparisons, were analyzed, by distinguishing three fields of research: knowledge base creation, the development of strategies for the integration of heterogeneous systems as well as the definition of inference rules, and the identification of methodologies for the visualization of qualification outcomes and curricula.

Specifically for knowledge base creation, the first steps to be performed for formalizing existing knowledge [Tao et al. 2005] have been presented, and then, studies that define metadata that supports the description of learning resources [Poyry et al. 2002, Poyry and Puustjarvi, 2003, Sampson and Fytros, 2008] were illustrated. Qualitative and quantitative comparisons exploiting Bloom's taxonomy [Spivey, 2007, Bourque et al. 2003, Starr et al. 2003, Hoffman, 2003] were also shown, and solutions exploiting the EQF guidelines for the description of qualifications [Pernici et al. 2006, Gatteschi et al. 2011a] were analyzed. Finally, strategies for improving the match between job offers and demands collected in different online recruitment services [De Meo et al. 2007] as well as the creation of ad hoc ontologies to express the relations among terms of a given domain [Hexin and Bin, 2006] or among the composing elements of curricula/job positions [Harzallah et al. 2002] and the use of existing semantic thesauri to annotate and compare job offers and demands [Gatteschi et al. 2011b] were presented and compared.

Approaches for the integration of heterogeneous systems and the definition of inference rules were presented by distinguishing between research activities investigating how ontologies could be exploited for aligning courses provided by an institution to a given standard [Ronchetti and Sant, 2007], studies proposing an ontology-based integration of classes provided by other institutions [Hackelbusch, 2006, Cubillos et al. 2006, Cubillos et al. 2007a, Cubillos et al. 2007b, Gatteschi et al. 2009], and research activities aimed at finding the best candidate for a given job [Hexin and Bin, 2006 and Gatteschi et al. 2011c].

For the visualization issue, studies that propose the representation of qualifications by means of two-dimensional [Gestwicki, 2008, Dexter and Davies, 2009] and three-dimensional [Hoffman, 2003, Sommaruga and Catenazzi, 2007] graphs, as well as research presenting whether tag clouds could be used in this context [Gatteschi et al. 2011a, Mirizzi et al. 2009 and Gatteschi et al. 2011b], were described.

The solutions developed showed an improvement in the training offer and job seeking and recruitment activities as well as in students' and workers' mobilities. Moreover, the adoption of European instruments, such as the EQF, leads to a more transparent comparison of qualifications.

Specifically, the exploitation of metadata turned out to be appropriate in contexts that are characterized by unstructured information; however, if the objective is the comparison of the content of a qualification (or curriculum), also an (ad hoc or already present on the internet) ontology reporting relations among concepts should be exploited. Moreover, in those cases in which the structure of elements to be compared could not be expressed by means of existing instruments, meta-ontologies linking them to a common framework should be developed, together with rules for matchmaking (which should possibly be applied to learning outcomes annotated by means of a shared ontology or a lexical dictionary such as DBpedia or WordNet). Finally, the choice of the type of approach to follow for showing results of matchmaking as well as the main aspects of a qualification/curriculum or job offer should prefer "standard" tag clouds or visual metaphors in the case of non-skilled users, or ad hoc tag cloud-based representations or UML graphs when there is a need to display a very large amount of information to expert actors.

Acknowledgment. The work in this paper was supported by the MATCH "Informal and non-formal competences matching device for migrants employability and active citizenship" project (http://match.cpv.org), ref. 510739-LLP-1-2010-1-IT-GRUNDTVIG-GMP CUP E17E11000140004.

References

[Anderson et al. 2001] Anderson, L.W., Krathwohl, D.R., Bloom, B.S.: A taxonomy for learning teaching, and assessing: a revision of Bloom's taxonomy of educational objectives. Longman (2001)

[Anderson et al. 2004] Anderson, T., Whitelock, D.: The Educational Semantic Web: visioning and practicing the future of education. Journal of Interactive Media in Education (JIME) 7, 1–15 (2004)

[Auer and Lehmann, 2007] Auer, S., Lehmann, J.: What have Innsbruck and Leipzig in common? Extracting Semantics from Wiki Content. In: Franconi, E., Kifer, M., May, W. (eds.) ESWC 2007. LNCS, vol. 4519, pp. 503–517. Springer, Heidelberg (2007)

[Aydin and Tirkes, 2010] Aydin, C.C., Tirkes, G.: Open source learning management systems in e-learning and Moodle. In: IEEE Education Engineering (EDUCON), pp. 593–600 (2010)

[Bittencourt et al. 2008] Bittencourt, I.I., Isotani, S., Costa, E., Mizoguchi, R.: Research Directions on Semantic Web and Education. Journal Scientia - Interdisciplinary studies in Computer Science 19(1), 59–66 (2008)

[Bizer et al. 2005] Bizer, C., Heese, R., Mochol, M., Oldakowski, R., Tolksdorf, R., Eckstein, R.: The Impact of Semantic Web Technologies on Job Recruitment Processes. In: International Conference of Wirtschaftsinformatik, pp. 17–26 (2005)

[Bloom, 1956] Bloom, B.S.: Taxonomy of educational objectives - the classification of educational goals - Handbook I: cognitive domain. Longman (1956)

[Bologna Declaration, 2001] Bologna Declaration Communiqué of the meeting of European Ministers in charge of Higher Education (May 19, 2001), `http://ec.europa.eu/education/policies/educ/bologna/bologna.pdf` (cited May 2011)

[Bourque et al. 2003] Bourque, P., Buglione, L., Abran, A., April, A.: Bloom's Taxonomy Levels for Three Software Engineer Profiles. In: Eleventh Annual International Workshop on Software Technology and Engineering Practice (STEP 2003), pp. 123–129 (2003)

[Cubillos et al. 2006] Cubillos, C., Lamberti, F., Demartini, C.: National Qualification Systems Integration Using Ontologies. IFIP, vol. 210, pp. 285–289 (2006)

[Cubillos et al. 2007a] Cubillos, C., Lamberti, F., Demartini, C.: Integration at Vocational Education and Training Level through Mapped Ontologies. In: Third International Conference on Convergence and Hybrid Information Technology (ICCIT 2008), vol. 1, pp. 117–122 (2008)

[Cubillos et al. 2007b] Cubillos, C., Lamberti, F., Demartini, C.: Integrating European Qualification Systems with OWL Ontologies. In: Electronics, Robotics and Automotive Mechanics Conference (CERMA 2007), pp. 752–757 (2007)

[De Meo et al. 2007] De Meo, P., Quattrone, G., Terracina, G., Ursino, D.: An XML-Based Multiagent System for Supporting Online Recruitment Services. IEEE Transactions on Systems, Man and Cybernetics, Part A: Systems and Humans 37(4), 464–480 (2007)

[Devedžić, 2005] Devedžić, V.: Intelligent Web-based computer-supported collaborative learning. In: Ghaoui, C., Jain, M., Bannore, V., Jain, L.C. (eds.) Knowledge-Based Virtual Education - User-Centred Paradigms, pp. 81–110. Springer, Berlin (2005)

[Devedžić, 2006a] Devedžić, V.: Computer-supported collaborative learning. In: Ghaoui, C. (ed.) Encyclopedia of Human Computer Interaction, pp. 105–111. Idea Group Reference, Hershey (2006a)

[Devedžić, 2006b] Devedžić, V.: Semantic Web and Education. Springer, New York (2006)

[Dexter and Davies, 2009] Dexter, H., Davies, I.: An ontology-based curriculum knowledgebase for managing complexity and change. In: 9th IEEE International Conference on Advanced Learning Technologies, pp. 136–140 (2009)

[ECVET, 2009] Recommendation of the European Parliament and of the Council on the establishment of a European Credit System for Vocational Education and Training (ECVET) (June 18, 2009), `http://eur-lex.europa.eu/LexUriServ/LexUriServ.do?uri=OJ:C:2009:155:0011:0018:EN:PDF` (cited May 2011)

[EQF, 2008] Recommendation of the European Parliament and of the Council on the establishment of the European Qualifications Framework for lifelong learning (April 23 2008), `http://eur-lex.europa.eu/LexUriServ/LexUriServ.do?uri=OJ:C:2008:111:0001:0007:EN:PDF` (cited May 2011)

[Gatteschi et al. 2009] Gatteschi, V., Lamberti, F., Salassa, F., Demartini, C.: An Automatic Tool Supporting Life-long Learning Based on a Semantic-Oriented Approach for Comparing Qualifications. In: IADIS International Conference on Cognition and Exploratory Learning in Digital Age (CELDA 2009), pp. 354–358 (2009)

[Gatteschi et al. 2011a] Gatteschi, V., Lamberti, F., Sanna, A., Demartini, C.: Using taxonomies and ontologies to define occupation and education-driven European qualifications. IADIS International Journal on Computer Science and Information Systems 6, 30–44 (2011)

[Gatteschi et al. 2011b] Gatteschi, V., Lamberti, F., Sanna, A., Demartini, C.: Using tag clouds to support the comparison of qualifications, résumés and job profiles. In: 9th IEEE International Conference on Emerging eLearning Technologies and Applications (ICETA2011), pp. 57–61 (2011)

[Gatteschi et al. 2011c] Gatteschi, V., Lamberti, F., Sanna, A., Demartini, C.: A ranking tool exploiting semantic descriptions for the comparison of EQF-based qualifications. Journal of Universal Computer Science 17(7), 1060–1077 (2011)

[Gestwicki, 2008] Gestwicki, P.: Work in Progress – Curriculum Visualization. In: Frontiers In Education Conference (FIE 2008), pp. T3E13–T3E14 (2008)

[Gruber, 1993] Gruber, T.R.: Toward principles for the design of ontologies used for knowledge sharing. In: Padua Workshop on Formal Ontology, pp. 1–23 (1993)

[Hackelbusch, 2006] Hackelbusch, R.: Handling Heterogeneous Academic Curricula: National qualification systems integration using ontologies. In: DEXA 2006, pp. 344–348 (2006)

[Harzallah et al. 2002] Harzallah, M., Leclere, M., Trichet, F.: CommOnCV: Modelling the competencies underlying a curriculum vitae. In: Proc. Int. Conf. SEKE, pp. 65–71 (2002)

[Hexin and Bin, 2006] Lv, H., Zhu, B.: Skill ontology-based semantic model and its matching algorithm. In: 7th International Conference on Computer-Aided Industrial Design and Conceptual Design CAIDCD 2006, pp. 1–4 (2006)

[Hoffman, 2003] Hoffmann, M.: Using Blooms Taxonomy of learning to make engineering courses comparable. In: EAEEIE 19th Annual Conference (2008), pp. 205–209 (2008)

[LOM, 2002] IEEE 1484.12.1 Final Draft Standard for Learning Object Metadata

[Mirizzi et al. 2009] Mirizzi, R., Di Noia, T., Di Sciascio, T.: A Semantic Web enabled System for Résumé Composition and Publication. In: IEEE International Conference on Semantic Computting (ICSC 2009), pp. 583–588 (2009)

[Netcraft, 2011] Netcraft Web Server Survey, `http://news.netcraft.com/archives/category/web-server-survey/` (cited May 2011)

[Ohler, 2008] Ohler, J.: The SemanticWeb in Education – What happens when the read-write web gets smart enough to help us organize and evaluate the information it provides? EDUCAUSE Quarterly Magazine 31(4), 7–9 (2008)

[Pernici et al. 2006] Pernici, B., Locatelli, P., Marinoni, C.: The eCCO System: An eCompetence Management Tool Based on Semantic Networks. In: Meersman, R., Tari, Z., Herrero, P. (eds.) OTM 2006 Workshops. LNCS, vol. 4278, pp. 1088–1099. Springer, Heidelberg (2006)

[Poyry et al. 2002] Poyry, P., Pelto-Aho, K., Puustjarvi, J.: The role of metadata in the CUBER system. In: Proceedings of the 2002 Annual Research Conference of the South African Institute of Computer Scientists and Information Technologists on Enablement Through Technology, pp. 16–18 (2002)

[Poyry and Puustjarvi, 2003] Poyry, P., Puustjarvi, J.: CUBER: a personalised curriculum builder. In: Proceedings of the 3rd IEEE International Conference on Advanced Learning Technologies, pp. 326–327 (2003)

[Protégé, 2000] Protégé user guide (2000), `http://protege.stanford.edu/doc/usersguide/` (cited May 2011)

[Ronchetti and Sant, 2007] Ronchetti, M., Sant, J.: Curriculum Management and Review: an ontology-based solution. In: World Conference on E-Learning in Corporate, Government, Healthcare, and Higher Education (ELEARN 2007), pp. 6476–6482 (2007)

[Saini and Ronchetti, 2003] Saini, P., Ronchetti, M.: Deriving ontology-based metadata for e-learning from the ACM Computing Curricula. In: IADIS International Conference e-Society 2003, pp. 892–898 (2003)

[Sampson and Fytros, 2008] Sampson, D., Fytros, D.: Competence Based Educational Metadata for Supporting Lifelong Competence Development Programmes. In: The 8th IEEE International Conference on Advanced Learning Technologies (ICALT 2008), pp. 288–292 (2008)

[Sommaruga and Catenazzi, 2007] Sommaruga, L., Catenazzi, N.: Curriculum visualization in 3D. In: 12th International Conference on 3D Web Technology, pp. 177–180 (2007)

[Spivey, 2007] Spivey, G.: A Taxonomy for learning, teaching, and assessing Digital Logic Design. Frontier. In: Education Conference – Global Engineering: Knowledge Without Borders, Opportunities Without Passports (FIE 2007), pp. F4G-9–F4G-14 (2007)

[Staab and Studer, 2004] Staab, S., Studer, R.: Handbook on Ontologies. International handbooks of Information Systems. Springer (2004)

[Starr et al. 2003] Starr, C.W., Manaris, B.Z., Stalvey, R.H.: Bloom's taxonomy revisited: specifying assessable learning objectives in computer science. SIGCSE (2008), 261–265 (2008)

[Tao et al. 2005] Tao, F.B., Millard, D., Davis, H., Woukeu, A.: Managing the Semantic Aspects of Learning using the Knowledge Life Cycle. In: The 5th IEEE International Conference on Advanced Learning Technologies (ICALT 2005), pp. 575–579 (2005)

[Tiropanis et al. 2009] Tiropanis, T., Davis, H., Millard, D., Weal, M.: Semantic Technologies for Learning and Teaching in the Web 2.0 Era. IEEE Intelligent Systems 24(6), 49–53 (2009)

[UML, 1996] Unified Modeling Language (UML), `http://www.uml.org/` (cited May 2011)

[UMLGraph, 2003] UMLGraph, `http://www.umlgraph.org/` (cited May 2011)

[W3C, 2004] W3C OWLWeb Ontology Language, `http://www.w3.org/TR/owl-features/` (cited May 2011)

[Wordle, 2009] Wordle, `http://www.wordle.net/` (cited May 2011)

[WordNet, 1985] WordNet, `http://wordnet.princeton.edu/` (cited November 2011)

Chapter 2: Reasoning for Personalization and Recommendation

The huge amount of information, in combination with the dynamic and heterogeneous nature of the Web, makes information retrieval a hard task for the average user. This situation is known as the “information overload problem". Informed decisions based on automatic reasoning processes and adapted to the user characteristics is the answer to this problem. Recommender systems proved to be an efficient way of combining reasoning and personalization by providing proactive and personalized suggestions. In such a system online users receive recommendations from other users with similar information needs and preferences. Getting the best of a recommender system requires a careful design of its technical characteristics (similarity estimation, reasoning, timely suggestions) along with implications of the actual work, Vafopoulos *et al.* provide an interesting article on this aspect. Extending and adapting the capabilities of recommender systems to digital TV systems is another area of interest which tries to minimize the digital divide. The TV, digital or not, is a more familiar medium to the older population rather than the Internet. Blanco-Fernández *et. al.* explore new ways of developing e-commerce services through digital TV. Modern e-commerce websites are multilingual environments which are further enriched by the comments of users. Saloun *et. al.* investigate ways of enhancing multilingual websites with automated navigation services, while at the end Paralic et al. present and compare experiences with two different approaches to the utilization semantic technologies for the personalized access to web services.

Article 2.1

Title: Recommendation Systems: Bridging Technical Aspects with Marketing Implications

Authors: Michalis Vafopoulos and Michalis Oikonomou

As the Web matures and becomes social and participatory, collaborative filters are the basic complement in searching online information about people, events and products. In Web 2.0, what connected consumers create is not simply content (e.g. reviews) but context. This new contextual framework of consumption emerges through the aggregation and collaborative filtering of personal preferences about goods in the Web in massive scale. More importantly, facilitates connected consumers to search and navigate the complex Web more effectively and amplifies incentives for quality. In this article a joint review of the basic stylized facts of relevant research in recommendation systems in computer and marketing studies is presented in order to share some common insights. The focus of analysis is on the fields of one-to-one marketing, network-based marketing Web merchandising and atmospherics and their implications in the processes of personalization and adaptation in the Web while Market Basket Analysis is investigated in context of recommendation systems.

Article 2.2

Title: Exploring New Ways for Personalized E-Commerce through Digital TV

Authors: Yolanda Blanco-Fernández, Martín López-Nores, José J. Pazos-Arias, and Manuela I. Martín-Vicente

The advent of new devices (e.g., Digital TV receivers, mobile phones or media players) and usage habits (e.g., social networking) is progressively rendering the classical search engines of the Internet insufficient to support users in their grasp of an ever-expanding information space. This fact has brought about a new paradigm of recommender systems, which aim at proactively discovering the items that best match the preferences, interests and needs of each individual at any time. In settings where the user is not focused on the kind of items that may be recommended to him/her (e.g. in delivering publicity while he watches TV programs), one could expect the recommender to select the best offers from among various providers, to look for the pieces of information that describe each item in the most complete or accessible way, to arrange the most suitable interfaces to place an order, etc. In other words, with the user around the marketplace, the recommender should be responsible for building the shop in which the user will feel most comfortable to browse the suggested items. Obviously, this is not a task for human developers, because no workforce would suffice to provide specific applications for all the different users, items and devices in all possible contexts.

Article 2.3

Title: Towards Automated Navigation over Multilingual Content

Authors: Petr Saloun, Zdenek Velart, and Jan Nekula

Web-based multilingual, adaptive and personalized systems are becoming standards on the web. This article describes the main features of such systems emphasizing to the domain of education where multilingual content is a key characteristic. As a case-study, XAPOS, a web-based adaptive personalized system is described in more detail. XAPOS's design is independent of the language of closed content and is used in a multilingual environment while XAPOS's language-independent navigation feature is based on the domain ontology and principles of semantic web.

Article 2.4

Title: Personalized and Adaptive Access to Services – The Semantic Web Services Approach

Authors: Marek Paralič, Peter Bednár, and Ján Paralič

Semantic web services are main concern of this article. The authors present and compare experiences with two different approaches to the utilization semantic technologies for the personalized access to services, where services are considered both web services accessible online and traditional services provided by business or government organizations. Based on state of the art of frameworks and ontologies for semantic description of web services they describe two different ways of achieving personalization for services. The authors follow their experiences from two projects, where applications in different areas, such as e-government, e-business and crisis management have been implemented.

Recommendation Systems: Bridging Technical Aspects with Marketing Implications

Vafopoulos Michalis and Oikonomou Michael

Mathematics Department, Aristotle University of Thessaloniki, Greece
ACM: H.3.3; J.4
WSSC: webscience.org/2010/D.2.4; webscience.org/2010/E.1.1.1
AMS: 90B60; 91D30; 62P25; 97M10; 97M70

Abstract. In 2010, Web users ordered, only in Amazon, 73 items per second and massively contribute reviews about their consuming experience. As the Web matures and becomes social and participatory, collaborative filters are the basic complement in searching online information about people, events and products.

In Web 2.0, what connected consumers create is not simply content (e.g. reviews) but context. This new contextual framework of consumption emerges through the aggregation and collaborative filtering of personal preferences about goods in the Web in massive scale. More importantly, facilitates connected consumers to search and navigate the complex Web more effectively and amplifies incentives for quality.

The objective of the present article is to jointly review the basic stylized facts of relevant research in recommendation systems in computer and marketing studies in order to share some common insights.

After providing a comprehensive definition of goods and Users in the Web, we describe a classification of recommendation systems based on two families of criteria: how recommendations are formed and input data availability. The classification is presented under a common minimal matrix notation and is used as a bridge to related issues in the business and marketing literature. We focus our analysis in the fields of one-to-one marketing, network-based marketing Web merchandising and atmospherics and their implications in the processes of personalization and adaptation in the Web. Market Basket Analysis is investigated in context of recommendation systems. Discussion on further research refers to the business implications and technological challenges of recommendation systems.

1 Preface

Searching, social networking, recommendations in various forms, blogging and micro-blogging have become part of everyday life whilst the majority of business applications have migrated to the Web. Understanding and modeling this enormous impact of the Web in macro (e.g. [1]) and micro scale (e.g. [2], [3]) has

I.E. Anagnostopoulos et al. (Eds.): Semantic Hyper/Multimedia Adaptation, SCI 418, pp. 155–180.
springerlink.com

become a major task for computer and social scientists. The trans-disciplinary field in this direction has been entitled "Web Science" and is focused in the significant reciprocal relationship among the social interactions enabled by the Web's design, the scalable and open applications development mandated to support them, and the architectural and data requirements of these large-scale applications [4], [5], [6].

The Web "curves" physical time and space by adding flexibility, universality [7], more available options [8], [9] and sources of risks [10]. At the current Web 2.0 era, Users can easily edit, interconnect, aggregate and comment text, images and video in the Web. Most of these opportunities are engineered in a distributed and self-powered level.

In particular, recommendation systems have become mainstream applications in the Web with massive User participation affecting an important part of offline and online industries. During the last twenty years, research and practice on recommendation systems is growing in an increasing pace. This massification creates new business opportunities and challenging research issues in software development, data mining, design of better algorithms, marketing, management and related issues. User and business demands are now setting part of the research agenda in recommendation systems literature. Recently, new research communities (e.g. network analysis) from diverse fields have started to involve in the research of recommendation systems in order to understand the economic behavior of online consumers and its implications to business process and competition.

Computer science literature and related fields are often enriched by bibliographic reviews on the advancements of recommendation systems ([11] is the most recent). To the best of our knowledge, it does not exist an effort to jointly review the technical and business aspects of recommendation systems. Thus, the objective of the present article is to overview the main aspects of relevant research in recommendation systems both in computer and marketing studies in order to create a bridge and facilitate the sharing of common insights.

The article is organized as follows. The first section is devoted in the description of the fundamental changes that the Web brings in the economy. Specifically, the role of recommendation systems is identified as an important part in the transition to more energetic and interdependent consumption patterns. The second section provides an overview of the technical aspects characterizing recommendation systems. After providing a targeted and comprehensive definition of goods and Users in the Web, we describe a classification of recommendation systems based on two families of criteria: how recommendations are formed and input data availability. The classification is presented under a common minimal matrix notation. The third section reviews the related issues of recommendation systems in the business and marketing literature. We focus our analysis in the fields of one-to-one marketing, network-based marketing, Web merchandising and atmospherics and their implications in the processes of personalization and adaptation in the Web. Market Basket Analysis is investigated in the context of recommendation systems. The final section discusses issues for further research.

2 Consumption in the Web

2.1 Introduction

Some economists expected that the Web would gradually lead to perfect information in consumption, acute price competition and pricing at the marginal cost followed by low dispersion [12]. The basic arguments were based on lower search and fixed costs, less product differentiation and "frictionless commerce" via the Web. There is not strong evidence that many things have changed in these directions in the markets of ordinary goods, since online prices are still dispersed, not much lower than offline (see for instance [13] and [14]) and many sectors continue to share oligopolistic characteristics.

But what actually changed, and not expected at all, was the emergence of new types of consumption and production, new service sectors (e.g. Software as a Service) and the transformation of existing industries (e.g. mass media). The resulting reconfigurations in the triptych of production-exchange-consumption stemmed from an update in the fundamentals of the economy that the Web brings. Basically, the Web is contributing one major new source of increasing returns in the economy: *more choices with less transaction costs in production and consumption.*

This source of value arises from the orchestration of digital and network characteristics of goods in the Web. More choices in consumption are ranging from larger variety of available goods, to online consumer reviews, recommendations and adaptive content. This updated mode of *connected consumption* allows consumers to make more informed decisions and provides them with stronger incentives to take part in the production and exchange of mainly information-based goods. On the other hand, the provision of more choices with less transaction cost in consumption is not always coming without compensation. The leading native business model in the Web is the forced joint consumption of online information and contextual advertisements in massive scale.

Turning in the *production* side, many business operations virtualized, went online and become less hierarchical, niche online markets and services emerged and traditional industries revolutionized. Decentralized Peer production through loosely affiliated self-powered entities is based on a broader baseline of input and output to create a larger range of possibilities for both producers and consumers [15]. Moreover, the recent emergence of "social commerce" as a consumer-driven online marketplace of personalized, individual-curated shops that are connected in a network, demonstrates the volatile boundaries among production, exchange and consumption in the Web.

2.2 More Energetic and Connected Consumption

Due to the rapid penetration of the Web in many technological platforms (e.g. mobile, TV) and social aspects, electronic commerce has become a major activity in ordinary business operations. Almost every firm in the developed world has online presence that describes or/and provides its goods to potential customers.

The migration of many business functions in the Web decreased operational costs, primarily, for service-oriented companies. Online commerce is one of the basic components of the Web economy and is gradually becoming an important sector for the entire economy. The Census Bureau of the Department of Commerce announced that the estimate of U.S. retail e-commerce sales for the first quarter of 2011 was $46.0 billion, an increase of 17.5% from the first quarter of 2010 while total retail sales increased 8.6% in the same period. E-commerce sales in the first quarter of 2011 accounted for 4.5% of total sales[1].

The expansion of online commerce has attracted many scholars from diverse disciplines such as Economics, Business and Operation Research, Computer and Information science, Law and others (for a review of e-commerce literature see [16] and [17]).

Trivially, the Web has enabled consumers to access round-the-clock services and to search and compare products, prices, catalogues, descriptions, technical specifications and so forth. Apart from searching and comparing the characteristics of goods and services in the Web, consumers can comment and be informed from others' consumers' purchases and comments. Consumption becomes more connected in the Web. In the rest of this subsection we describe the basic characteristics of connected consumption as a broad economic phenomenon related to recommendation systems.

Basically, positive *network effects* characterize a good when more usage of the good by any User increases its value for other Users. These effects are also called *positive consumption or demand side externalities*. As consumers become more connected in the Web ecosystem, the network effects are gradually based on the mutual benefits of consumption, [18]. *Connected consumption* defines a new form of direct complementarity among consumers. When Users consume goods through the Web, reveal and contribute private information about their preferences and expectations, which is beneficial to other consumers if aggregated and made public. These publicly aggregated consumption patterns and comments are valuable in two ways: (a) indirectly, by reducing search and transaction costs (e.g. tags, playlists, collaborative filtering) and (b) directly, by increasing consumption gains (e.g. discovery of complementary goods in co-purchase networks [19]). Actually, what connected consumers create is not simply content (e.g. reviews) but *context*. This new contextual framework of consumption emerges through the aggregated personal preferences about goods in the Web in massive scale. More importantly, facilitates connected consumers to search and navigate the complex Web more effectively and amplifies incentives for quality. *But how so many and heterogeneous consumers around the globe can coordinate their preferences and expectations?*

In the world of Coase, a small number of consumers can effectively coordinate their preferences through informal agreements and formal contracts to capture the benefits of network effects [20]. However, the coordination of large number of consumers requires high transaction costs. Hayek [21] argued that the price system acts as a coordination device that synchronizes substantial numbers of producers

[1] `http://www.census.gov/retail/mrts/www/data/pdf/ec_current.pdf`

and consumers. Spulber [18] extended Hayek's analysis of "spontaneous order" to include many other market mechanisms for accomplishing coordination at large. These coordination devices include mass media and marketing, mass communications and observation of other consumers.

In the Web era, search engines, social networks and recommendation systems of online retailers (e.g. Amazon, BestBuy) are the most prominent examples of mass coordination devices of consumers' preferences. Search engines and social networks are general coordination devices that include the full spectrum of preferences. Recommendation systems of the Web merchants are focused in increasing the amount of sales by synchronizing purchasing patterns and adapting content provision in individual Users.

It is beyond the scope of the present article to review and analyze the various dimensions of Web economy and commerce, but rather to focus on a specific part of it: the marketing implications of the technical aspects characterizing recommendation systems.

3 Recommendation Systems in the Web: The Technical Aspect

3.1 Introduction

Recommendation systems are basic aspects of the current collaborative Web era that complement search engine algorithms in information discovery. Today, almost every Web commerce business uses information filtering techniques to propose products for purchase like a "virtual" salesperson.

Actually, recommendation systems are information filters that exploit user's characteristics (e.g. demographics) and preferences (e.g. views and purchases) to form recommendations or to predict user's future behavior. Commonly, recommendations in the Web are made automatically based on either individual or collective preferences (collaborative filtering) and are presented as hierarchical lists or schemes. For recent surveys on the technical aspects of implementing and analyzing recommendation systems you may refer to [11] and [22].

In the rest of the section we provide a minimal descriptive framework of existing research in recommendation systems that focus on understanding the main characteristics of related functions in the Web. The proposed classification builds on the previous attempts in related research (see for instance Adomavicius and Tuzhilin [23]) and extends the underlying categories by taking into consideration the criteria of input data source and availability. In addition, the concepts of "items" and "products" are specified to the more relevant concept of "Web Goods".

3.2 Web Goods and Users

Recommendation systems have emerged to elaborate efficient searching of goods in the Web through personalized and collective evaluation by the Web Users.

Before going into the details of recommendation systems let us define what kind of goods are available in the Web.

First, Web Goods has been defined as sequences of binary digits, identified by their assigned URI and hypertext format, and affect the utility of or the payoff to some individual in the economy [24]. Their market value stems from the digital information they are composed from and a specific part of it, the *hyperlinks*, which link resources and facilitate navigation over a network of Web Goods. Web goods can be further elaborated in the following categories. *Pure* Web Goods are the primary focus of the Web Science research [25] because they are defined to include goods that are basically exchanged and consumed in the Web and are not tightly connected to an ordinary good or a service (pre-) existing in the physical world. For instance, a blog entry that comments the market of used cars is a pure Web Good, but a car sales advertisement is not. According to a production incentives-based categorization, Web Goods are discriminated into *commercial* (e.g. sponsored search results) and *non-commercial* (e.g. Wikipedia entries). In contrast to commercial, non-commercial Web Goods are produced outside the traditional market mechanisms of price and property and are based on openness, Peer production and qualitative ex post reward schemes.

In recommendation systems literature, Web Goods are commonly referred as "items" or "products". In the present article we interchangeably use the term of "Web Goods" with the established terms because it better describes the realistic spectrum of goods and services available in and through the Web.

Web Users (or simply Users) produce and consume Web Goods. The emergence of the participatory Web highlighted the decisive role of Web Users in the collaborative creation of online content (refer to Vafopoulos [24] which provides a simple and comprehensive categorization of Web Users based on motivations and economic impact of their actions in the Web ecosystem).

3.3 The Main Classifications of Recommendation Systems

In the present article we adopt the minimal descriptive definition of recommendation systems initiated by Berkovsky et al [26] in order to enable the comparative analysis of the technological characteristics with the emergent marketing implications. For more formal and detailed definition of recommendation systems refer to [11].

Initially, it is assumed the existence of N Users with n distinct features, which may request recommendations for M Web Goods with m distinct features. All possible User and Web Good pairs are described by a $n + m$ dimensional space. In the simplest case, a single feature as unique identification describes Users and Web Goods, resulting a two-dimensional space. The $N \, x \, M$ User-Web Good rating matrix represents the ratings given by the Users to Web Goods. These ratings could be formed explicitly or implicitly in a predefined scale. Explicitly is considered in the sense that are directly contributed by Users and not by the recommendation system (implicit).

Table 1 List of symbolic representations of the main variables in recommendations systems' framework

U	$1, 2, \ldots, N$ Users with n distinct features
I	$1, 2, \ldots, M$ Web goods with m distinct features
$User_{feat}$	the user features (e.g. age, location, income)
$User_{id}$	a unique identifier of the Users
$Web\ Good_{feat}$	Web Good features (e.g. ID, price, availability)
$Web\ Good_{id}$	a unique identifier of the Web Goods
$Rating$	the ratings given by the Users to the Web Goods
R_{gen}	the general Recommendation function
R_{CF}	the Recommendation function in collaborative filtering
R_{CB}	the Recommendation function in content based filtering
Exp_{CA}	Context-aware experience
$Context_{feat}$	the context features (e.g. personal attitudes and tasks)
	Data Models
$Rating$	$Rating \sim (numerical, ordinal)$
$Binary$	$Rating \sim (0, 1)$
$Unary$	$Rating \sim (\emptyset, 1)$

The general Recommendation function (R_{gen}) is described as follows:

$$R_{gen}: User_{feat} \times Web\ Good_{feat} \rightarrow rating \quad (1)$$

Since in most cases (1) is not defined for all possible $n + m$ Users-Web Goods pairs (sparsity problem), a completion rule is needed to fill the missing values. Therefore, the main focus of analysis of recommendation systems is twofold: (a) to estimate the ratings of Web Goods that have not been rated by the Users and (b) to provide methodologies and techniques that will facilitate the formation of recommendations (Table 1 contains the symbols used in this section).

Missing values of the not-yet-rated Web Goods can be assessed by either empirical validation of specific heuristic forms of the recommendation function (1) (data-driven approach) or by estimating the recommendation function that optimizes statistical performance criteria (e.g. MSE) (model-driven approach). For a comprehensive review according to the rating estimation approach refer to [23].

In the present article, recommendation systems are categorized according to two sets of different criteria: how recommendations are created and what kinds of data are available, and a unified analytical framework is provided under common symbolism.

3.3.1 How Recommendations Are Formed

Regarding to *how recommendations are formed,* recommendation systems can be classified into the following three categories [27]:

a. Content-based recommendations

The User will be recommended Web Goods similar to her past preferences [28]. In the case of content-based recommendations the two-dimensional matrix R_{CB} is given by the following representation:

$$R_{CB}\text{: User}_{id} \text{ x Web Good}_{feat} \rightarrow \text{rating} \quad (2)$$

where User_{id} is a unique identifier of the Users, Web Good_{feat} refers to a feature space that represents the Web Good's features and rating reflects the User's evaluation for the Web Good's features [26].

b. Collaborative recommendations

The User will be recommended Web Goods that groups of Users with similar tastes preferred in the past (e.g. co-purchase network of Web Goods). In the case of collaborative filtering the two-dimensional matrix R_{CF} is becoming:

$$R_{CF}\text{: User}_{id} \text{ x Web Good}_{id} \rightarrow \text{rating} \quad (3)$$

Cacheda et al [29] in their recent work compare different techniques of collaborative filtering by identifying their main advantages and limitations.

c. Hybrid recommendations

In the hybrid approach, content-based and collaborative methods are orchestrated in forming recommendations (for a survey see [30]).

Adomavicius and Tuzhilin [23] contributed a more detailed classification of recommendation systems research by analyzing the statistical methodology followed in each of the above main three categories. In particular, they discriminated user-based (or memory - or heuristic - or neighborhood-based) and model-based recommendation techniques for each one of the content-based, collaborative and hybrid approaches. User-based filters consider that each User participates in a larger group of similarly behaving individuals and therefore, physical or Web Goods frequently viewed, liked or purchased, by group members of the group, are the main input for recommendation algorithms [31]. User-based algorithms are heuristics that classify items based on the entire collection of previously rated Web Goods by the Users. The model-based approach *"analyzes historical information to identify relations between different items such that the purchase of an item (or a set of items) often leads to the purchase of another item (or a set of items), and then use these relations to determine the recommended items."* [31]. The most popular type of model-based recommendations in Web commerce is referred to the literature as the "item-based top-N recommendation algorithms" (example is presented in subsection 4.4.4). These algorithms exploit the similarities among various Web Goods to define the set of them to be recommended (for an updated survey on collaborative filtering refer to [32] and [11]).

3.3.2 Data Sources and Data Availability

The second main category of recommendation systems is based on *data sources* and *data availability*. Regarding the data source, four different types of User's feedback are identified: no feedback, explicit, implicit and hybrid feedback. Explicit feedback is formed by direct input of Users regarding their preferences for specific items. For instance, Amazon reviews; Netflix star ratings and similar high quality data can be used in collaborative filtering algorithms. In cases where explicit feedback is not available or inadequate for building efficient collaborative systems, implicit feedback is employed. Implicit feedback data basically include browsing, usage patterns, purchase history and social network analysis. For instance, usage patterns in hypermedia systems are employed to enable adaptation to the individual User's needs (for a review see [33]). The aforementioned item-based top-N recommendation algorithms are based on implicit feedback mechanisms.

According to Yifan Hu et al [34], collection of implicit data is characterized by the following four main characteristics:

1. The option for Users to express no negative feedback is usually absent.
2. Data is inherently noisy.
3. The numerical values of implicit feedback declare *confidence* and not *preference* as in the case of explicit feedback.
4. Evaluation of recommendation systems based on implicit feedback require updated statistical measures that account for new features such as Web Good availability, competition and dynamic feedback.

Hybrid feedback recommendation systems are jointly exploiting explicit and implicit feedback from Users (see for instance [35]).

Analysis of recommendation systems can be also indexed on the basis of *data availability*. According to Bodapati [36] three different types of models are analyzed by the relevant literature, namely: the *ratings*, the *binary* and the *unary* data models. Specifically, the above classification refers to the availability of data in the $N \times M$ User-Web Good rating matrix. In the first case of ratings data model, each User provides explicit feedback by reporting a vote for a subset of Web Goods on a numerical (e.g., 1-6) or ordinal (e.g., like, indifferent or dislike) scale (Table 1). The binary data is considered to be a truncated version of the first model since Users express either a positive or a negative feedback. Purchasing a Web Good or awarding it a rating that meets some threshold could identify positive feedback, commonly recorded as 1. On the contrary, the User declares negative feedback (recorded as 0) if she expresses the intention not to purchase the Web Good or if her rating falls below the threshold. Finally, the *unary* data model is a restricted version of the binary data model because only positive valences are observed. Statistical analysis of item-based top-N recommendations is commonly based on the unary data model.

3.3.3 Context-Aware Recommendation Systems

In the introductory part of this article we argue that, through recommendation and feedback systems, what connected consumers create is not simply content (e.g. reviews) but *context.* This new contextual framework of consumption emerges through the aggregated personal preferences about goods in the Web and enables connected consumers to search and navigate more effectively and amplifies incentives for quality in the production of online content.

The investigation in different aspects of contextual information is gaining attention in fields not only technical such as the Semantic Web, data mining, information retrieval and computing, but also in economics and business studies. There are many diverse definitions of context. Context in recommendation systems analysis could be defined to consist of five concrete aspects: environment, personal attitudes, tasks, social and spatiotemporal information [37].

Recently, [38] Adomavicius and Tuzhilin contributed a thorough review of context-aware recommendation systems. They highlight that context-aware recommendations are characterized by complexity and interactivity and they initiate three different algorithmic paradigms for incorporating contextual information into the recommendation process.

In order to capture the various dimensions of feedback and context McNee et al [39] generalized all possible forms of rating to evaluation and Berkovsky et al [26] extended the general Recommendation function R_{gen} to include the experience (Exp) of User for a Web Good. Initially, *"an experience is defined as an evaluation function that maps a pair, the user that had the experience and the item experienced by the user, to an evaluation."* [26]. On this basis, the additional third dimension of context shapes the context-aware experience of the User in a recommendation system. Formally, is represented as follows:

$$Exp_{CA}: User_{feat} \times Web\ Good_{feat} \times Context_{feat} \rightarrow evaluation \quad (4)$$

The inclusion of contextual information into the recommendation process creates new opportunities in personalizing and adapting more efficiently online content through existing and innovative business practices. In the next section, we discuss the marketing implications of recommendation systems based on the understanding of core functional aspects that this section has built.

4 Recommendation Systems in the Web: The Marketing Aspect

4.1 Introduction

In the current Web, almost every e-commerce business uses information filtering techniques to propose products for purchase like a "virtual" salesperson. Salespersons in the physical world are responsible of making product recommendations to customers, which are integrated and aligned with the firm's marketing strategy. According to the Chartered Institute of Marketing, *Marketing* is the management process, which fulfills the following objectives: identifying, anticipating and satisfying customer requirements profitably[2].

Marketing in the Web (or internet marketing or e-marketing) is intuitively defined as the process of achieving the aforementioned objectives of traditional Marketing through mainly the Web ecosystem [40]. Therefore, recommendation systems in Web 2.0, beyond their role as online "virtual" salespersons, extend firm's marketing strategy by providing a hypermedia two-way channel between producers, distributors and customers. In particular, recommendation systems contribute in the fulfillment of marketing objectives by:

- *Identifying customer requirements*
  - facilitating massive, more detailed and cheaper data acquisition
  - extending one-to-one marketing analysis
- *Anticipating customer requirements*
  - enriching statistical modeling of customer's behavior
  - extending Market Basket Analysis
- *Satisfying customer requirements*
  - providing more informed, personalized and adaptive recommendations
  - implementing one-to-one marketing analysis
  - facilitating better merchandising and atmospherics

4.2 One-To-One Marketing, Personalization and Adaptation in the Web

One-to-one marketing (also referred as personalized marketing) instead of targeting an entire group of customers, as in traditional marketing, is designed to increase the revenue of a business by servicing each customer individually and fitting its needs perfectly [41], [42]. Thus, the function of one-to-one marketing is twofold:

- understand each customer's needs individually and
- recommend products that suit the customer's needs

In particular, one-to-one marketing is defined by four principles [43], namely: (1) identify customers, (2) differentiate each customer, (3) interact with each customer and (4) customize products for each customer.

Despite the fact that one-to-one marketing has been employed by researchers and practitioners before the Web, the advances on digital and Web technologies accelerated its expansion. First to mention the contribution of new technology on one-to-one marketing were Gillenson and Sherrell [44]. They also underlined that

---

2 http://www.cim.co.uk/resources/understandingmarket/definitionmkting.aspx

although super markets dominate the shopping behavior, customers still want to be treated as individuals. They highlighted that *"One-to-one marketing activities are characterized by a desire to interact individually with the most profitable customers of a firm. By learning the needs and desires of the most profitable customers and responding to those desires, companies can build an intensely loyal and profitable clientele."* In the Web era, the massive employment of recommendation systems enables the realization of one-to-one marketing not only to "the most profitable customers" as Gillenson and Sherrell [44] suggests, but for all Web customers. Mainly, one-to-one marketing employs content-based and hybrid recommendations.

One-to-one marketing in the Web converges to the processes of personalization and adaptation, which have extensively investigated in Computer science and related fields. Web personalization could be considered as the process in which the content, the layout and the functionality of a website are being changed dynamically, according to User's features in order to provide recommendations that would fit her. For an extensive discussion concerning various definitions and approaches on personalization refer to [45], [46], [47] and [42].

Kim [42] distinguishes the modes of personalization in two different aspects: information in the Web and one-to-one marketing. However, we suggest that participatory Web commerce through recommendation and feedback mechanisms unifies these aspects. In practice, Web 2.0 merchants attempt to jointly deliver successful marketing mixtures with personalized and adaptive information. Going a step further, online content adaptation could be analyzed as system-driven personalization (not to be confused to adaptability which is User-driven personalization).

4.3 Web Merchandising and Atmospherics

Merchandising and store atmospherics are the basic aspects of satisfying the customer requirements, which are included in the third axe of marketing objectives.

In traditional marketing of physical stores, these fields are responsible for efficient placing of products on the "shelf" and creating the appropriate store atmosphere to attract and sustain new customers. The emergence of click-and-mortar commerce resulted important transfers and transformations in the business functions of merchandising and store atmospherics. We briefly discuss the relevant changes to recommendation systems in the Web.

Merchandising *"consists of the activities involved in acquiring particular products and making them available at the places, times, prices and in the quantity to enable a retailer to reach its goals"* [48]. In the Web, there is no physical shelf, retailer or merchandiser but instead websites to present and dispose products (in the form of Web Goods, as have been defined in subsection 3.2). These products are stored in stock centers, homes or databases (in the case of pure Web Goods). *Web Merchandising* focuses on how to make available products in the Web. Online merchandisers are responsible for product collection and display, including promotions, cross-selling and up-selling. Studies in Web Merchandising could be divided in four areas [49]: (1) product assortment, (2) merchandising cues, (3) shopping metaphor and (4) Web design features.

Merchandising cues are techniques for presenting and/or grouping products to motivate purchase in online stores. A good example of merchandising cues is recommendation systems apart from traditional promotion methods [50]. Web design features share similar functionalities and analysis with Web atmospherics.

Web Atmospherics is the conscious designing of Web environments to create positive effects in Users in order to increase favorable customer responses. Just like retailers provide important information through atmospherics in conventional stores, online retailers also create an atmosphere via their website, which can affect shoppers' image and experience in the online store [51]. For a more extended overview on Web atmospherics refer to [52], [53] and [54]. For example, Nanou et al [55] investigate the effects of recommendations' presentation on customer's persuasion and satisfaction in a movie recommender system and they concluded that the most efficient presentation method is based on the "structured overview" and the "text & video" interfaces.

Recommendation systems could be employed not only for the product's placement, promotion and related functions but to improve the online environment and atmosphere via dynamic and adaptive features to the User's and product's characteristics. Mass merchants in the Web (e.g. Amazon) use various forms of recommendations (both content-based and collaborative) to build the main part of their store's architecture, functionality and adaptation. Indicatively, the webpage of an item in Amazon [3] contains recommendations of the following types: "Frequently Bought Together", "What Other Items Do Customers Buy After Viewing This Item?", "Customers Who Bought This Item Also Bought", "Customers Viewing This Page May Be Interested in These Sponsored Links", "Product Ads from External Websites", "Customer Reviews", "Related Items", "Tags Customers Associate with This Product", "Customer Discussions", "Listmania!", "Recently Viewed Items" and "Recent Searches".

To conclude, the transition from traditional marketing to marketing in the Web is not a trivial task. Recommendation systems contribute in this transition as follows:

- Marketing objectives sustain their core principles but extend their field of implementation in a more interactive communication process between business and customers and among customers themselves.
- In this updated communication process, recommendation systems play a prominent role by aggregating individual preferences and enabling massive one-to-one marketing.
- Recommendation systems also facilitate personalization and adaptation of Web commerce, which drastically affect Web merchandising and atmospherics.

---

[3] The considered example refers to Apple MacBook Pro MC725LL/A 17-Inch Laptop `http://www.amazon.com/Apple-MacBook-MC725LL-17-Inch-Laptop/dp/B002C74D7A/ref=sr_1_5?s=pc&ie=UTF8&qid=1309243292&sr=1-5`

We suggest that one-to-one marketing, recommendation systems and adaptation in the Web are crucial functions that should be further explored, both in Marketing and Web studies, in order to enhance Web merchandising and atmospherics.

4.4 Market Basket Analysis

4.4.1 Introduction

There is a classic example of Market Basket Analysis stating, "beers and diapers are often being purchased together in the same basket". In the Web, beers and diapers have been expanded to online music, movies and various types of services. The new marketing strategy mix has to anticipate changes in purchase behavior and customer requirements. Market Basket Analysis is a prominent tool in this effort since a growing number of research communities are involving in understanding consumption patterns in Web commerce.

Apart from traditional database and data mining-oriented research on Market Basket Analysis, investigators in network analysis and econometrics have recently contributed rich insights in this field. Network analysis is mainly based on crawling publicly available data from recommendation systems in the Web (e.g. Amazon).

4.4.2 Definitions, History and Applications

Let us first describe the basic definitions and techniques of Market Basket Analysis.

Market Basket is the set of items purchased by a customer during one single shopping occasion [56]. During a shopping trip, the customers are in a "pick-any"-situation because they have the option to choose no item, one or any other number of items from each category [57]. Market Basket data are binary data (e.g. an item added or not added into basket) organized in sets of items bought together by customers (often called transactions [58]).

Market Basket Analysis (MBA) studies the composition of shopping baskets in order to identify customer's purchase behavior [56]. It is also known as *association rule mining*, which is a method of identifying customer purchasing patterns by extracting associations from stores' transactional databases [59].

A mathematical definition of MBA is provided by Chen et al [59]: *"Given two non - overlapping subsets of product items, X and Y, an association rule in form of X/Y indicates a purchase pattern that if a customer purchases X then he or she also purchases Y. Two measures, support and confidence, are commonly used to select the association rules. Support is a measure of how often the transactional records in the database contain both X and Y, and confidence is a measure of the accuracy of the rule, defined as the ratio of the number of transactional records with both X and Y to the number of transactional records with X only."*

MBA is occupied for mainly marketing purposes. In particular, could be helpful in designing and implementing:

- cross and up-selling strategies[4]
- promotion strategies and discounts
- loyalty programs
- store atmospherics

Chib et al [60] summarize the motivation of MBA in the following arguments:

- Create improved estimates of brand-choice elasticities with respect to marketing mix variables, properly accounting for not just the direct impact but also the indirect impact on brand-choice via category purchases [61].
- Facilitate the understanding of what factors drive category purchase incidence and what impact marketing-mix variables have at the brand level on category purchases.
- Describe the isolating correlations amongst various product categories within the shopping basket in order to identify which categories are complements and which are substitutes.

Before the advent of the Web, MBA was enhanced by the technological evolutions on transactional systems in commerce (e.g. barcode implementations, RFID etc.) (see for example [62] and [56]). Electronic transactions provided the first stream of massive Market Basket data. Aggarwal et al [62] were the first to propose an influential algorithmic way for data mining in large transactions databases. One of the tools commonly used to perform MBA is Affinity analysis. Affinity analysis is a technique that discovers co-occurrence relationships among transactions performed by specific individuals or groups. For a literature review on MBA and related techniques refer to Mild and Reutterer [56] who classify relevant literature depending on the followed statistical approach (exploratory or explanatory analysis).

The second stream of input data came from Web commerce transactions including clickstreams, log files and other browsing data captured through voluntary and compulsory collection processes.

Hao et al [63] argue that MBA has become a key success factor in e-commerce and Kantardzic [64] states *"A business can use knowledge of these patterns to improve the Placement of these items in the store or the layout of mail-order catalog page and Web pages."*

4.4.3 Market Basket Analysis and Recommendation Systems

MBA and recommendation systems in Web commerce share the same input transaction data but employ different techniques and address their results to

---

[4] Cross-selling and up-selling are strategies of providing existing customers the opportunity to purchase additional or more expensive items, respectively.

different end users. MBA identifies customer's purchasing patterns by extracting associations from the data to inform the decisions of marketing managers, while recommendation systems provide relative information to Web Users.

Traditional MBA in purchasing data of physical stores is an exclusive property of storeowners and contains less information about the customer's purchasing behavior than in online stores. Specifically, in Web commerce every purchase is assigned with a unique time stamp of occurrence, Users' reviews and evaluations are often contributed and purchasing behavior could be inter-connected to general browsing patterns and website visits. These extra features of commerce data analysis in the Web offer a comparative advantage to online merchants and raises concerns of excessive market power and personal data abuse through selling to third parties and profiling without permission.

What is also changing in Web commerce is that other entities than the store owner/administrator can commence (partial) MBA by collecting online recommendations.

For example, Amazon, the biggest Web merchant, is based on a successful item-based collaborative filtering system[5] providing a wide range of general and personalized recommendations. Specifically, the list of "most customers that bought this item, also bought" (BLB) recommended products presents related items that were co-purchased most frequently with the product under consideration.

BLB recommendations form the store's co-purchase network and can be represented as a directed graph in which nodes are products and directed links connect each product with its recommended products. In such setup: *"The virtual aisle location of a product is determined, in part collectively by consumers rather than being chosen based on fees paid by manufacturers, or explicit strategic considerations by the retailer"* [65].

4.4.4 Network Analysis of Market Basket Data

The idea of examining the Amazon BLB co-purchase network was initiated by Krebs [66] who proposed the analysis of emergent patterns of connections that surround an individual, or a community of interest, based on book purchases. Dhar et al [67] extended considerably the analysis by assessing the influence of BLB networks on demand and revenue streams in Web commerce. They also contributed new experimental verification about the significance of visible item recommendations on the long tail of commerce in the Web.

Oestreicher-Singer and Sundararajan conducted a series of research efforts to analyze time series of co-purchase item networks incorporating methods from economic theory, econometrics and computer science. By repeatedly crawling the same items they create time series data for the books under consideration. In their initial research Oestreicher-Singer and Sundararajan [68], [69], they employed a depth-first crawler to collect from Amazon approximately 250,000 distinct books during the period 2005-6.

---

[5] For more detailed descriptions of collaborative filtering please refer to subsection 2.3.1.

Their main focus was to infer demand levels for each item and to evaluate if the network structure influences individual items. This influence was measured by adapting the PageRank algorithm to account for weighted composite graphs. The variation in the demand distribution across categories was estimated by computing the Gini coefficient [70] for each category. Their conjecture was found to be in accordance to the "long tail" for demand phenomenon in ecommerce [71]. Their results also include the following:*"(1) an increase in the variance in the extent to which the network influences products in a within a category increases the category's demand inequity. (which makes intuitive sense in the context of our theory of the network "flattening" demand), (2) The number of products in a category is positively associated with demand inequity, and (3) the average demand within a category is associated with an increase in the category's demand inequity."*

Using a similar to [68], [69], dataset Oestreicher-Singer and Sundararajan [72] econometrically identified the influence that visible co-purchase networks have on demand of individual items. Based on a simple set of conditions, which imply minimal empirical restrictions on the network structure concluded that visible co-purchase networks more than triples the average influence that complementary items have on demand. They also estimated that the magnitude of this social influence is higher for more popular and more recently published books. On the contrary, pricing, secondary market activity and assortative mixing across product categories are related by counter-intuitive ways with network position of individual items. Furthermore, in a newer investigation [65] they found among others that within a books category an increase in the influence of the recommendation network is consistently related with a more even distribution of both revenue and demand and when the recommendations are internal to a category itself, *"the redistribution of attention they cause compensates demand more within the category, rather than redirecting demand to a popular book in a different category."*

These results have practical implications for Web commerce since it is becoming clear that connected consumption exists online.

In 2009, Carmi et al [73] extend [72] to address new research questions related to the diffusion of exogenous shocks in the Amazon co-purchase network. In particular, they estimated how far these shocks propagate, how long they survive and if they affect demand of neighboring items and network structure. The initial dataset was augmented to include two years more data and book reviews about the books of Oprah Winfrey that appeared on the Oprah.com and "Sunday Book Review" section of the New York Times. They identified a consistent relationship between the shape of the diffusion curve and the level of clustering in the co-purchase network.

They also concluded that two subsets of reviewed books exist: "those whose demand increase is substantially higher than the total increase for its neighbors, and those for which the total increase in demand for the neighbors is an order of magnitude higher."

Going a step further from the econometric identification of impact that the visible co-purchase links have to demand [65], Dhar et al [67] assessed whether co-purchase item networks contain useful predictive information about sales. In particular, they estimated a simple auto-regressive model, where demand in the next period is modeled as a linear combination of demands in previous periods. By analyzing an extensive dataset, which covers a diverse set of books spanning over 400 categories over a period of three years with a total of over 70 million observations they concluded that changes in demand for each item can be predicted more accurately using network information.

Recently, in a relevant experimental study, Vafopoulos et al [19] crawled a set of 226,238 products from all the thirty Amazon's categories, which form 13,351,147 co-purchase connections. They introduce the analysis of local (i.e. dyads and triads) and community structures for each category and the more realistic case of different product categories (Market Basket Analysis). Their main results concerning the purchasing behavior of Web consumers are the following:

- The cross-category analysis revealed that Amazon has evolved into a book-based multi-store with strong cross-category connections.
- Co-purchase links not only manifest complementary consumption, but also switching among competitive products (e.g. the majority of consumers switch from Kaspersky to Norton Internet security suite).
- Top selling products are important in the co-purchase network, acting as hubs, authorities and brokers (or "mediators") in consumer preference patterns.
- Ostensibly competitive products may be consumed as complements because of the existence of compatibility and compatible products that facilitate their joint consumption.

Let us focus on the community analysis of the co-purchase network of products. For a given network, a community (or cluster, or cohesive subgroup) is defined to be a sub-network whose nodes are tightly connected, i.e. cohesive. Since the structural cohesion of the nodes can be quantified in several different ways, many different formal definitions of community structures have been emerged [74]. The analysis of community structures offers a deeper understanding for the underlying functions of a network. Figure 1 shows a part of the software products co-purchase network, where different colors indicate different community membership. Different product communities have been identified based on the spin glass community detection algorithm [75].

Analysis indicated that the seemingly competitive products of Apple and Microsoft are in reality consumed as if they were complementary. Microsoft (nodes with purple color) and Apple (nodes with orange color) product communities are "mediated" by compatibility like VMware Fusion, Parallels Desktop and compatible products like Office for Mac.

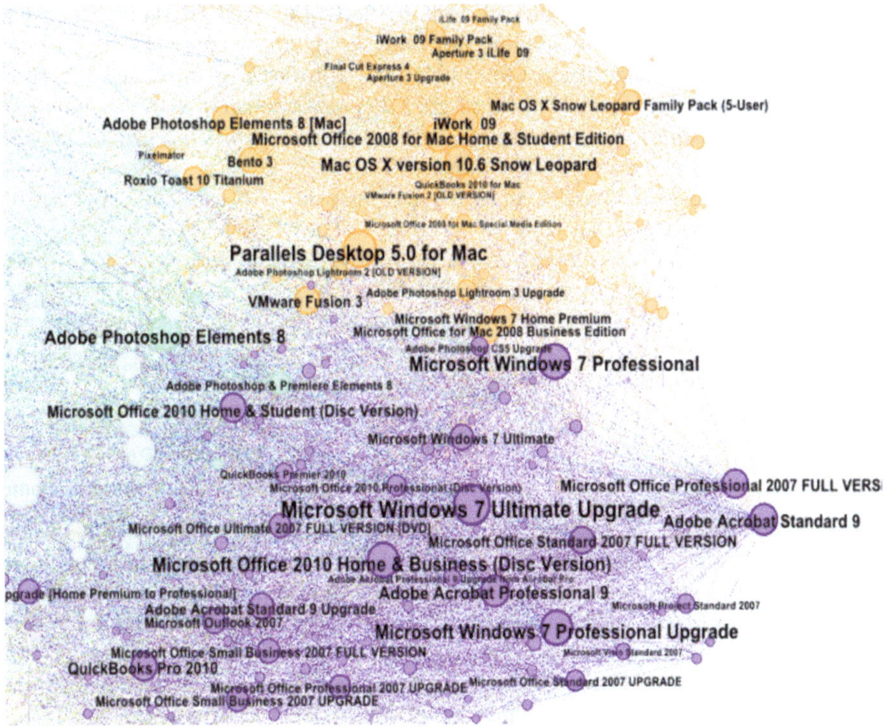

Fig. 1 Microsoft and Apple software programs are consumed as complements, because of compatibility (e.g. Parallel Desktops) and compatible (e.g. MS Office for Mac) products.

4.5 Network-Based Marketing

In parallel, apart of e-marketing studies has been recently emerged the field of network-based marketing. Network-based marketing refers to a collection of marketing techniques that take advantage of links between consumers to increase sales and should not be confused with network or multilevel marketing [76]. It could be also found in the literature as word-of-mouth and buzz marketing [77] and viral marketing [78]. The main focus is to measure how product adoption propagates from consumer to consumer through recommendation systems [78] and customer feedback mechanisms [79], [80].

Contrastingly to traditional marketing studies, network-based marketing models interdependent consumer preferences through explicit and implicit links among consumers. According to Hill et al [76] statistical research in network-based marketing includes six main fields: (1) econometric modeling, (2) network classification modeling, (3) surveys, (4) designed experiments with convenience samples, (5) diffusion theory and (6) collaborative filtering and recommendation systems.

Recommendation systems are relevant to network-based marketing because share the same objective to exploit the underlying knowledge residing in the stored data that are related to customer behavior (for a review in related literature refer to [81]). As Hill et al [76] highlight *"Recommendation systems may well benefit from information about explicit consumer interaction as an additional, perhaps quite important, aspect of similarity."*

Leskovec et al [78] studied an extensive snapshot of Amazon's person-to-person recommendation network of products. They modeled propagation of recommendations and the cascade sizes according to a simple stochastic model and concluded that product purchases follow a 'long tail' and that on average recommendations are not very effective at inducing purchases. Based on Bayesian network analysis identified communities, product, and pricing categories for which viral marketing is considered to be efficient.

5 Discussion

Making good product recommendations in the Web is not just matter of fast algorithms, but also a marketing task. Moreover, selling products and services in the Web has become a complex issue with equally important technical and marketing aspects, because Users, apart from searching and comparing the characteristics of products, can comment and be informed from others' Users' purchases and comments. In Web 2.0, Users reveal and contribute private information about their preferences and expectations, which is beneficial to other consumers if aggregated and made public. Consumption becomes more connected in the Web. This fact calls for new ways of investigating related Web phenomena and behavior.

In this article, we review recommendation systems, not only as technical artifacts, but also as parts of the more general problem of studying online purchasing behavior. The main focus is to highlight useful connections among diverse research efforts, which share some common tasks and challenges. In particular, we discuss the specific fields of one-to-one marketing, network-based marketing, Web merchandising and atmospherics and their implications in the processes of personalization and adaptation in the Web. The transformation of traditional marketing methodologies in the Web ecosystem is a multifold task, in which recommendation systems could contribute because (a) Marketing objectives sustain their core principles but extend their field of implementation in a more interactive communication process between business and customers and among customers themselves, (b) in this updated communication process, recommendation systems play a prominent role by aggregating individual preferences and enabling massive one-to-one marketing and (c) recommendation systems also facilitate personalization and adaptation of Web commerce, which drastically affect Web merchandising and atmospherics.

Market Basket Analysis is analyzed in the context of recommendation systems. The discussed issues in related literature are micro-economic issues, which directly refer to what types of recommendations are appropriate in achieving certain tasks. But participatory and interactive Web commerce raises a series of issues related to the Web economy in the macro level. As more and more companies are participating in the Web commerce, finding and analyzing consumption patterns is an essential key to their success. Data is the "king" and trust the "queen" in Web commerce and if combined with navigational patterns and social networking, provide to online mass merchants a strong comparative advantage, not only against their direct competitors in the Web but also against to the "brick-and-mortar" retailers.

At the same time, these massive amounts of personal and market data raise concerns about *privacy* and *excessive market power*. To the best of our knowledge, there is no yet scientific investigation in the economic or law literature concerning the excessive market power of Web merchants, which steams from data exploitation. As Clemons and Madhani [82] admit: *"Some digital business models may be so innovative that they overwhelm existing regulatory mechanisms, both legislation and historical jurisprudence, and require extension to or modification of antitrust law."*

An emergent challenge for recommendation systems will be the case of extensive application of Web 3.0 technologies in Web commerce (e.g. Good Relations ontology [83]). The easier exploration and comparison of Web Goods from Users will enable further competition and lower price dispersion. Semi-automatic contracting and business rules formation [84] have the potential to extend recommendation systems in a wider range of functionality.

The research agenda of economists, computer and information scientists is filling up with issues coming from Web commerce and in our point of view, this is going to happen for many years. Experts in algorithms, statistics and business intelligence will experiment with alternative recommendation and feedback systems (e.g. context-aware systems) but it seems that their efforts should account more seriously for data privacy concerns. An alternative and more generic approach may be to extend the Web architecture to support ex ante information transparency and accountability rather than ex post security and access restrictions [85]:

"Consumers should not have to agree in advance to complex policies with unpredictable outcomes. Instead, they should be confident that there will be redress if they are harmed by improper use of the information they provide, and otherwise they should not have to think about this at all".

Consequently, future challenges in recommendation systems will not be purely technical or business-oriented. They will involve issues like privacy, trust and provenance in semantic and ubiquitous Web environments, market competition and regulation.

References

[1] Albert, R., Barabási, A.: Statistical mechanics of complex networks. Reviews of Modern Physics 74(1), 47–97 (2002)

[2] Vafopoulos, M., Theodoridis, T., Kontokostas, D.: Inter-Viewing the Amazon Web Salespersons: Trends, Complementarities and Competition. In: 15th Panhellenic Conference on Informatics (PCI), pp. 299–303 (2011)

[3] Anagnostopoulos, I., Stavropoulos, P.: On the feasibility of applying capture recapture experiments for web evolution estimations. Applied Mathematics Letters 24(6), 1031–1036 (2011)
[4] Hendler, J., Shadbolt, N., Hall, W., Berners-Lee, T., Weitzner, D.: Web science: an interdisciplinary approach to understanding the web. Communications of the ACM 51(7), 60–69 (2008)
[5] White, S., Vafopoulos, M.: Web Science: expanding the notion of Computer Science. In: 43rd ACM Technical Symposium on Computer Science Education (2011)
[6] Vafopoulos, M.: Web Science Subject Categorization (WSSC). In: Proceedings of the ACM WebSci 2011, pp. 1–3 (2011)
[7] Vafopoulos, M.: Being, space and time in the Web. Metaphilosophy (forthcoming, 2012)
[8] Vafopoulos, M., Gravvanis, G., Platis, A.: The personal Grid e-workspace (g-work), Grid Techn., pp. 209–234. WIT Press (2006)
[9] Vafopoulos, M.: A Roadmap to the Grid e-Workspace. In: Szczepaniak, P.S., Kacprzyk, J., Niewiadomski, A. (eds.) AWIC 2005. LNCS (LNAI), vol. 3528, pp. 504–509. Springer, Heidelberg (2005)
[10] Vafopoulos, M.: Information Society: the two faces of Janus. Artificial Intelligence Applications and Innovations, 643–648 (2006)
[11] Desrosiers, C., Karypis, G.: A comprehensive survey of neighborhood-based recommendation methods. In: Recommender Systems Handbook, pp. 107–144 (2011)
[12] Ellison, G., Ellison, S.F.: Lessons about Markets from the Internet. The Journal of Economic Perspectives 19(2), 139–158 (2005)
[13] Brynjolfsson, E., Hu, Y., Smith, M.: Consumer Surplus in the Digital Economy: Estimating the Value of Increased Product Variety at Online Booksellers. Management Science 49(11), 1580–1596 (2003)
[14] Baye, M., Morgan, J., Scholten, P.: Information, search, and price dispersion. Economics and Information systems 1, 323–373 (2006)
[15] Benkler, Y.: The wealth of networks: How social production transforms markets and freedom. Yale University Press, New Haven (2006)
[16] Ngai, E., Wat, F.: A literature review and classification of electronic commerce research. Information & Management 39(5), 415–429 (2002)
[17] Wang, C.C., Chen, C.C.: Electronic Commerce Research in Latest Decade: A Literature Review. International Journal of Electronic Commerce 1(1), 1–14 (2010)
[18] Spulber, D.: Consumer Coordination in the Small and in the Large: Implications for Antitrust in Markets with Network Effects. Journal of Competition Law and Economics 4(2), 207 (2008)
[19] Vafopoulos, M., Theodoridis, T., Kontokostas, D.: Inter-Viewing the Amazon Web Salespersons: Trends, Complementarities and Competition. In: 15th Panhellenic Conference on Informatics (PCI), pp. 299–303 (2011)
[20] Coase, R.: The Problem of Social Cost. Journal of Law & Economics 3, 1–44 (1960)
[21] Hayek, F.: Spontaneous ('grown') order and organized ('made') order. Markets, hierarchies and networks: the coordination of social life, 293 (1991)
[22] Jannach, D., Zanker, M., Felfernig, A., Friedrich, G.: Recommender Systems: An Introduction. Cambridge Univ. Pr., New York (2010)
[23] Adomavicius, G., Tuzhilin, A.: Toward the Next Generation of Recommender Systems: A Survey of the State-of-the-Art and Possible Extensions. Knowledge Creation Diffusion Utilization 17(6), 734–749 (2005)

[24] Vafopoulos, M.: Modeling the Web Economy: Web Users and Goods. In: WebSci 2011, Koblenz, Germany, June 14-17 (2011)
[25] Hendler, J., Shadbolt, N., Hall, W., Berners-Lee, T., Weitzner, D.: Web science: an interdisciplinary approach to understanding the web. Communications of the ACM 51(7), 60–69 (2008)
[26] Berkovsky, S., Kuflik, T., Ricci, F.: Mediation of user models for enhanced personalization in recommender systems. User Modeling and User-Adapted Interaction 18(3), 245–286 (2007)
[27] Balabanović, M., Shoham, Y.: Fab: content-based, collaborative recommendation. Communications of the ACM 40(3), 66–72 (1997)
[28] Morita, M., Shinoda, Y.: Information filtering based on user behavior analysis and best match text retrieval. In: Proceedings of the 17th Annual International ACM SIGIR Conference on Research and Development in Information Retrieval, pp. 272–281 (1994)
[29] Cacheda, F., Carneiro, V., Fernández, D., Formoso, V.: Comparison of collaborative filtering algorithms: Limitations of current techniques and proposals for scalable, high-performance recommender systems. ACM Transactions on the Web (TWEB) 5(1), 2 (2011)
[30] Burke, R.: Hybrid recommender systems: Survey and experiments. User Modeling and User-Adapted Interaction 12(4), 331–370 (2002)
[31] Deshpande, M., Karypis, G.: Item-based top-n recommendation algorithms. ACM Transactions on Information Systems 22(1), 143–177 (2004)
[32] Su, X., Khoshgoftaar, T.M.: A survey of collaborative filtering techniques. Advances in Artificial Intelligence 2009, 4 (2009)
[33] Kobsa, A., Koenemann, J.: Personalized Hypermedia Presentation Techniques for Improving Online Customer Relationships. The Knowledge Engineering Review 16(2), 111 (2000)
[34] Hu, Y., Koren, Y., Volinsky, C.: Collaborative filtering for implicit feedback datasets. In: Eighth IEEE International Conference on Data Mining, ICDM 2008, pp. 263–272 (2008)
[35] Resnick, P.: Recommender systems. Communications of the ACM (1997)
[36] Bodapati, A.V.: Recommendation Systems with Purchase Data. Journal of Marketing Research 45(1), 77–93 (2008)
[37] Samaan, N., Karmouch, A.: A mobility prediction architecture based on contextual knowledge and spatial conceptual maps. IEEE Transactions on Mobile Computing 4(6), 537–551 (2005)
[38] Adomavicius, G., Tuzhilin, A.: Context-aware recommender systems. In: Recommender Systems Handbook, pp. 217–253 (2011)
[39] McNee, S., Lam, S., Konstan, J.: Interfaces for Eliciting New User Preferences in Recommender Systems. In: Brusilovsky, P., Corbett, A.T., de Rosis, F. (eds.) UM 2003. LNCS (LNAI), vol. 2702, pp. 178–187. Springer, Heidelberg (2003)
[40] Chaffey, D., Ellis-Chadwick, F., Mayer, R., Johnston, K.: Internet Marketing Strategy Implementation and Practice. Prentice Hall (2009)
[41] Hung, L.: A personalized recommendation system based on product taxonomy for one-to-one marketing online. Expert Systems with Applications 29(2), 383–392 (2005)
[42] Kim, W.: Personalization: Definition, status, and challenges ahead. Journal of Object Technology 1(1), 29–40 (2002)

[43] Pitta, D.: Marketing one-to-one and its dependence on knowledge discovery in databases. Journal of Consumer Marketing 15(5), 468–480 (1998)
[44] Gillenson, M., Sherrell, D.: Information Technology as the enabler of one-to-one marketing tutorial. Communications of the AIS 2 (September 1999)
[45] Blom, J.: Personalization: a taxonomy. In: CHI 2000: Extended Abstracts on Human Factors in Computing Systems, pp. 313–314 (April 2000)
[46] Cingil, I., Dogac, A., Azgin, A.: A broader approach to personalization. Communications of the ACM 43(8), 136–141 (2000)
[47] Germanakos, P., Mourlas, C.: Adaptation and Personalization of Web-based Multimedia Content. Media, 1–29 (2000)
[48] Berman, B., Evans, J.R.: Retail management: a strategic approach. Prentice Hall, Upper Saddle River (1998)
[49] Lee, J., Podlaseck, M., Schonberg, E., Hoch, R., Gomory, S.: Analysis and Visualization of Metrics for Online Merchandising. In: Masand, B., Spiliopoulou, M. (eds.) WebKDD 1999. LNCS (LNAI), vol. 1836, pp. 126–141. Springer, Heidelberg (2000)
[50] Lee, J., Podlaseck, M., Schonberg, E., Hoch, R.: Visualization and analysis of clickstream data of online stores for understanding web merchandising. Data Mining and Knowledge Discovery 5(1), 59–84 (2001)
[51] Machleit, K.A., Eroglu, S.A., Mantel, S.P.: Online retail atmospherics: empirical test of a cue typology. Journal of Consumer Psychology 9(1), 29–42 (2000)
[52] Manganari, E.E., Siomkos, G.J., Vrechopoulos, A.P.: Store atmosphere in web retailing. European Journal of Marketing 43(9/10), 1140–1153 (2009)
[53] Siomkos, G., Vrechopoulos, A., Manganari, E.: Web-Atmospheric Effects on Online Consumer Behavior: A Review of the Literature. In: Proceedings of IADIS International Conference e-Commerce, pp. 333–337 (2006)
[54] Vrechopoulos, A., O' Keefe, R., Doukidis, G., Siomkos, G.: Virtual store layout: an experimental comparison. Journal of Retailing (2004)
[55] Nanou, T., Lekakos, G., Fouskas, K.: The effects of recommendations' presentation on persuasion and satisfaction in a movie recommender system. Multimedia Systems 16, 219–230 (2010)
[56] Mild, A., Reutterer, T.: An improved collaborative filtering approach for predicting cross-category purchases based on binary market basket data. Journal of Retailing and Consumer Services 10(3), 123–133 (2003)
[57] Boztug, Y., Hildebrandt, L.: A market basket analysis conducted with a multivariate logit model. SFB 649 Discussion Papers (2005)
[58] Aggarwal, C.C., Procopiuc, C., Yu, P.S.: Finding localized associations in market basket data. IEEE Transactions on Knowledge and Data Engineering 14(1), 51–62 (2002)
[59] Chen, Y., Tang, K., Shen, R., Hu, Y.: Market basket analysis in a multiple store environment. Decision Support Systems 40(2), 339–354 (2005)
[60] Chib, S., Seetharaman, P., Strijnev, A.: Analysis of multi-category purchase incidence decisions using IRI market basket data. Advances in Econometrics 16, 57–92 (2002)
[61] Chiang, J.: A simultaneous approach to the whether, what and how much to buy questions. Marketing Science 10(4), 297–315 (1991)
[62] Aggarwal, C.C., Wolf, J.L., Yu, P.S.: A new method for similarity indexing of market basket data, vol. 28(2), pp. 407–418. ACM, New York (1999)

[63] Hao, M., Dayal, U., Hsu, M.: Visualization of directed associations in e-commerce transaction data. Data Visualization (2001)
[64] Kantardzic, M.: Data mining: concepts, models, methods, and algorithms. Wiley-Interscience (2003)
[65] Oestreicher-Singer, G., Sundararajan, A.: Recommendation Networks and the long tail of electronic commerce. Working Paper (2009)
[66] Krebs, V.: The Social Life of Books: Visualizing Communities of Interest via Purchase Patterns on the WWW (1999)
[67] Dhar, V., Oestreicher-Singer, G., Sundararajan, A., Umyarov, A.: The gestalt in graphs: Prediction using economic networks. NYU Working Paper No. CEDER-09-06 (2009)
[68] Oestreicher-Singer, G., Sundararajan, A.: Network Structure and the Long Tail of ECommerce Demand. In: Proc. 27th Internat. Conf. Inform. Systems, Milwaukee WI (2006)
[69] Oestreicher-Singer, G., Sundararajan, A.: Linking Network Structure to Ecommerce Demand: Theory and Evidence from Amazon.com's Copurchase Network. In: 34th Telecommunications Policy Research Conference, pp. 1–14 (2007)
[70] Yitzhaki, S.: Relative deprivation and the Gini coefficient. The Quarterly Journal of Economics 93(2), 321–324 (1979)
[71] Anderson, C.: The long tail. Hyperion, New York (2006)
[72] Oestreicher-singer, G., Sundararajan, A.: The Visible Hand of Social Networks in Electronic Markets. Electronic Commerce Research (2008)
[73] Carmi, E., Oestreicher-Singer, G., Sundararajan, A.: Spreading the Oprah Effect: The Diffusion of Demand Shocks in a Recommendation Network. In: ICIS 2009 Proceedings, p. 78 (2009)
[74] Boccaletti, S., Latora, V., Moreno, Y., Chavez, M., Hwang, D.U.: Complex networks: Structure and dynamics. Physics Reports 424(4-5), 175–308 (2006)
[75] Reichardt, J., Bornholdt, S.: Statistical mechanics of community detection. Physical Review E 74(1), 016110 (2006)
[76] Hill, S., Provost, F., Volinsky, C.: Network-Based Marketing: Identifying Likely Adopters via Consumer Networks. Statistical Science 21(2), 256–276 (2006)
[77] Rosen, E.: The anatomy of buzz: How to create word-of-mouth marketing. Crown Business (2002)
[78] Leskovec, J., Adamic, L., Huberman, B.: The dynamics of viral marketing. ACM Transactions on the Web 1(1), 5-es (2007)
[79] Dellarocas, C.: The digitization of word of mouth: Promise and challenges of online feedback mechanisms. Management Science, 1407–1424 (2003)
[80] Dellarocas, C.: Efficiency and robustness of binary feedback mechanisms in trading environments with moral hazard.MIT Sloan School of Management Working Paper no. 4297-03 (2003)
[81] Getoor, L., Taskar, B.: Introduction to statistical relational learning. The MIT Press (2007)
[82] Clemons, E.K., Madhani, N.: Regulation of Digital Businesses with Natural Monopolies or Third-Party Payment Business Models: Antitrust Lessons from the Analysis of Google. Journal of Management Information Systems 27(3), 43–80 (2010)
[83] Hepp, M.: Goodrelations: An ontology for describing products and services offers on the web. Knowledge Engineering: Practice and Patterns, 329–346 (2008)

[84] Boley, H., Tabet, S., Wagner, G.: Design rationale of RuleML: A markup language for semantic web rules. In: International Semantic Web Working Symposium (SWWS), pp. 381–402 (2001)

[85] Weitzner, D., Abelson, H., Berners-Lee, T., Feigenbaum, J., Hendler, J., Sussman, G.J.: Information accountability. Communications of the ACM 51(6), 82–87 (2008)

Exploring New Ways for Personalized E-Commerce through Digital TV*

Yolanda Blanco-Fernández, Martín López-Nores, José J. Pazos-Arias, and Manuela I. Martín-Vicente

Department of Telematics Engineering, University of Vigo
EE Telecomunicación, Campus Universitario s/n, 36310 Vigo (Spain)
{yolanda,mlnores,jose,mvicente}@det.uvigo.es

Summary. The evolution of information technologies is consolidating *recommender systems* as essential tools in e-commerce. To date, these systems have focused on discovering the items that best match the preferences, interests and needs of individual users, to end up listing those items by decreasing relevance in some menus. In this paper, we propose extending the current scope of recommender systems to better support trading activities, by automatically generating interactive applications that provide the users with personalized commercial functionalities related to the selected items. We explore this idea in the context of Digital TV advertising, with a system that provide personalized commercial functionalities, gathering contents from multiple sources and bring together semantic reasoning techniques, SWRL rules and new architectural solutions for web services and mashups.

1 Introduction

The advent of new devices (e.g., Digital TV receivers, mobile phones or media players) and usage habits (e.g., social networking) is progressively rendering the classical search engines of the Internet insufficient to support users in their grasp of an ever-expanding information space. Basically, it is no longer realistic to think that the users will always bother to visit a site and enter queries describing what they are looking for. This fact has brought about a new paradigm of *recommender systems*, which aim at proactively discovering the items that best match the preferences, interests and needs of each individual at any time. Such systems are already working behind the scenes in many e-commerce sites [12, 4], as well as in personalized TV programming guides [34, 2], news systems [21], web navigation assistants [36] and so on.

The problem we address in this paper is that recommender systems have hitherto limited themselves to providing the users with lists of the items that best match the

* This work has been partially funded by the Ministerio de Educación y Ciencia (Gobierno de España) research project TIN2010-20797 (partly financed with FEDER funds), and by the Consellería de Educación e Ordenación Universitaria (Xunta de Galicia) incentives file CN 2011/023 (partly financed with FEDER funds).

I.E. Anagnostopoulos et al. (Eds.): Semantic Hyper/Multimedia Adaptation, SCI 418, pp. 181–201.
springerlink.com

information stored in their profiles. This approach works well when the user's aims and attention are sufficiently close to the recommended items. For instance, if the user has finished watching a TV program, he may welcome a list of the most interesting forthcoming contents over the different channels available; likewise, when he has just entered an online bookshop, it makes perfect sense to face him with a list of potentially interesting books. In such contexts, which we classify under the metaphor *the user inside the shop*, the actions triggered by the user selecting one item from the list are either straightforward (zap to the new channel) or given by the provider in question (the bookshop), so the recommender system has completed its task. However, in settings where the user is not focused on the kind of items that may be recommended to him (e.g. in delivering publicity while he watches TV programs), one could expect the recommender to select the best offers from among various providers, to look for the pieces of information that describe each item in the most complete or accessible way, to arrange the most suitable interfaces to place an order, etc. In other words, with *the user around the marketplace*, we believe the recommender should be responsible for building the shop in which the user will feel most comfortable to browse the suggested items. Obviously, this is not a task for human developers, because no workforce would suffice to provide specific applications for all the different users, items and devices in all possible contexts.

We have realized the aforementioned vision by enhancing the recommender system of MiSPOT [18], originally designed to identify the best items to advertise within TV programs. The new version unleashes the possibilities of the Digital TV technologies for e-commerce, by automatically generating interactive applications (henceforth, *i-spots*) that provide the users with personalized commercial functionalities related to the selected items. To contextualize this work, Sect. 2 includes an overview of previous research about (i) recommender systems and (ii) provision of interactive applications through the TV. Then, Sect. 3 describes the elements involved in the assembly of i-spots, whereas Sect. 4 focused on the role *web services*, *web mashups* and *semantic reasoning* processes play in their automatic composition. Section 5 provides an example drawn from a system prototype, while Sect. 6 details architectural challenges and opportunities raised by the specifics of Digital TV. Finally, Sect. 7 summarizes the conclusions from this research.

2 Related Work

Research in recommender systems is hectic nowadays, in an attempt to address the many new questions raised by the growing number of practical applications. Next, we provide an overview of the milestones in recommenders history and semantic technologies, and thereafter focus on the issue of automatically composing personalized interactive applications, which remains practically unexplored in literature.

2.1 Related Work in Recommender Systems

Given a set of items, the goal of a recommender system is to identify the most suitable ones according to the information stored in a user's profile. When designing a recommender system, several concerns must be taken into account:

- First, it is necessary to choose a specific profile representation technique, which is a key decision since the recommender's success will greatly depend on the ability to represent accurately the user's current interests. Several approaches have been adopted for this purpose, ranging from indexed vectors of features, matrixes of ratings and sets of demographic features, to decision trees, classifiers based on neural networks and semantic networks.
- Second, a recommender system cannot work until the user profile has been created. Besides, the system must know as much as possible about the particular interests of each user. For that reason, a suitable technique to model an accurate initial profile is clearly required in personalization tools. To this aim, researchers have adopted many diverse approaches ranging from user-driven manual input to semi-automatic procedures such as training sets or stereotyping.
- Third, to generate and maintain the user profile, a recommender system requires fresh information about his/her personal preferences. This requirements justifies the adoption of relevance feedback tecniques to gather information about the user's tastes and interests. The feedback can be either given explicitly by the own users, or implicitly inferred by the recommender system from their actions and previous interactions. In both scenarios, the recommender needs techniques to adapt the users' profiles as their interests and preferences evolve over time. Such techniques typically consider that the most recent observations inferred from the relevance feedback are better than older ones, so that the goal is to adapt the user profiles to new interests and forget old preferences. Some approaches involves manual procedures (based on adding the new information to the user profile), whereas more sophisticated methods adopt time windows, gradual forgetting functions and even natural selection for ecosystems of agents [1].
- Lastly, a recommender system requires a filtering strategy to match the information modeled in the user profile against the available items. Initially, many recommender systems dealt with demographic filtering, which uses descriptions of people to learn the relationship between a single item and the type of users who like it. The user profiles are created by classifying users as per *stereotypes*, which are mechanisms that provide general descriptions for a set of similar users [27]. Specifically, stereotyping creates a user model that is initialized by classifying each user in (stereotypical) representations driven by features of classes of individuals. Typically, the data used in the classification is demographic and the user is asked to fill out a registration form including a wide diversity of data (city, age, sex, lifestyle, etc). As described in the systems proposed in [23, 30, 14, 15], stereotypes have also been widely adopted in diverse filtering strategies for the selection of the most appropriate recommendations for each user.

The first strategies merely looked at demographic information (e.g. age, gender or marital status) to recommend items that had interested other users with similar data. The results so obtained tend to be imprecise and fail to reflect changes of the user preferences over time (because personal data are often stable for long periods). This problem was addressed by *content-based filtering*, that looks for items similar to others that gained the user's interest in the past [1, 5]. This strategy is easy to adopt, but bears a problem of *overspecialization*: the recommendations tend to be repetitive for considering that a user will always appreciate the same kind of items. Furthermore, the minimal data available about new users makes the first results highly inaccurate.

To tackle these problems, the scientific community came up with *collaborative filtering*, that proceeds by evaluating not only the profile of the target user (the one who will receive the recommendations), but also those of users with similar interests (his/her neighbors) [29, 23]. This approach can solve the lack of diversity in the recommendations, but faces problems like the *sparsity* when the number of items is high (which makes it hard to find users with similar evaluations for the same items) or the treatment given to users whose preferences are dissimilar to the majority (the *gray sheep*). There exist *hybrid approaches* that attempt to neutralize the weaknesses and combine the strengths of content-based and collaborative filtering, e.g. recommending items similar to the ones listed in the user's profile, but considering two items similar if the individuals who show interest in the one tend to be interested in the other [3].

Regardless of the filtering strategy, it is noticeable that most of the recommender systems have relied on syntactic matching techniques, that relate items by looking for common words in their attached metadata. Even though there exist plenty of different approaches, they all miss much knowledge during the personalization process, because they are unable to reason about the meaning of the metadata. A syntactic approach is also a source of overspecialization, because the recommendations so computed can only include items very similar to those the users already know [1].

To go one step beyond in personalization quality and diversity, research is now focused on applying techniques from the Semantic Web, that allow to gain insight into the meaning of words.

2.2 Related Work in Semantic Technologies

As the Semantic Web is an initiative that uses metadata to provide explicit meaning to the information and services available on the Web, an automated processing is possible that prevents the users from searching manually appealing resources among those offered in a specific domain. The technological cornerstone for that purpose are the *ontologies*, conceptualizations that provide a commonly-accepted description of the domain of interest and whose application meets most of the challenging requirements related to the recommender system field:

- Firstly, the provision of a common semantic base (vocabulary) fights the interoperability problem that comes along with the integration of heterogeneous data sources.

- Secondly, an ontology provides a formal basis which is the prerequisite for adoption of semantic reasoning mechanisms aimed at discovering knowledge starting from the explicit relationships formalized in ontologies.
- Lastly, thanks to these mechanisms, ontologies pave the road for valuable services in the realm of a recommender system, ranging from information search to the provision of fully automatic personalization capabilities.

Since ontologies describe and interrelate items and their attributes by means of class hierarchies and properties, many authors have enhanced the traditional filtering strategies with the aforementioned semantic reasoning mechanisms, which are aimed at discovering the items that best match the preferences of each user by reasoning about their semantic descriptions.

- In [12], Hung proposed a recommender system for one-to-one marketing based on a taxonomy of products, revealing the advantages of such semantics when it came to providing instant online recommendations and identifying potential customers upon release of a new product.
- Similarly, in [22], Middleton *et al.* explored a novel ontological approach to user profiling within semantics-based recommender systems, addressing the problem of recommending academic research papers. Actually, the ontology-driven user modeling has been also adopted for development of clustering techniques in the formation of users' neighborhood and even for the creation and management of stereotypes [37, 7, 28].
- Another recommender system that relies on ontologies is Bibster, a peer-to-peer (P2P) system aimed at researchers that share bibliographic metadata [8]. The goal is to provide the users with personalized access to the metadata available in the P2P network. For this purpose, the authors use an ontology for representing the knowledge of the application domain. Bibster proposes a semantic similarity metric in order to compare the user's expertise (inferred from the bibliographic metadata stored in his/her local repository) to the data that other users want to share. This comparison process is based on the hierarchical relationships stated in the ontology and some properties explicitly represented in it. This fact (common in existing semantics-recommender system) limits greatly the amount of knowledge that can be discovered about the users' preferences, which is typically coded in hidden relationships inferred from the attributes of the available items. In fact, richer inferential capabilities would help to overcome the drawbacks of traditional content-based and collaborative approaches, which is the premise of the reasoning-driven filtering included in our strategy.

2.3 Related Work in Automatic Assembly of Personalized Interactive Applications

None of the mentioned recommender systems has generalized the possibility of linking the advertisements with interactive applications —no matter how simple— that would let the users browse details of the corresponding items, purchase online, subscribe to the notification of novelties, etc. At the most, they could link the advertisements to one from among a set of manually-developed applications, which is certainly insufficient.

A solution for the automatic generation of interactive applications can be inspired by the extensive literature of web services composition, that builds upon a set of standards (e.g. WSDL or OWL-S) to characterize web services in terms of inputs and outputs, some conceptualization of their internal processing or purpose, etc. There already exist various engines that can automatically provide the back-end logic of many e-commerce services on the Internet [11, 31, 35, 17].

- Specifically, [11] describes one of the first distributed algorithms for automatic service composition based on the desired input to and output from the process. The algorithm makes use of the content-based publish/subscribe model, with service inputs modeled as subscriptions, and outputs as advertisements. Service interfaces are mapped to publish/subscribe messages so that publish/subscribe matching is used to evaluate service compatibility.
- In [31], authors deal with automated composition of web services by Artificial Intelligence planning techniques. Specifically, they propose a system that plans over sets of OWL-S descriptions using SHOP2 and then executes the resulting plans over the Web. The resulting system executes information-providing web services during the planning process.
- The work presented in [35] describes a technique called interface-matching automatic composition for generation of complex web services which is based on capturing the users' expected outcomes when a set of inputs are provided. The approach results in a sequence of services whose combined execution achieves the users' goals.
- In [17], Lin *et al.* describe an automated generation mechanism for service composition which is based on assessing structural properties formalized in multiple ontologies with a very similar structure.

As regards the front-end, there have been various attempts to automate the generation of user interfaces from web service descriptions (e.g. WSUI, WSXL or WS-GUI) but they have been discontinued. The important point for this paper, though, is that all of this research has implicitly targeted the World Wide Web as accessed from personal computers, therefore paying no attention to the specifics of the TV. Thus, there is nearly nothing in literature about (i) providing specialized user interfaces, (ii) assembling deferred/local interaction services when there is no return channel available for bidirectional communication, or (iii) harnessing the bandwidth and multicasting possibilities of the broadcast networks. In general, research is also scarce in considering user preferences during the composition process, to drive the decision of what providers to link, what types and pieces of information to display, etc.

3 Cooking I-Spots: Ingredients

We conceive the i-spots as a sort of specialized mashups for DTV receivers, that bring together multimedia contents (e.g. pictures, videos or maps) from various sources and deliver functionality through pieces of software characterized as web services. The elements that come into play in the i-spot generation process, to be detailed in the following sections, are as follows:

- **A domain ontology.** The semantic reasoning features of our system require an ontology to formally represent the concepts and relationships of the e-commerce domain. The creation of such an ontology is problematic due to the high degree of specificity (that leads to a very large number of concepts) and the need for timely maintenance, owing to the continuous innovations that take place in the domain of products and services. Actually, we did not intend to create an ontology covering all possible types of items, but rather to use one that could be easily extracted from some of the classification standards available for industrial products and services. To this aim, we looked at standards like UNSPSC[1], eCl@ss[2], eOTD[3] and the RosettaNet Technical Dictionary[4], which reflect some level of community consensus and contain multiple definitions of hierarchically-organized concepts. Finally, we chose eCl@ss as the main input for creating the domain ontology, due to the reasons of completeness, balance and maintenance discussed in [10]. More specifically, we have borrowed from eClassOWL —the OWL ontology for products and services developed by M. Hepp [9]— many concepts referred to categorizations of commercial products, and we have also defined new properties and classes to accommodate some missing features. The ontology was populated by retrieving information from multiple online retailers. A core part of the ontology we use is a hierarchy that classifies items and their attributes as shown in the excerpt of Fig. 1, which includes a documentary about the *Bergamo Alps*, *vinegar* and *wine* brands from the Italian city of *Modena*, and a *sports car* also manufactured in Modena.
- **User profiles.** We capture the preferences of each user in a data structure that stores demographical information (e.g. gender, age, marital status, income, etc) plus *consumption* and *viewing histories* to keep track of items purchased and TV programs watched in the past. Items and programs are linked to a number between -1 and 1 that represents the degree of interest (DOI) of the user in them. This number may be explicitly provided by the user, or inferred from ratings given to other related concepts in the domain ontology (see [18] for the details).
- **Filtering logic.** In order to decide whether a given item or a related piece of information might be appealing to a user, our system runs a hybrid filtering strategy against his/her preferences and the semantic annotations of the item

[1] http://www.unspsc.org

[2] http://www.eclass-online.com

[3] http://registry.eccma.org/eotd

[4] http://www.rosettanet.org

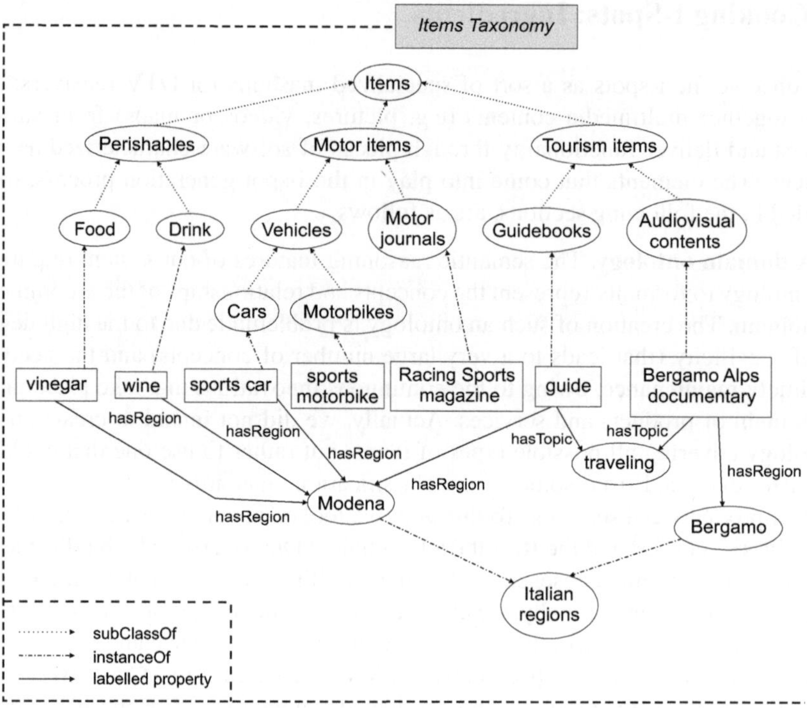

Fig. 1 A micro-excerpt from the items taxonomy of our domain ontology.

in the domain ontology (further details will be given in Sect. 4.1). For example, the reasoning can relate the *Racing Sports* magazine and a *sports car* of Fig. 1 through the ancestor *Motor items*; likewise, a given *guidebook* and *Bergamo Alps documentary* would be recognized as similar items because they share the attribute *traveling*, and also because they are bound to two *Italian regions* (*Modena* and *Bergamo*, respectively).

- **Functional elements.** The functionality of the i-spots in the scenarios we are considering results from combining software elements that may provide either *local interactivity* (dealing exclusively with contents delivered in the broadcast stream) or *deferred interactivity* (storing information to forward when a return channel becomes available). We characterize all of them as *semantic web services* as per the OWL-S standard [20], which involves three interrelated subontologies: *Profile* ("*what the service does*" for purposes of discovery), *Process Model* ("*how the service works*" for invocation and composition) and *Grounding* (a realization of the process model into detailed specifications of message formats and protocols).

  OWL-S provides an extensive ontology of functions where each class corresponds to a class of homogeneous functionalities (see [20]). Leaning on that ontology, we built a *services taxonomy* that represents the capabilities of the elements that may be included in an i-spot. Our domain ontology merges the items and services taxonomy so that the semantic reasoning features can relate the available functionalities to the items stored in the user profiles. A micro-excerpt from the services taxonomy is depicted in Fig. 1, with each item linked to different types of services by *hasTypeofService* properties. For example, the guidebook about Modena is associated to a type of service (denoted by *OWL-S Service* #1) that may be used to buy one copy while browsing information about the main tourist attractions of Modena; it is also linked to a service (*OWL-S Service* #2) that may be used to purchase a travel to this Italian region with possibilities to book accommodation and flight. Similarly, the sports car is linked to a service (*OWL-S Service* #3) that may be used to simply describe its features, and to a second one (*OWL-S Service* #4) that may offer the chance to arrange an appointment with a dealer.

- **Templates.** The i-spots are realized over templates that provide the Java code (as per the MHP standard [32]) that glues together different functional elements and pieces of content. This includes the programming of the user interfaces and the SOAP logic needed to interact with web services [33]. Actually, a same type service as described above can be offered through multiple templates, which differ from each other in the interactive elements shown to the user. This fact is represented in Fig. 2, where the elements of each template are identified by semantic concepts formalized in our ontology (e.g. *Map*, *Dealer* or *Calendar*).
- **Semantics-enhanced UDDI registry.** To enable discovery of the functional elements to include in a given template, we use a semantics-enhanced *registry* based on the UDDI standard [33]. The registry is accessed via a proxy module running in the receivers, that provides a unique interface to retrieve elements regardless of whether they are received through broadcast or through the return channel. Apart from that, our registry works just like the one presented in [13], which takes advantage of the expressiveness of OWL-S to match the capabilities offered against the requested ones.

4 Cooking I-Spots: Procedure

Our process of i-spot assembly is activated when the user clicks on some region of the scene appearing in the content she is viewing [19]. This scene is associated with several keywords that are represented as semantic concepts in our domain ontology. Starting from these concepts, the recommender system of MiSPOT identifies (and shows to the user) a set of potentially appealing items. Once the user has selected one of these items, it is necessary to choose: (i) which type of OWL-S web service best match the user preferences, and (ii) which is the most convenient template to assemble the i-spot. Finally, the selected template must be populated by retrieving the contents included in each one of its interactive elements. The aforementioned processes are detailed through this section.

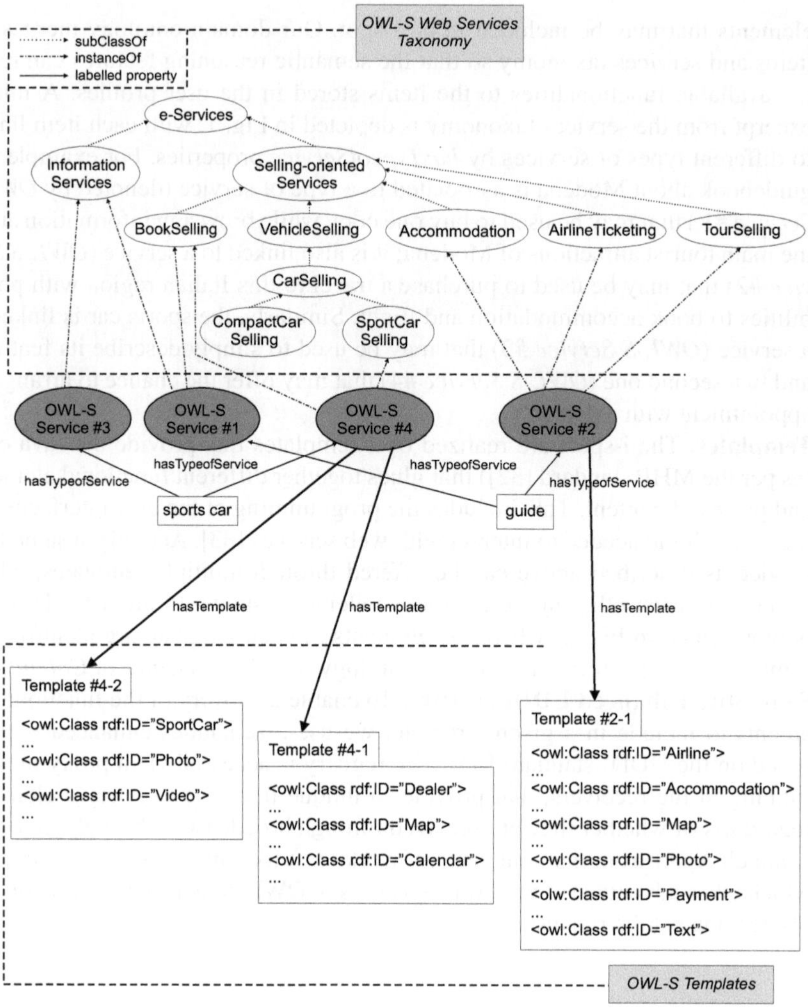

Fig. 2 A micro-excerpt from the taxonomy of semantic web services and templates.

4.1 Selecting Potentially Appealing Items

Our personalization procedure starts out from the assumption that the program viewed by the user is related to his/her current interests (otherwise he/she would not have chosen it). For this reason, the items that MiSPOT selects as relevant for the user are related to the content he/she is viewing. Specifically, MiSPOT explores only the items in the ontology that are directly connected (via properties) to the user-activated concepts (hereafter *activated items*). The *filtering* module of our system incorporates content-based filtering and collaborative filtering as recommendation strategies, so that MiSPOT can suggest both items similar to others that the user liked in the past, and those have interested other users with similar preferences (i.e. his/her neighbors).

In the rest of this section, we explain how our two filtering strategies decide whether a given *activated item* $\mathscr{I}$ is appealing (and therefore it would be recommended) to a user $\mathscr{U}$ (whose profile is $\mathscr{P}$).

4.1.1 Content-Based strategy

Our content-based strategy consists of matching the considered *activated item* against the user's preferences. This *matching level* is computed by averaging the values of similarity between $\mathscr{I}$ and the items stored in $\mathscr{U}$'s profile, weighted by their respective DOIs. Intuitively, the resulting *matching level* grows with the similarity between $\mathscr{U}$ and the favorite items for $\mathscr{U}$ (i.e. those with the highest ratings in the user's profile). To measure such similarity, we have defined a metric that quantifies the strength of the relationships that can be inferred —from the knowledge captured in the domain ontology— between $\mathscr{I}$ and $\mathscr{U}$'s preferences. Specifically, our *semantic similarity metric* considers not only the explicit relations defined by the hierarchy of classes, but also others which are hidden behind the attributes of the items. Therefore, we talk about two similarity criteria, that we refer to as *hierarchical similarity* and *inferential similarity*, which are finally weighted and combined by means of a factor $\alpha \in [0,1]$.

- The notion of hierarchical similarity has appeared in many previous approaches (see for example [6, 25, 26]), and consists of valuing the relationship between two nodes of the ontologies by the existence and specificity of a common ancestor in a hierarchy of classes. In our approach, the value of hierarchical similarity between $\mathscr{I}$ and each item i in U's profile grows with the depth of their *lowest common ancestor* (LCA) and also with its proximity to $\mathscr{I}$ and i in the items taxonomy. The depth of a node is given by the number of hierarchical links traversed to reach the node from the root of the hierarchy; thereby, the hierarchical similarity between two nodes is 0 if they do not have other common ancestor than the root class. For example, in Fig. 1, the *sports car* is more similar to the *sports motorbike* than to the *Racing Sports magazine*, because the *LCA* between the two sports vehicles is more specific (deeper) than the *LCA* between the *sports car* and the *motor magazine* (i.e. depth(*Vehicles*)=2 > depth(*Motor items*)=1).
- The notion of inferential similarity consists of measuring similarity by looking at relationships between the semantic attributes of the items compared. In this regard, two items are considered similar if they share some attributes (*common attributes*), or if they have attributes belonging to the same class in some hierarchy (*sibling attributes*). For example, the *guidebook* depicted in Fig. 1 and *Bergamo Alps documentary* are similar because both share the attribute *traveling*, and also because they are bound to two *Italian regions* (*Modena* and *Bergamo* are sibling attributes).

  To calculate the value of inferential similarity between $\mathscr{I}$ and i, in addition to the existence of common or sibling attributes, we consider the level of interest of

$\mathcal{U}$ in such attributes (i.e. his/her DOI indexes in those attributes). Consequently, inferential similarity between both items is high when they share many attributes and when these attributes are highly rated in $\mathcal{U}$'s profile.

After computing the *matching level* between $\mathcal{U}$ and the *activated item* $\mathcal{I}$ (by combining the semantic similarity values and DOI indexes), our content-based strategy recommends this item to the user if the resulting value is greater than a given threshold. Otherwise, the item $\mathcal{I}$ is considered again by our collaborative strategy.

4.1.2 Collaborative Strategy

Our collaborative strategy starts out by delimiting the neighborhood of the user $\mathcal{U}$. Before describing this process, it is necessary to highlight some concerns about the architectural design of our system. As we will explain in Sect. 6, the architecture of MiSPOT enables to work both with permanently-enabled return channels and with scenarios where the users' preferences can only be stored locally. In the first case, the users' profiles are stored in dedicated servers, whereas in the second one, such profiles are not available and collaborative systems must handle stereotypes instead. In other words, the goal of the collaborative filtering will be either to find the users' profiles most similar to a given one, or to identify which are the stereotypes in which the user fits best.

For that purpose, we first create a *rating vector* containing the DOIs of the item classes most appealing or most unappealing to the user (identified by DOIs close to 1 and −1, respectively). Next, we look for either the user profiles or stereotypes that contain DOIs for at least 70% of those classes, and create their respective rating vectors. Finally, we compute the Pearson-*r* correlation between the rating vector of each partial stereotype and the vector corresponding to the user $\mathcal{U}$.

The user's neighborhood is formed by the users' profiles/stereotypes that yield correlation values greater than a given threshold γ. Once the neighbors have been identified, we predict the level of interest of the user in the *activated item* by considering both his/her preferences and the interest of his/her neighbors in $\mathcal{I}$. Specifically, if a neighbor has rated this item, we use his/her DOI index; otherwise, we predict his/her level of interest by computing the *matching level* between $\mathcal{I}$ and the neighbor's preferences. As a result, the interest value predicted for $\mathcal{I}$ is greater when this item is very appealing to the selected neighbors and these are strongly correlated with $\mathcal{U}$'s preferences.

Analogously to what we commented in the content-based strategy, the *activated item* is finally recommended to the user when the predicted interest is greater than a given threshold.

4.2 Selecting Type of OWL-S Service

The generation of an i-spot is triggered when the user shows interest in one of the items recommended to him/her. Then, it is necessary to decide which type of service best matches the user preferences, and which is the most convenient template to create one i-spot that provides the service. The selected template must be finally populated by retrieving contents for each one of its interactive elements.

In order to select the type of service to provide, we adopt an approach similar to the one presented in [24], which consists in using SWRL rules to relate user preferences and context information in a personalization system. We have defined a set of SWRL rules to associate personal information of the users (such as hobbies and income) with the kind of services classified into the services taxonomy. The user preferences are antecedents in the rules, while the service types appear in their consequents. For example, by the rule #1 shown below, we infer that a user with sufficient income who is fond of traveling may appreciate services that permit to book a tour (instead of services that, for instance, only describe the tourist attractions of a given destination). Likewise, rule #2 makes it possible to conclude that a user with high income will likely be interested in services for the selling of luxury items (against services that, for instance, only provide information about them).

SWRL Rule #1

user (?u) ∧ swrlb:notEqual (income (?u), "LOW") ∧
swrlb:equal (hobby (?u), "TRAVEL") ⇒
appealingTypeOfService (?u,?ws) ∧ TourSellingService (?ws)

SWRL Rule #2

user (?u) ∧ swrlb:equal (income(?u), "HIGH") ⇒
appealingTypeOfService (?u,?ws) ∧ LuxuryItemsSellingService (?ws)

Such inferences are made by a *Description Logic* (DL) reasoner module that computes a relevance factor for each type of service associated to the item chosen by the user. Obviously, the factor is greater when the type of service appears in the consequent of a rule whose antecedent is highly rated among the user's preferences.

4.3 Selecting Template

Having selected a type of service, the next step is to decide which one of the templates associated to it may be most interesting for the user. In other words, the goal is to identify the most suitable interactive elements to assemble an i-spot about the item selected by the user. This decision is driven by both the user preferences (e.g. the kind of elements the user has accessed in previous i-spots) and parameters such as the computational power, input/output capabilities or return channel availability. Therefore, for example, we refrain from using templates with maps in the case of users who have never fiddled with such artifacts in previous i-spots, while we limit the amount of text to be shown in the case of small screen devices or with users who have never fully read lengthy descriptions of other items.

Following the selection of a template to shape the i-spot, it is time to look in the UDDI registry for services categorized in the taxonomy under the selected service type, and also for contents to place in all the elements of the selected template. With this information, an OWL-S/UDDI matchmaker maps the OWL-S Service Profile into the corresponding WSDL representations of the services offered in our semantic registry. As a result, the registry provides a set of services to retrieve the contents to be lodged in the template. For example, assembling an i-spot via *Template #2-1* shown in Fig. 2 requires to discover selling services that offer information about airlines and hotels providers, as well as maps and photos of rooms.

To conclude, it is necessary to invoke the OWL-S services discovered in the preceding step, by exploiting the WSDL descriptions of message formats and in/out arguments provided by the service groundings. This way, our i-spot composition logic retrieves contents that are pre-filtered considering the user preferences, as we will describe next.

4.4 Populating the Template

Analogously to what we commented for the selection of types of OWL-S service, the population of the templates is also based on an approach similar to the one presented in [24], where the authors used SWRL rules to relate user data and context information in a personalization system. In our case, as noted in Section 3, the antecedents of the rules involve demographic data and levels of familiarity of the user with certain semantic concepts, whereas the consequents involve service types or sources of information.

All the inferences that lead to a personalized i-spot are made by a *Description Logic* (DL) reasoner module, that computes a relevance factor for each source of information $\mathscr{S}_i$ associated to an item $\mathscr{I}$ chosen by a user $\mathscr{U}$. Intuitively, the factor is greater when $\mathscr{S}_i$ appears in the consequent of a rule whose antecedent includes terms that match the data stored in $\mathscr{U}$'s profile. The matching is computed differently for demographic features and semantic concepts:

- On the one hand, we measure how much $\mathscr{U}$'s data match the demographic features associated to $\mathscr{I}$ by computing a value between 0 and 1 in a way that varies with the type of feature. For example, *age* takes values lower than 5 and greater than 80, so a difference of more than 75 years yields a value 0, whereas coinciding values yield 1 and all other possibilities get intermediate values. In contrast, with features like *gender* or *marital status*, we get 1 if the values in $\mathscr{U}$ and the SWRL rules coincide, and 0 if they do not. We conclude that $\mathscr{U}$ meets a demographic condition if the value is greater than a configurable threshold (0.67 in our preliminary tests).
- On the other hand, to calculate levels of familiarity, we lean on the *semantic similarity* metric described in Sect. 4.1. The level of familiarity of the user $\mathscr{U}$

with a semantic concept $\mathscr{C}$ results from accumulating the semantic similarities between $\mathscr{C}$ and the concepts rated in $\mathscr{U}$'s profile, weighted by their DOIs and by a factor that measures the *relative consolidation* of those DOIs in the profile. The latter factor guarantees that the greatest contributions correspond to the concepts (remember, classes or attributes) linked to most TV programs or items rated by $\mathscr{U}$ —in turn, this ensures that the DOI values are not distorted by negative ratings given to programs or items that $\mathscr{U}$ might have found unsatisfactory for other reasons than their topic. Thus, for example, if a user is a fond viewer of a thematic channel devoted to motor sports, we always take *Mechanics* as a highly informative concept compared to others.

We decide that the level of familiarity of $\mathscr{U}$ with a semantic concept is '*high*' when the computed sum is greater than 0.67, '*medium*' if it is between 0.34 and 0.66, and '*low*' if it is lower than 0.33.

5 A Sample Scenario

For a brief example of the personalization features offered by the new version of the MiSPOT system, we shall consider the case of Paul, a childless man in his early 20s with significant income (demographic data), who lives close to London and is subscribed to the *Racing Sports* magazine (consumption history). We assume that Paul is currently watching a documentary about Modena, using a high-end Digital TV receiver permanently connected to a cable network, and that the filtering logic presented in [18] has selected the *sports car* manufactured in that Italian city as a relevant item for a recommendation.

If Paul accepts the recommendation, clicking a button in his remote control tells our system to assemble an i-spot about the sports car. The procedure goes on as follows:

- To begin with, our DL reasoner finds that Paul's demographic information matches the conditions of SWRL Rule #4 (Section 4.2), which leads to selecting a car selling service (*OWL-S Service* #4) instead of one that would simply provide information about the sports car, due to his high level of income.
- As a consequence of the computing and communication capabilities of Paul's high-end receiver, the system selects a template with demanding interactive elements, namely #4 − 2 from Fig. 2.
- Next, asking the UDDI semantic registry, we obtain the OWL-S services needed for retrieving the contents of the selected template.
- When it finally comes to populating the template, the DL reasoner finds SWRL Rule #1 (Section 3) to be applicable for Paul, due to his fondness for motor-related issues. This leads to retrieving car specifications from the specialized site *www.autoguide.com*.

As shown in the snapshots of Fig. 3, one tab of the i-spot assembled for Paul displays the aforementioned information about the car specifications. Another one provides a collection of pictures from Flickr and videos from *Youtube*. The last tab provides

Fig. 3 Snapshots of the i-spot assembled for Paul.

an interactive map to locate dealers of the car brand (Maserati) around the region where he lives, plus a calendar to arrange an appointment for a test drive.

6 Architectural Design

We have designed the MiSPOT system so that it is possible to arrange the logic explained in the previous sections in many different ways, in order to accommodate a range of scenarios defined by the continuous, intermittent or null availability of a return channel, the greater or lower bandwidth of the broadcast networks, the computational power of the receivers, privacy requirements and various other factors. To this aim, we have extended the solutions presented in [18] for the recommendation of items, to embrace also the generation of i-spots. The supported possibilities match either of the following schemes:

- The *server-side scheme* is intended for scenarios with permanently-enabled return channels (e.g. with cable networks) and when the legal conditions are met to store the users' profiles in a centralized repository. As depicted in Fig. 4, the personalization tasks are entirely carried out by dedicated servers, which may be powerful enough to apply complex reasoning processes (with the content-based and collaborative semantic metrics) over huge amounts of data about items, functional elements, service providers and so on. The receiver simply downloads a full-fledged application when the user decides to launch an i-spot for a given item.

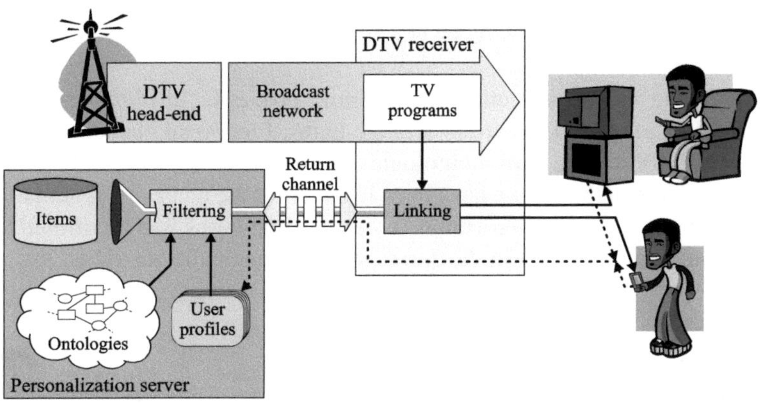

Fig. 4 The common design of server-based schemes to personalization in DTV.

- The *receiver-side scheme* enables personalization whenever the user's preferences can only be stored locally, so it is the receiver that must take charge of both the recommendation of items and the generation of i-spots. This is facilitated by a pre-filtering that takes place in servers, driven by a set of stereotypes that characterize the interests and needs of the potential audience of the TV programs being broadcast. The pre-selected material (corresponding to the items that may be most interesting in average) is delivered to all the receivers through broadcast, containing functional elements, pieces of content to populate templates and excerpts from the domain ontology, that will be matched against the profile of each user. The matching uses the same content-based and collaborative metrics as above, though the latter considers stereotypes rather than the profiles of other users (which are kept private). In this scheme, we have introduced a module that acts as a proxy to the semantic UDDI registry, in order to provide a unique interface to retrieve functional elements regardless of whether they are received through broadcast or through the return channel. This module prioritizes the material received one way or the other depending on the speed of the return channel and the latency times of the broadcast networks. We had shown in [18] that the item recommendations provided by the client-side scheme are less accurate than the ones provided by the server-side scheme, because the latter can involve much more knowledge about users and items. Likewise, concerning the enhancements presented in this paper, the i-spots assembled by the client-side scheme may not bring into play functional elements and pieces of content that would appear otherwise. The quality of the semantic metrics employed in the pre-filtering phase and in the final filtering is therefore crucial, thus ruling out heuristic or syntactic approaches.

As shown in Fig. 5, we have also devised a *pruning* procedure to reduce the amount of information to be handled by the receivers. This procedure consists of cutting off metadata from the domain ontology to leave only the most relevant concepts about the pre-selected items. As a result, we get *partial ontologies* of a manageable size for the receivers to work with, plus *partial stereotypes* to support the filtering.

Following the pre-selection and pruning processes, a *planning* module in the DTV head-end takes care to arrange items, functional elements, templates, pieces of information, partial ontologies and partial stereotypes in the broadcast emissions. When those data have been loaded into the receivers, it is finally possible to run the filtering algorithms to decide what items will be offered to each individual user, and to build the corresponding personalized i-spots.

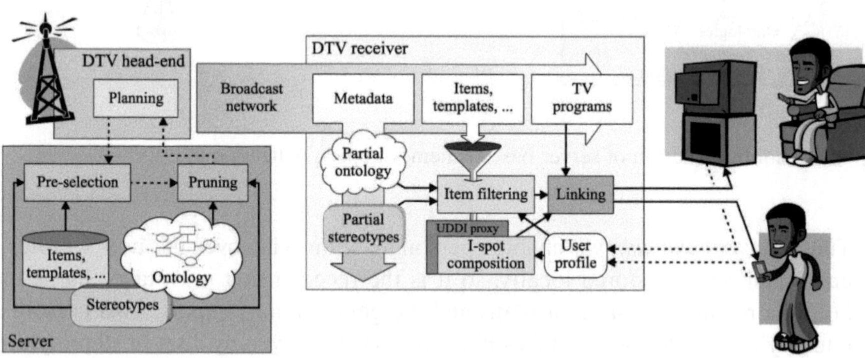

Fig. 5 Architecture of the receiver-side scheme of MiSPOT recommender.

7 Concluding Remarks

Traditional recommender systems do not fully meet the needs of e-commerce settings in which the users are not focused on the kind of items that may be offered to them. Instead of merely providing a list of potentially interesting items, the MiSPOT recommender presented in this paper takes advantage of its internal reasoning about items and user profiles to automatically compose interactive applications offering personalized commercial functionalities. Applied to the specific domain of digital TV, this paper paves the road to generalizing the provision of personalized interactive applications, which have been noticeable for their absence despite being one of the major assets of the new technologies. The architectural design of our system allows to deal with a wide range of communication possibilities, ranging from continuous availability of a return channel to intermittent or null access to it. The general contributions from this work have to do with introducing semantic reasoning processes in the composition logic through SWRL rules and similarity metrics, and with fostering the convergence of web services, TV contents and networks.

References

1. Adomavicius, G., Tuzhilin, A.: Towards the next generation of recommender systems: a survey of the state-of-the-art and possible extensions. IEEE Transactions on Knowledge and Data Engineering 17(6), 739–749 (2005)
2. Blanco-Fernández, Y., Pazos-Arias, J., López-Nores, M., Gil-Solla, A., Ramos-Cabrer, M.: AVATAR: An improved solution for personalized TV based on semantic inference. IEEE Transactions on Consumer Electronics 52(1), 223–231 (2006)
3. Burke, R.: Hybrid Recommender Systems: Survey and Experiments. User Modeling and User-Adapted Interaction 12(4), 331–370 (2002)
4. Cho, Y., Kim, J.: Application of Web usage mining and product taxonomy to collaborative recommendations in e-commerce. Expert Systems with Applications 26, 233–246 (2004)
5. Dias, M., Locher, D., Li, M., Elderedy, W., Lisboa, P.: The value of personalized recommender systems to e-business: A case study. In: Proc. of the 2008 ACM Conference on Recommender Systems, pp. 291–294 (2008)
6. Ganesan, P., Garcia-Molina, H., Widom, J.: Exploiting hierarchical domain structure to compute similarity. ACM Trans. Inf. Syst. 21(1), 64–93 (2003)
7. Gawinecki, M., Mateusz, K., Paprzycki, M.: Ontology-based stereotyping in a travel support system. In: Proceedings of the XXI Fall Meeting of Polish Information Processing Society, pp. 73–85 (2005)
8. Haase, P., Ehrig, M., Hotho, A., Schnizler, B.: Personalized Information Access in a Bibliographic Peer-to-Peer System. In: AAAI Workshop on Semantic Web Personalization, pp. 1–12 (2004)
9. Hepp, M.: Products and services ontologies: a methodology for deriving OWL ontologies from industrial categorization standards. International Journal on Semantic Web & Information Systems 2(1), 72–99 (2006)
10. Hepp, M., Leukel, J., Schmitz, V.: A quantitative analysis of product categorization standards: Content, coverage and maintenance of eCl@ss, UNSPSC, eOTD, and the RosettaNet Technical Dictionary. Knowledge and Information Systems 13(1), 77–114 (2007)
11. Hu, S., Muthusamy, V., Li, G., Jacobsen, H.A.: Distributed automatic service composition in large-scale systems. In: Proc. of the 2nd International Conference on Distributed Event-Based Systems, Rome, Italy (July 2008)

12. Hung, L.: A personalized recommendation system based on product taxonomy for one-to-one marketing online. Expert Systems with Applications 29, 383–392 (2005)
13. Kawamura, T., de Blasio, J.A., Hasegawa, T., Paolucci, M., Sycara, K.P.: Public Deployment of Semantic Service Matchmaker with UDDI Business Registry. In: McIlraith, S.A., Plexousakis, D., van Harmelen, F. (eds.) ISWC 2004. LNCS, vol. 3298, pp. 752–766. Springer, Heidelberg (2004)
14. Kobsa, A., Koenemann, J., Pohl, W.: Personalized Hypermedia Presentation Techniques for Improving Online Customer Relationships. The Knowledge Engineering Review 16(1), 111–155 (2001)
15. Krulwich, B.: LifeStyle Finder: Intelligent User Profiling Using Large-Scale Demographic. AI Magazine 18(2), 37–45 (1997)
16. Kvaloy, T.A., Rongen, E., Tirado-Ramos, A., Sloot, P.: Automatic Composition and Selection of Semantic Web Services. In: Sloot, P.M.A., Hoekstra, A.G., Priol, T., Reinefeld, A., Bubak, M. (eds.) EGC 2005. LNCS, vol. 3470, pp. 184–192. Springer, Heidelberg (2005)
17. Lin, N., Kuter, U., Hendler, J.: Web Service composition via problem decomposition across multiple ontologies.In: Proc. of the IEEE Congress on Services, Salt Lake City, USA (July 2007)
18. López-Nores, M., Pazos-Arias, J.J., García-Duque, J., Blanco-Fernández, Y., Martín-Vicente, M.I., Fernández-Vilas, A., Ramos-Cabrer, M., Gil-Solla, A.: MiSPOT: Dynamic product placement for digital TV through MPEG-4 processing and semantic reasoning. Knowledge and Information Systems (in press), doi:10.1007/s10115-009-0200-8
19. López-Nores, M., Rey-López, M., Pazos-Arias, J.J., García-Duque, J., Blanco-Fernández, Y., Gil-Solla, A., Díaz-Redondo, R.P., Fernández-Vilas, A., Ramos-Cabrer, M.: Spontaneous interaction with audiovisual contents for personalized e-commerce over Digital TV. Expert Systems with Applications 36, 4192–4197 (2009)
20. Martin, D.L., Paolucci, M., McIlraith, A.A., Burstein, M.H., McDermott, D.V., McGuinness, D.L., Parsia, B., Payne, T.R., Sabou, M., Solanki, M., Srinivasan, N., Sycara, K.P.: Bringing Semantics to Web Services: The OWL-S Approach. In: Cardoso, J., Sheth, A.P. (eds.) SWSWPC 2004. LNCS, vol. 3387, pp. 26–42. Springer, Heidelberg (2005)
21. Maybury, M., Greiff, W., Boykin, S., Ponte, J., McHenry, C., Ferro, L.: Personalcasting: Tailored broadcast news. In: Personalized Digital Television. Targeting programs to individual users. Kluwer Academic Publishers (2004)
22. Middleton, S.: Capturing knowledge of user preferences with recommender systems. PhD thesis, Department of Electronic and Computer Science. University of Southampton (2003)
23. Montaner, M., López, B., De la Rosa, J.: A taxonomy of recommender agents on the Internet. Artificial Intelligence Review 19(4), 285–330 (2003)
24. Plas, D., Verheijen, M., Zwaal, H., Hutschemaekers, M.: Manipulating context information with swrl. Report of A-MUSE project (2006)
25. Rada, R., Mili, H., Bicknell, E., Blettnet, M.: Development and application of a metric on semantic nets. IEEE Trans. Syst., Man, and Cybern. 19(1), 17–30 (1989)
26. Resnik, P.: Semantic similarity in a taxonomy: An information-based measure and its application to problems of ambiguity in natural language. J. Artif. Intell. Res. 11(4), 95–130 (1999)
27. Rich, E.: User modeling via stereotypes. International Journal of Cognitive Science 3(1), 329–354 (1979)
28. Rosco, R.D., Rogers, M.K., Liles, S.: Breaking Down Stereotypes: A Call for an Ontological Model. In: Proceedings of International Conference on System Sciences, pp. 271–278 (2007)

29. Schafer, J., Konstan, J., Riedl, J.: E-commerce recommendation applications. Data Mining Knowledge Discovery 5(1), 115–151 (2001)
30. Shani, G., Rokach, L., Meisles, A., Gleizer, Y., Ben-Shimon, D.: A stereotypes-based hybrid recommender system for media items. In: 22nd AAAI Conference on Artificial Intelligence (AAAI 2007), pp. 60–75 (2007)
31. Sirin, E., Parsia, B., Wu, D., Hendler, J.A., Nau, D.S.: HTN planning for Web Service composition using SHOP2. Journal of Web Semantics 1(4), 377–396 (2004)
32. DVB: The Multimedia Home Platform.ETSI standard TS 102 812 (2003), `http://www.mhp.org`
33. Walsh, A.E.: UDDI, SOAP, and WSDL: The Web Services Specification reference book. Pearson Education (2002)
34. Zhang, H., Zheng, S., Yuan, J.: A personalized TV guide system compliant with MHP. IEEE Transactions on Consumer Electronics 51(2), 731–737 (2005)
35. Zhang, R., Arpirar, I.B., Aleman-Meza, B.: Automatic composition of semantic Web Services. Information Systems Journal, 38–41 (2003)
36. Zhu, T., Greiner, R., Häubl, G.: An effective complete-web recommender system. In: Proceedings of the 12th International World Wide Web conference, Budapest, Hungary (May 2003)
37. Ziegler, C., Lausen, G., Konstan, J.: On exploiting classification taxonomies in recommender systems. AI Comnmunications 21(2-3), 97–125 (2008)

Towards Automated Navigation over Multilingual Content

Petr Šaloun[1], Zdenek Velart[1], and Jan Nekula[2]

[1] VSB-Technical University of Ostrava, 17. listopadu 15, 70833 Ostrava, Czech Republic
petr.saloun@vsb.cz, zdenek.velart@gmail.com
[2] University of Ostrava, 30. dubna 22, 70103 Ostrava, Czech Republic
jan.nekula@osu.cz

Abstract. Web-based adaptive and personalized systems are becoming standards on the web. The chapter describes the main features of selected web-based adaptive systems in the domain of education, and summarizes the architecture of such systems. One of the representatives of web-based adaptive personalized systems is described in more detail, XAPOS, designed and developed by the authors. XAPOS's design is independent of the language of closed content and was used in a multilingual environment. XAPOS's language-independent navigation feature is based on the domain ontology and principles of semantic web. We conducted a controlled experiment in an educational course of programming language Lisp in universities in Ostrava, Czech Republic, and Ankara, Turkey; results of two groups of users navigated using our algorithm as well as the standard one, and are compared and discussed in detail.

1 Introduction

Web-based information systems are standard parts of the Internet. Most of them offer more functionality than just content. Such extended services are focused on the presented content, user, or on both. From one point of view, the goal is to sell more goods, and from the other it is to provide a content recommendation fitting a user's interest. More generally, all successful extensions and enhancements have to give some benefit for both the content provider and the user.

Content adaptation in the "wild web" area is a difficult task, partially successfully solved using means for implicit user feedback such as adaptive proxy, acting as an intermediate between a user and the web [2]. An easier task, in comparison with the wild web content adaptation, seems to be closed content adaptation and/or personalized navigation over the information space. A lot

I.E. Anagnostopoulos et al. (Eds.): Semantic Hyper/Multimedia Adaptation, SCI 418, pp. 203–229.
springerlink.com

of its subtasks are more general in the adaptation area on which this chapter, and our research, is focused.

One branch of current web-based systems are personalized adaptive systems. Content adaptation and personalization is still a matter of worldwide research focus. This chapter gives the state of the art as well as our approach to personalization and adaptation of navigation over a multilingual closed content. The chapter extends our paper [25] presented on SMAP 2010 conference in Limassol, Cyprus, in December 2010.

In this chapter we give an overview in the area of navigational adaptation, guiding and personalization, mostly used in web-based educational systems. We employ representation formalisms of the semantic web such as ontology or topic maps. Our contribution in this area is realized by XAPOS – which is an experimental adaptive personalized web-based system. We describe XAPOS as well as discuss the closed content ranking and navigation techniques, more deeply described in case studies with our experimental results given at the end of chapter. A newly realized idea here is automated categorization of added new content and its interconnection to the existing content.

2 Adaptive Web-Based Educational Systems – State of the Art

The amount of available information accessible to a user is constantly growing. New web sites are springing to life every day, news is coming from every direction, and social networking is the word of the day. A possible way to improve user experience is to employ personalization and adaptation, see [19] for an example.

Personalization can be achieved by personalization of navigation of available information (links), or personalization of the presented information content. The very basics of the field of adaptation and personalization can be found in [3, 4]. Brusilovsky defines two groups of techniques for adaptive presentation and adaptive navigation.

Adaptive presentation employs techniques for altering the content such as insertion, removal, or offering alternate content:

inserting and removing – fragments of the content are added or removed on the fly;

altering – a longer more informative fragment can be presented to the user against a shorter version of the fragment;

sorting – fragments are sorted according to their relevance to the user;

dimming of fragments – fragments can be visually altered according to their relevance. Irrelevant fragments can have for example gray color, making them less visible;

strechtext – only a small portion or name of the fragment is shown to the users. On click the whole fragment is expanded and shown.

Adaptive navigation concerns with presentation of links to the content and navigation:

direct guidance – the user is directly guided through the problem domain, for example using the next and previous buttons;
sorting – links are sorted for the users according their relevance;
hiding – links are presented to the user, but have the same color as the surrounding text;
disabling – links are rendered not clickable, remaining only as part of the text;
removal – links are removed from the text;
annotation – links are presented visually or textually differently according their relevance, for example interesting links can be painted green;
creation of links – new links can be created by the system according the gathered data from the users;
map adaptation – graphical representation of local or global problem domain. The user can navigate directly to their desired goal by clicking on representation of the fragment in the link map.

De Bra and Houden [6] generalized previous heuristic approaches in 1999 designing AHAM, naming parts of a general adaptive system focused on an individual user dealing with models of a user, application domain and approaches of the content delivery to the user. Last research adds adaptive proxy filtering and reordering which uses all previous approaches and seems to be quite fruitful, see [2, 27]

Many of these techniques are used in most of the systems with which we interact daily – search engines, help systems, recommendation systems and many others. Beside commercial systems there are those which are non-commercial, mostly focused on universities and education. Our application domain is education, and therefore our focus is on this domain.

2.1 Navigation in Information Space

Navigation is the most crucial task of any web-based system. The goal is to provide a user with information which is useful and will provide direction to the target. For this purpose an adaptive system needs to collect data about its users. The system collects data either in the form of knowledge, learning style or the past visited educational material of an individual user. The system employs one or more approaches and their combination for handling users' navigational metadata.

Concept-based navigation employs a concept as an elementary unit of information. A concept can have multiple relations with other concepts. A set of concepts with their relationships is called concept space. It represents metadata information about a particular problem domain. A concept could play one of two roles in navigation – outcome or prerequisite. Concepts first introduced in educational material are called outcomes. Concepts used in the educational material, but introduced earlier are prerequisites. Other approaches [10, 11] use as an elementary unit a topic, and focus on the navigation based on topic maps, and creation of ontologies from topics.

In order that the adaptation could work, the systems need to know about the presented material, and users, gather information about them, and be able to utilize all such information. Adaptive system models are [6]:

Domain model – represents the domain or the scope in which application operates. The model contains concepts, their relations and the content. Many systems use domain ontology to describe the domain, content repository and content structure ontology for describing the relations across the content.

User model – stores user's characteristics – preferences, knowledge and other attributes – which are gathered by the system and used for the adaptation.

Adaptation model – specifies methods how the adaptation is performed. The model is defined as a set of rules which are applied to the presented content. The presentation of the information has impact on the user model.

Because of its complexity and impact on personalization, the user model and the gathering of user characteristics deserves a closer look. A user is represented in the user model by various user characteristics which were gathered by the system itself, or obtained for example from a survey. These characteristics are used to adapt the content, presentation and navigation. Many different approaches use different terminology such as user attributes, features, characteristics or properties. Attempts in terminology standardization can be found in the User Model Meta-Ontology [32].

Several approaches regarding how the user characteristics change define *permanent characteristics* as characteristics which are independent of the currently used system, and *variable characteristics* as which can be changed independently of the system or characteristic directly tied to users' experience with the system [20]. Others use divisions of users' characteristics into *static* and *dynamic* groups. Static characteristics are often obtained from some user survey or questionnaire, whereas dynamic characteristics are gathered by the systems itself observing a user's behavior. User characteristics can be distinguished by the domain dependency [15] where *domain-dependent* characteristics are those that relate to the application domain and *domain-independent* are unrelated to the application domain and therefore can be shared by a common user model across domains.

Cold start problem is a situation when the user encounters the system for the first time. This problem, as well as many others, is solved more or less successfully by some user model approaches:

Overlay model – an instance of the domain model (represented by concepts and their relations) is kept for every user where the knowledge of the users is tracked for every concept. Estimation of the concept knowledge can be *binary, qualitative* or *quantitative* [3]. The system doesn't necessarily need to keep the whole copy of the domain model for every user. A differential user model [17] reflects only a part of the domain in which a user might be interested. The disadvantage of the overlay model is its initialization (the cold start problem, in other words) – creation of the user copy of the domain model. No information on the user knowledge is known at this point, so some initial values for the concepts must be set – zero, or for example some mean value based on how the knowledge is tracked.

Stereotype model [15, 20, 22] – stereotype is a collection of frequently occurring characteristics of users. Users are associated with one or more stereotypes which reflect their knowledge, social background, experience, interests, and others. Stereotypes can be ordered hierarchically and characteristics inherited. The designers need to identify stereotypes which will be used by the system. Kobsa in [18] identifies three important tasks which are needed for successful identification of stereotypes: – User subgroup identification; – Identification of key characteristics; – Representation in (hierarchically ordered) stereotypes. Kay in [17] suggested approaches for identifying stereotypes – *handcrafted*, based on designers' observations, and *empirically-based*, which is on collected data about the users. The disadvantage of this model is that a user can inherit inappropriate or unneeded characteristics because the model focuses not on an individual user but on a group. The definition of states of stereotype has statistically based reasoning [16]. On the other hand a big advantage of the model is overcoming a cold start problem. According to the supplied information the user can be associated with selected stereotypes and inherit the characteristics.

Shared model brings connections between the previously mentioned model across systems and domains. Data gathered by one system is available for others. Therefore the user does not have to provide the same information more times [1]. The advantage is that user information is stored logically in one place and when the user encounters new systems, the information can be used for initialization (cold start problem). The disadvantage is usage of the same terminology for elements in the model and security issues when not all systems need to know all information about the user.

2.2 Selected Adaptive Systems

Many adaptive systems are based on the models described above. Some others use a totally different approach. The common goal of these systems is to provide a meaningful adaptation for the user.

PersonalReader Framework [12] – is a service-based environment for implementing and accessing personalized services. The framework utilizes three types of services – connector service for passing the request and results between the user interface and personalization services; – personalization

service for adaptation functionality such as recommending learning objects, pointing to more detailed information, quizzes, exercises and other;
– visualization service for interpreting the result of the personalization service to the user and to create the actual interface.

TANGRAM [14] – is a web-based educational system. It uses ontologies for representing learning object content structure, learning object content type and educational path ontology for specifying pedagogical relations among domain concepts. Student must complete a questionnaire and exam that determine learning style and knowledge level. Based on these deduced results and the ontologies, the system builds a visual representation of educational content adapted to the user in the form of a tree of links using link annotation and link-hiding techniques.

GALE [9] – is an extensible adaptive hypermedia engine which tries to offer all possible forms of adaptation for hypermedia applications based on individual user models. GALE architecture is generic and extensible. Part of it includes authoring tools. GALE stretch locality (typical adaptive limitation of the content) allows the use of an adaptive application from within different Learning Management Systems, and sharing user model information through a common user model framework. Open corpus adaptation is the way how GALE allows use of decentralized applications.

NavEx [31] – the system applies adaptive annotation on links providing visual cues with additional information about the content behind the links. These "traffic lights" help the user to choose the most relevant and proper link to follow. NavEx is focused on source code examples of programming languages. NavEx uses two sets of prerequisite and outcome concepts for every example. The example is recommended to the user by "traffic lights" if the user has mastered the prerequisite concepts.

ALEF [27] (Adaptive LEarning Framework) is framework for creating adaptive and highly interactive personalised web-based educational systems. It incorporates principles of Adaptive Web-based Learning 2.0 and modular web-based educational systems. ALEF adopt a scheme for Adaptive Web-based Learning 2.0, so reusability, easy maintenance and manageability of the educational content and metadata are guaranteed. The framework and its underlying models are designed with respect to the course authoring that is intended to be as flexible as possible. The domain model is simple enough to allow efficient usage of methods for automatic domain metadata extraction based on lightweight semantics acquisition, see [26] for example.

WebImp [13] – this system provides an effective navigation within the content of the website. WebImp does automatic estimation of the user's interest in a web page they visit is proposed. This estimation is used for recommendation of web portal pages (through presenting adaptive links) that the user might like.

For the purpose of testing our ideas we developed an adaptive web-based educational system XAPOS. The system is based on the AHAM reference model implemented in AHA! [5, 7, 8], the GALE's predecessor. AHAM defines three basic parts which are implemented as follows.

User model – we store basic information about the student as name, login information and information needed for navigation. Every student has a defined *achieved knowledge set* (known concepts) and *visited learning object set*, where timestamps of visits to the learning objects is stored. Our model also includes a system log such as information about login or logout of the student.

Domain model – in our case a course represents a unit which bundles all the needed information. The course has a defined concept set which describe the problem domain. The course has also defined learning objects set which represents educational material and examples of the course. In our system a mixed learning object can be represented by one or more HTML pages with linked multimedia material. Educational example as a learning object is represented by an XML definition containing the educational example in itself, and its inputs and expected outputs.

Adaptation model – in XAPOS we implemented our novel algorithm for concept set evaluation and previously described navigation scheme. The evaluation of the concept set is carried only in the preparation phase of the course or when some changes to the concept set occur.

An example of our system GUI is shown in Fig. 1. The menu is presented to the user in the left part (1). Every menu item has a mouse over annotation describing the learning object. In part 2, the whole course menu is shown together with a language selection. In the right part, a related menu is shown (3). Related learning objects and suitable educational examples are shown here. Menu history is also located here. The educational material of the current learning object is situated in the middle (4).

3 Concept Space Based Navigation

Our approach for the navigation of a student through the learning object space uses concept space. A concept is an element of information in a problem domain. A concept space, referenced also as ontology, represents a set of interconnected concepts defining a particular problem domain. The interconnection creates a multidimensional space based on a number of relation classes.

A learning object represents a closed unit of the educational course. Learning object usually consists of educational material in a form of text (HTML, plain text), multimedia, educational examples and, optionally, a test. In our

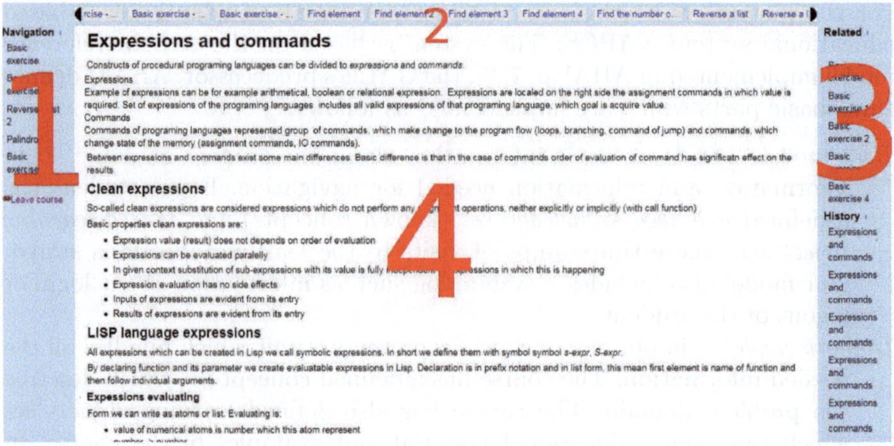

Fig. 1 XAPOS user's interface

case we distinguish two types of learning objects – mixed learning object and educational example learning object. A mixed learning object can contain educational material and educational examples bundled together. An example learning object contains the educational example only. The learning objects are treated differently in terms of the recommendation of similar learning object and categorization of educational examples. In the terms of navigation we treat these two types of learning objects the same.

The principle of our navigation scheme is based on the premise that relations that exist in the concept space also exist in the learning object space between the corresponding learning objects. Every learning object has defined a set of concepts from concept space which describe it (Fig. 2). The navigation scheme stands on the offer of suitable learning objects to the student according his actual knowledge. We define student's actual knowledge as a set – *achieved knowledge set*. The set contains concepts already known by the student. A suitable learning object is selected according the actual student's achieved knowledge set and concepts of learning objects.

Our approach to concept space rating [24, 25] uses an evaluated concept space for purposes of navigation. We need to evaluate every concept in the concept space in relevance to its graph-position – we evaluate every existing relation to other concepts. The calculated value expresses the relations-based complexity of the concept in relation to others. For this purpose we designed an algorithm to evaluate such a value. Our algorithm is based on the idea of the PageRank algorithm [21]. It calculates, for every concept, a single value based on concept's relations. We distinguish two groups of relations

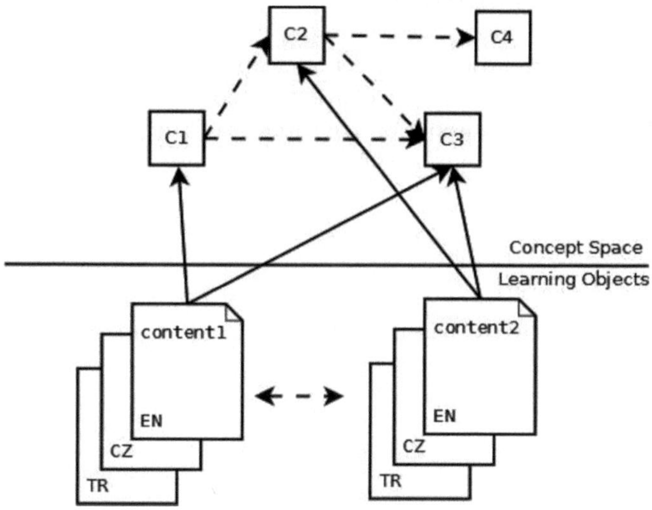

Fig. 2 Concept space and learning object space

concept-concept – structural one and dependency one. A structural relation describes the structure of the concept space - a mostly hierarchical one; the typical example of this is*subclassOf* relation.

A dependency relation denotes dependencies in the meaning of knowledge and complexity of a concept – typical examples are *uses*, *definedBy*.

First we give basic notation used through the chapter:

concept c – a basic knowledge item. The concept has defined several attributes:

- rating – is the calculated rating of the concept;
- related – a set of concepts that link to c with one of structural relations;
- countr – is the count of outbound structural relations of the concept c;
- depend – defines a set of concepts that c has dependency to;

$setC$ is a set of all concepts ($c_i \in setC$);

r denotes a relation;

- rating – defines a rating of a relation;

$setR$ is a set of all relations ($r_i \in setR$);

N is the number of concepts in the set $setC$, $N = |setC|$;

d is damping factor used in PageRank algorithm.

3.1 Elementary Ratings

As the first step the relations between concepts are analyzed. For every dependency relation $r_i \in setR$ in the concept space, there is a rating ($r_i.rating$) calculated based on the number of occurrences of the type of relation in the concept space in ratio to number of concepts.

$$r_i.rating = \frac{count(r_i)}{N}$$

A concept rating ($c_i.rating$) defines a rating of the concept in the meaning of educational complexity based on the relations to other concepts[1]. The equation is following:

$$c_i.rating = \frac{1-d}{N} + d \times \left(\sum_{c_j \in c_i.related} \frac{c_j.rating}{c_j.countr} + \sum_{c_k \in c_i.depend} c_k.rating \times r_k.rating \right)$$

The part

$$\sum_{c_j \in c_i.related} \frac{c_j.rating}{c_j.countr}$$

calculates a partial rating of the concept based on concept's structural relations. For every concept c_j, the current concept is connected. A rating is calculated out of concepts actual rating ($c_j.rating$) and the number of structural relations of the concept ($c_j.countr$). This part is adopted from the PageRank algorithm.

The second part

$$\sum_{c_k \in c_i.depend} c_k.rating \times r_k.rating$$

calculates a partial rating from the concepts that the current concepts depend on. The rating of the concept ($c_k.rating$) is multiplied by the rating of the relation itself ($r_k.rating$). This allows us to incorporate an explicit dependency into the equation of the rating.

The final rating of the concepts is then calculated from the sum of previously defined two partial ratings with the use of the dumping factor and total number of concepts in the concept space.

[1] Note, that described concept rating was tested and results are suitable for limited content adaptation, see section 6.

Concept rating algorithm written in the form of programming metalanguage follows:

while *numIterations* > 0 do
  foreach c_i in *setC*
    $rating = 0$
    foreach c_j in $c_i.related$
      $rating \mathrel{+}= \frac{c_j.rating}{c_j.countR}$
    end foreach
    foreach c_k in $c_i.depend$
      $rating \mathrel{+}= (r_k.rating \times c_k.rating)$
    end foreach
    $$c_i.rating = \frac{1-d}{N} + d \times rating$$
  end foreach
  $numIterations - -$
end while

Several starting attributes need to be defined for the deterministic run of the concept rating algorithm:

- Basic *rating* of a concept is defined as $\frac{1}{N}$.
- *Damping factor d* – parameter which can be set between 0 and 1. Usually the value is set to 0.85.
- *Number of iterations* is based on the number of concepts. For a default value we choose a tenth of the concept space size ($\frac{N}{10}$).

Figure 3 represents a part of the programming language C concept space. The evaluated values for several concepts (not all concepts are in figure) are shown in Table 1. These calculated values reflect the layout of the concept space. As we can see, the spread of the values is not a big one. The concept space for programming language C has over a thousand concepts (1100 exactly) with a relatively small number of relations between them. A concept space with a higher number of relations reflects the complexities of concepts with a higher degree of variance. For example for our concept space for Lisp programming language which only contains 105 concepts and a higher ratio of relations between them the algorithm calculated values between 0.0036 and 2.635.

The concept ratings of programming language C concept space given in Table 1 are suitable for adaptive navigation. If the number of relations among concepts increases then the concept ranking could increase too – specifically, a rating greater than 1 means that the concept space contains more relations than the concepts themselves. Such a situation appears in the subset of programming language Lisp educational course with no affect for adaptive navigation, which is the goal of our research.

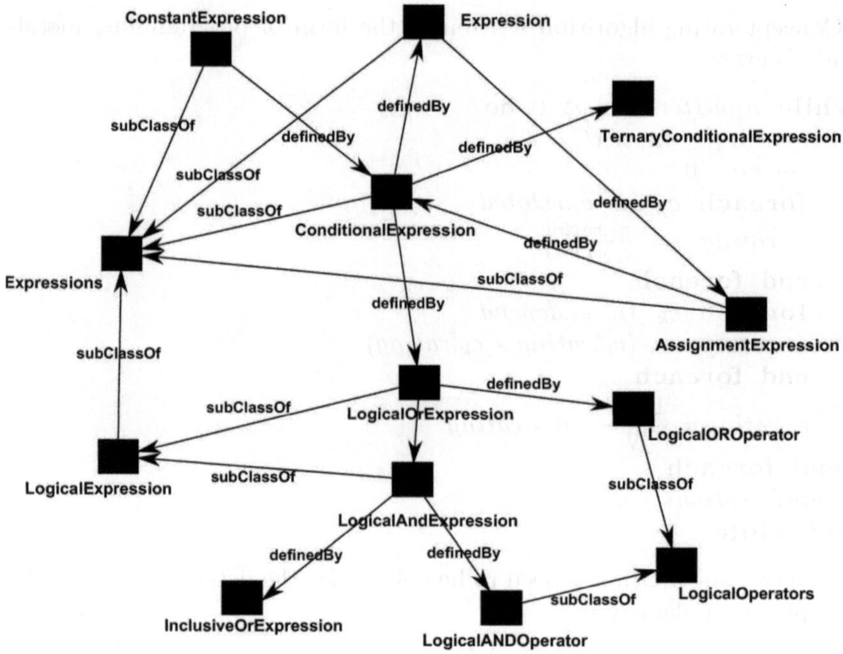

Fig. 3 Part of the programming language C concept space

Table 1 Calculated ratings of programming language C concept space

Concept	rating
WEOF	0.0035
UnaryExpression	0.0023
ConditionalExpression	0.0009
Expression	0.0009
InclusiveOrExpression	0.0007
ConstantExpression	0.0006
LogicalAndOperator	0.0003
LogicalOrOperator	0.0003
LogicalExpression	0.0002
TernaryConditionalOperator	0.0002
Symbols	0.0001

Note that it is necessary to run the concept rating algorithm several times, since relations among concepts take part in the rating. Our initial estimation of the number of iterations was $\frac{N}{10}$ to achieve a stabilized concept rating. Experimental evaluation proves, that the estimation is correct. A higher amount of iterations has a minimal influence on the concept rating. Experimental data showing selected concepts and their ranking are given in Table 2. As concept space of programming language C contains 1100 concepts, the necessary number of iteration $\frac{N}{10} = 110$ is highlighted in Table 2, moreover no difference among 10, 20 and 100 iterations and a small difference between 5 and 10 iterations are shown.

Table 2 Selected concepts (prog. lang. C) rating after a given # of iterations

# of iterations	ExpressionStatement	UnaryExpression	CaseStatement
1	0.000861	0.002422	0.001672
2	0.000509	0.002025	0.000948
5	0.000376	0.001306	0.000648
10	0.000369	0.001269	0.000637
20	0.000369	0.001269	0.000637
100	0.000369	0.001269	0.000637
$\frac{N}{10} = 110$	0.000369	0.001269	0.000637

4 Navigation over Learning Objects and over Educational Examples

Our navigation scheme utilizes the previously defined and rated concept space. A student is presented with a course containing the concept space as course definition and learning object space as educational material. A limited set of learning objects is presented in the menu to the student. They are ordered according to the student's actual knowledge and available learning objects.

Every student visiting the course has two sets stored in his profile – *visited learning objects set* and *achieved knowledge set* where the concepts the student successfully learned are stored. When the student visits, a learning object timestamp is stored in the visited learning object set. The set and its values are later used for path reconstruction or categorization of students according their navigation style.

For the individual navigation an overlay model is used. A set of visited objects, achieved knowledge set, tasks and test evaluations are stored for an individual user.

The recommendation algorithm follows (see Fig. 4).

1. The student chooses a learning object from the recommended learning objects in the menu which is presented to them. A menu usually contains more than one item. The items are ordered according the calculated ranking.

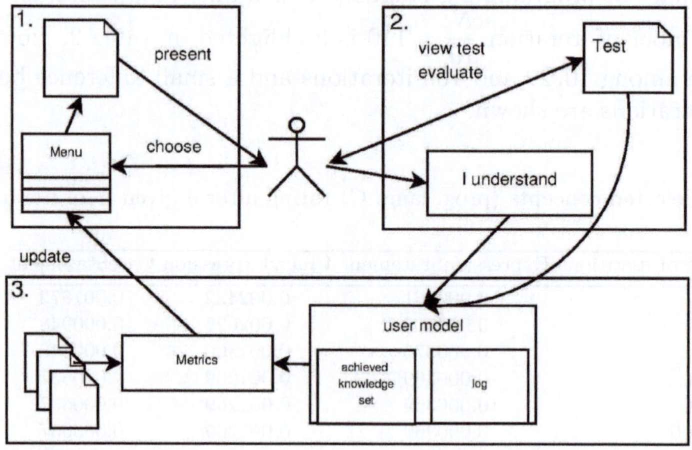

Fig. 4 Navigation scheme

2. According to the chosen menu item, the student is presented with the educational content. The educational content can be either a simple page, or set of pages with educational examples and other multimedia content. A timestamp is also stored in the visited learning objects set for the learning object. The timestamp represents a starting time of the learning object visit.
3. When the student finishes the educational material from the presented learning object there are two possible ways to proceed, see Fig. 5.

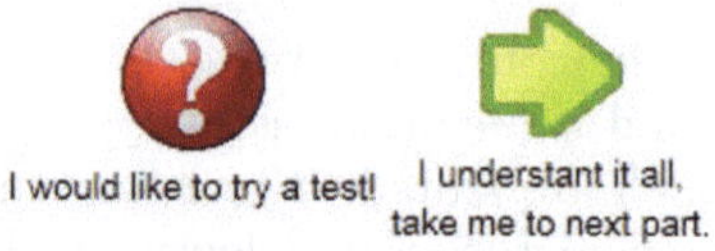

Fig. 5 A choice at the end of a learning object

- Student is certain about the gained knowledge and chooses to continue with the next recommended learning object or any other learning object. The achieved knowledge set is updated with all the concepts from the current learning object, Fig. 4 gives explanation.
- Student takes an available knowledge test. According the correctness of the answers concepts are added or removed from the achieved knowledge set. For scheme "update after test" evaluation see Fig. 6.

A second timestamp for the learning object is stored. It represents an ending time of the concepts visit.

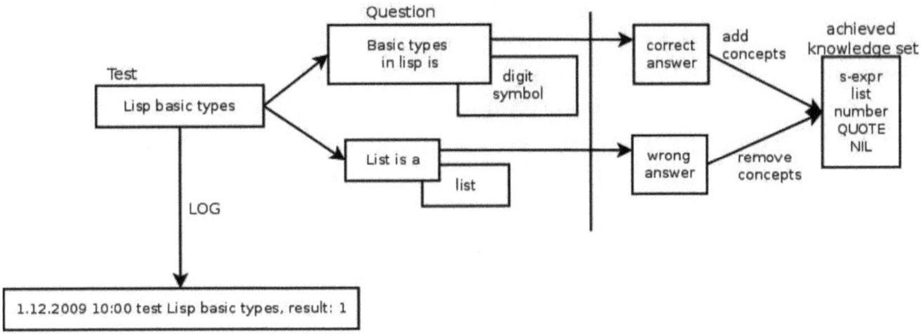

Fig. 6 Scheme "update after test" evaluation

4. The student's menu is updated according his actual achieved knowledge set and available learning objects.
   We defined several metrics for selecting suitable learning objects:

   Unvisited concepts ratings – unknown concepts are used in the calculation of the rating only – a complement between the achieved knowledge set and learning object's concept set is created.
   All concepts rating – all concepts included in the learning objects concept set are used.
   Similarity metrics – learning object are selected according the similarity – based on the concept set – to the current one.

   The selected learning objects are ordered according their ascending calculated value. The navigation continues with step 1.

5 Categorization of Educational Examples

NavEx is used for GUI navigation in the programming language C examples domain. There are only C source examples. We propose navigation over educational materials including both lecture texts as well as the examples. If student needs more examples, he can choose another one from the educational text content.

New educational examples are added to the educational course. They are analyzed to find concepts related with their educational content. Moreover, they are compared on the level of concepts with existing learning objects. Then we link examples with similar learning objects; we call it *categorization of educational examples* – the reason is ability to give additional examples to the user if more practice is needed to understand the learning object which is actually being taught.

Examples are processed in a different way compared to other learning content – no natural language processing is necessary, as we know, as the content of the educational example is programming language source code.

The result of the educational examples categorization is enrichment of an educational example with a defined set of concepts from the concept set. By assigning a concept set to the educational example we are able to link the educational example to existing learning objects. The complexity is additional value to the ordering in the same category examples. Our educational examples are from programming language C course. The educational examples come in different languages – comments, identifier names. We have five language sets of examples:

CZ – Czech without accents – 58 examples;
SK – Slovak, without stop-words and terms register – 20 examples;
EN – English without lemmatization – 20 examples;
EN-L – English with lemmatization – 20 examples;
ALL – CZ + SK + EN – 98 examples.

This splitting is only for the reader's knowledge. This information is not explicitly needed during the processing.

The algorithm for evaluation of educational examples has three steps:

1. Source code cleaning.
2. Examples – concepts pairing.
3. Similarity analysis and assignment to existing learning objects.

5.1 Source Code Cleaning

We need to transfer the source codes of the examples in a form which can be automatically processed. As a suitable representation an extended vector model was chosen. The items in the model have attributes such as cardinality, relevancy in the meaning of the measure of importance, and others. The steps in the preprocessing phase are explained below.

5.1.1 Text Analysis

The first step is the natural language processing of the example and recognition of tokens. Cleaning source code means execution of lemmatization on the example which creates set of lemmas and removes stop-words. After cleaning we have final set of tokens from the example. Because, as part of the experiment, we would like to test if we are able to use any educational example even without a set of stop-words and lemmatization, examples in SK will not be stripped of stop-words. We would like to prove that no stop-words will be marked as keywords. For the same purpose we have EN and EN-L set of examples to compare the results with and without lemmatization.

5.1.2 Creating Vector Representation

From the tokens we create vectors which contain cardinality of occurrences of the tokens. The following step is a transformation of the vector model into the extended vector model. To achieve that we add a relevance $rlvc_{i,j}$ of the token t_i in the source code sc_j. For the calculation of relevance we are using two approaches:

- creating a terms register,
- processing the semantics of the example.

In our experiment into the terms register we include keywords from the concept space. Thus we created a list of essential keywords which we are using in the evaluation. When a token is equal to a term from terms register the relevance equals cardinality of the term multiplied by a constant. For testing purpose we set this constant to 2. Semantics of the example we processed as an evaluation of source code complexity.

The minimum, maximum and average numbers of found tokens in the experiment are in the Table 3. In the table all found lemmas are in the first group. In the second group are all tokens after the cleaning process. In the third group are numbers of tokens which equal to the terms in the register.

Table 3 Lemmas and tokens count in language set analyzed source

		CZ	SK	EN	EN-L	ALL
	MIN	14	18	16	17	14
Lemmas	AVG	43	49	41	40	44
	MAX	115	142	92	89	142
	MIN	9	13	7	8	7
Cleaned tokens	AVG	34	38	25	20	33
	MAX	92	116	57	53	116
	MIN	4	7	4	4	4
Tokens in register	AVG	11	13	9	10	11
	MAX	24	21	23	23	24

5.2 Educational Examples – Concepts Pairing

After the initial phase we are able to determine if the found tokens are the keywords from the concept space. First we weight the tokens to the educational examples. For the evaluation, if a token weight is enough for bounding to a concept, we define a lower–bound value.

5.2.1 Calculating Degree of Membership

For the calculation of weight tokens to the example we use *tf–idf* [23] metrics. The metrics calculate a weight $w_{i,j}$ of the token t_i in the particular source code sc_j with respect to its cardinality in the set of all educational examples SC. Calculation of the weight $w_{i,j}$ follows:

$$w_{i,j} = \frac{rlvc_{i,j}}{\sum_t rlvc_{t,j}} \times idf_i$$

Where $rlvc_{i,j}$ is relevance of the token t_i in the source code sc_j and idf_i is inverse frequency of the token in all educational examples:

$$idf_i = log \frac{|SC|}{|\{sc_j : t_i \in sc_j\}|}$$

where SC is the set of all educational examples. In Table 4 an example of calculated relevance and weights of tokens in particular source code sc_j is shown.

Table 4 The relevance versus weight in source code sc_j

token t_i	$rlvc_{i,j}$	$w_{i,j}$
int	0.21521	0.00266
main	0.33335	0.00299
void	0.16646	0.01513
printf	0.14542	0.00681

5.2.2 Definition of the Minimal Weight of Token

We define a lower bound weight of tokens to be significant enough to be used for connecting the educational example with concepts. The lower bound value can be set using three approaches:

Static: from all the terms in the terms register the lowest relevance is chosen. This approach ensures that all terms in terms register are used. A disadvantage is a huge range of weight. The minimum, maximum and average numbers of the relevant tokens are in Table 5. In Table 6 you can see the numbers of tokens in educational examples, with most of the tokens having a weight bigger then the lower bound value. In same table you can see range (minimum and maximum value) of weight for these tokens.

Table 5 Numbers of relevant keywords

	CZ	SK	EN	EN-L	ALL
MIN	1	1	1	1	1
AVG	3.18	2.6	3.85	3.85	4.25
MAX	7	6	8	8	10

Table 6 Values range of weight

	CZ	SK	EN	EN-L	ALL
Number of tokens	8	6	8	8	10
MIN	0.02611	0.04860	0.04416	0.04416	0.01473
MAX	0.23681	0.07501	0.06536	0.06536	0.23045

Dynamic: a maximal number of terms which will be used for assignment. This approach ensures that a relatively similar number of terms will be used for each educational example. The maximal number should not be set to high, because, as well as with the static way, the degree of membership range will be extensive.

Dynamic with defined range: as well as in the previous dynamic approach, a maximal number of terms is defined. Moreover, a percentage range for the relevance variability of terms is defined.

5.2.3 Concept Assignment

Terms which match the previously defined minimal weights are used for assigning the concepts from the concept space. There are two possibilities:

- Term exists in the concept space. Corresponding concept is assigned to the educational example.
- Term does not exist in the concept space. The term can be added to the particular concept in the concept space or an entirely new concept can be defined.

5.3 Educational Examples Similarity Analysis and Categorization to Existing Learning Objects

With the educational examples assigned a concept set we are able to determine which educational examples correspond to which learning objects. The similarity analysis is calculated between educational examples and learning objects and between different educational examples. A method for evaluating the similarity is described here.

Learning object LO has relation to x concepts. Similarity coefficient k_s is defined as $\frac{1}{x}$. Coefficient of difference was defined as $k_d = \frac{k_p}{5}$. Similarity s is calculated as:

$$s = k_s \times |CS| - k_d \times |CD|$$

where CS is set of the same concepts and CD is set of different concepts.

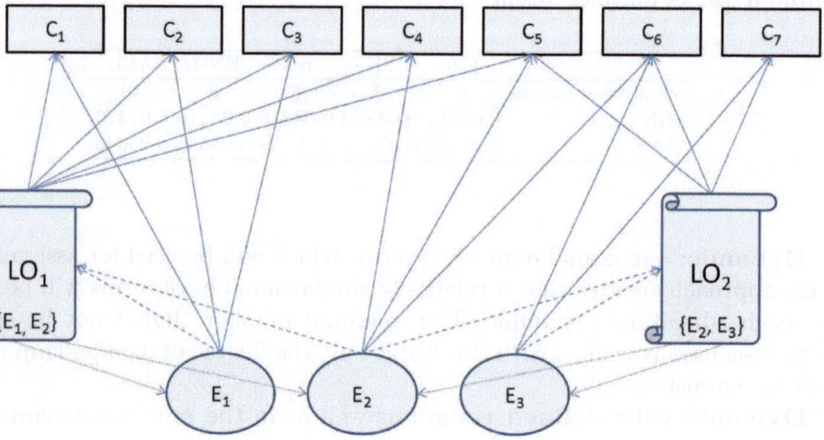

Fig. 7. Categorization of educational examples to learning objects based on concepts – graph view

Table 7 Calculation of similarity s for the educational example in Fig. 7

example	LO_1	LO_2
E_1	$s_{LO_1E_1} = 0.2 \times 3 - 0.04 \times 2 = \ \ 0.52$	$s_{LO_2E_1} = 0.2 \times 0 - 0.04 \times 6 = -0.24$
E_2	$s_{LO_1E_2} = 0.2 \times 2 - 0.04 \times 4 = \ \ 0.24$	$s_{LO_2E_2} = 0.2 \times 2 - 0.04 \times 2 = \ \ 0.32$
E_3	$s_{LO_1E_3} = 0.2 \times 0 - 0.04 \times 7 = -0.28$	$s_{LO_2E_3} = 0.2 \times 2 - 0.04 \times 1 = \ \ 0.36$

In the educational example shown in Fig. 7 and Table 7 we can see that the most suitable educational example for learning object LO_1 is E_1. We can also see that educational example E_3 in this case is unrelated to learning object LO_1.

6 Case Study – Lisp Course

Our course of programming language Lisp consists of 11 learning objects and 105 concepts in the concept space. The rating calculation (used for navigational support) was done only once when the concept space was loaded. Educational content is available in Czech, English, and Turkish languages.

35 Czech users and 89 Turkish users carried out the experiment. 23 volunteers did not finish it – they are not included in the evaluation. The maximal length of the controlled experiment was expected to be 60 minutes. The users were split into two groups; each was navigated by a different rating algorithm in order to compare them. The algorithms were PageRank and Ranking.

Achieved user's knowledge, measured as concepts known by the user, was the source of our results. The knowledge was evaluated from the tests located at the end of the learning objects, from the final test, and from the activity log.

First let us give some basic statistics on the course usage during the experiment. The number of the visited learning objects by the users varies between 11 visits, which is exactly one visit to each learning object, and 74 visits.

The average number of the visited pages was 19.4 visits per student (Ranking CZ: 22, Ranking TR: 18, PageRank CZ: 14, PageRank TR: 19). Median for all data is 15 visits (Ranking CZ: 21, Ranking TR: 15, PageRank CZ: 12.5, PageRank TR: 15).

The learning object was visited between 1.2 (Ranking CZ: 1.4, PageRank CZ: 1, Ranking TR: 1, PageRank TR: 1.15) to 2.9 (Ranking CZ: 2.54, PageRank CZ: 2, Ranking TR: 3.05, PageRank TR: 3.02) times by a user. The median value of the number of visits is 1.6 (Ranking CZ: 2.11, PageRank CZ: 1.36, Ranking TR: 1.57, PageRank TR: 1.64). Differences between users with different languages and algorithms were minor.

Figure 8 (Czech students dashed line, Turkish solid line) shows user navigation in the course graph. The thickness of the connection between two learning objects represents the number of transitions between these two objects (connections with minimal value of visits were omitted for better presentation). An interesting fact is the repeated visits immediately to the same learning object, which represents 18% of all visits. This is mainly because the test is situated at the end of the learning object. Users' desire to have the

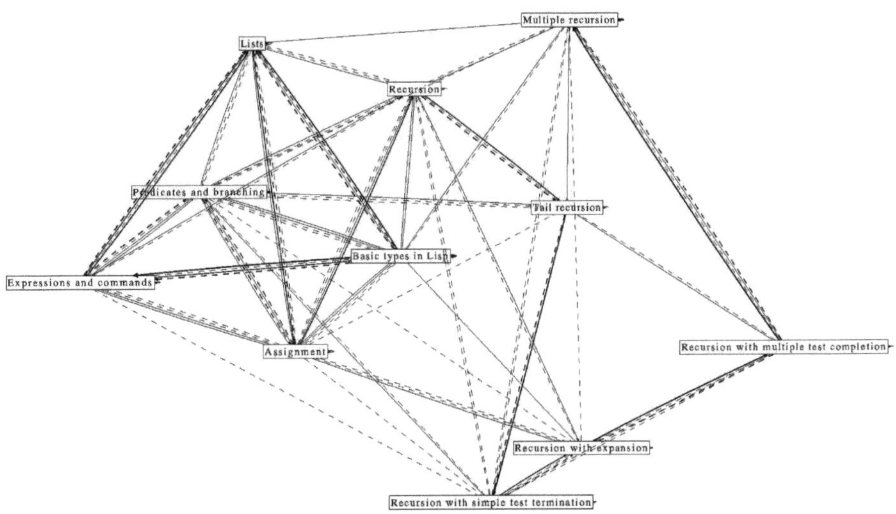

Fig. 8 Paths through the course

best possible score from the test leads to repeated visits to the same learning object.

Interestingly, the learning objects with the highest repeated visits were "Basic types" and "Multiple recursions". These two learning objects, plus "Expressions" and "Lists", represent more than 40% of all visits to the course. "Basic types" is the first learning object of the course which is presented to the user without the knowledge of the problem domain and contains basic knowledge. The high value suggests that there can be a problem either with the educational material of the learning object, or with the test, or that students were not acquainted with the navigation and system.

6.1 Time Evaluation and Final Test

Interesting results can be observed when evaluating the logs with regards to the time spent on learning objects. We clear data by avoiding extreme values, when students just clicks on the link or has no activity for a long time, 5 % of data was thrown out because of clearing.

Evaluation of the cleared data gives an average time spent in one learning object per user as 2.9 minutes, and median time as 1.8 minutes. As we can see in Fig. 9 (times in seconds, median values emphasized), the time varies from several seconds spent on a learning object up to hundreds of seconds. Low values (below 10 seconds) indicate that the user just "passed" the learning object – such values were 7% of all values. Such data cannot be simply avoided – some of these values appear from the repeated the visits of learning objects when the users wanted to try to do the test again – the system shows the test at the end of the learning object, thus logging the users access to the learning object itself. Some values were from users with overall low time spent in the course, and some from users when they only checked some information in the learning object.

The final test is a good indicator when evaluating users' performance in the course. The users were allowed to try the final test more than once without any penalty. Several users used this opportunity to test their knowledge during the course, some users missed the final test, and therefore an average user took 1.4 final tests. The results show that only 2% of the tests were with result 0, by examining the test log we discovered that 90% of these users tried the test again with better results.

The results also show that the possibility to try the test more than once was used by almost a half (40%) of the attempting users to achieve better results in the final test. Furthermore 10% of the users tried the final test more than two times.

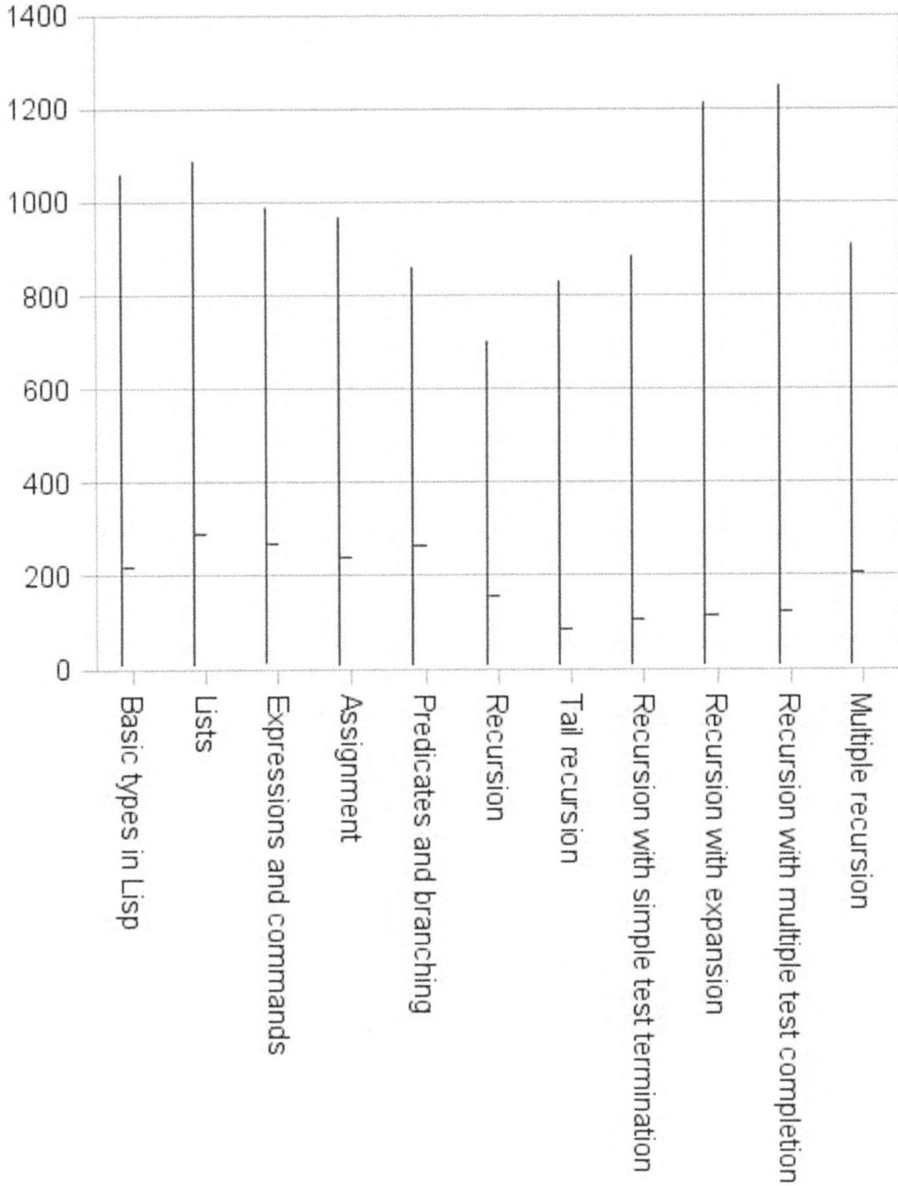

Fig. 9 Time spent on learning objects

6.2 Groups of Users

To better understand how users respond to the proposed navigation, we performed detailed evaluation of users' log and sessions. We were looking for patterns of the navigation as were described in [28]:

- *Sequential* – a user moves from one learning object to the next according to the proposed navigation – this pattern is typical for users who follow the system recommended navigation.
- *Repetition* – a user visits the same object several times immediately after the first visit – this happens mostly in situations where the user unsuccessfully tries the test at the end of the learning object and immediately returns to the same learning object to broaden his knowledge, and tries the test once more.
- *Go-back* – a user decides to return to an already visited learning object – this happened in situations where the user wants to "check" some information in the previous learning object or to read the educational material again.
- *Skipping* – a user decides to move ahead to a learning object which has not yet been visited and not yet been offered by the system.

The path that users take through the course can be covered by more than one pattern. Every transition between two learning objects is analyzed and classified as an instance of the defined pattern of the navigation. The results we obtained show that almost every student sometimes diverts from the proposed navigation – only 5% of students followed unconditionally. In Table 8, results for four categories are shown – the first category represents a situation where the user used a pattern of the navigation at least more than once. The other categories represent situations where the user uses a pattern of the navigation for 1/3, 1/2, and 2/3 of their stay in the course. The results are shown for all students and separately for Czech and Turkish students.

A traditional goal of adaptive navigation is to guide a user in the most effective way – in our case using the sequence pattern. However, the evaluation showed that a significant number of users return to a previous learning object (navigation pattern *Repetition* used by 49% of users and pattern *Go-back* by 60% of users more than once). The observed behavior from one point of view indicates a willingness of the users to be navigated by the system, but on the other hand it shows limitations either on the user's side or the course side. A course-side limitation can be caused by poorly described educational material, or too much information at once, or lack of additional educational examples. Recognized patterns of the navigation can become an additional source of information for the system about the user. For example if user's pattern of navigation is analyzed as Repetition, keywords in the following content can be annotated with their definition.

Table 8 Patterns of the navigation results

pattern	students	1/10	1/3	1/2	2/3
	ALL	94	79	59	42
sequential	CZ	78	74	41	29
	TR	100	81	65	46
	ALL	49	5	3	1
repetition	CZ	22	0	0	0
	TR	59	7	4	1
	ALL	60	28	7	0
go-back	CZ	67	44	22	0
	TR	57	23	1	0
	ALL	38	6	4	3
skipping	CZ	41	22	15	11
	TR	37	0	0	0

If we look at the results as a whole we can see that Turkish students were more closely following recommended path through the system – almost a half of the students (46%) used sequential pattern of the navigation, and other patterns were used sporadically.

7 Conclusion and Future Work

We describe a few adaptive systems and their features. Moreover, we describe in more detail our approach, its implementation in XAPOS system and results of the experiments testing both areas, area of solving categorization of educational examples currently adding into XAPOS and their interconnection with existing learning objects there, and area of adaptive navigation testing with Czech and Turkish groups of students, utilizing the multilingual abilities incorporated in XAPOS as well.

As future work we will improve the navigation ordering of educational examples – currently based on concepts and their concept-concept relations. In the future we plan to utilize a genuine feature of educational examples, such as difficulty to do more precisely mutual comparison of educational examples to better order them in the navigation of the user. The automated ontology domain building also belongs to the future work, together with a user-friendly interface to support correction of imperfection in machine-based raw results.

Our approach to educational examples categorization can be used for addition of new learning objects into the course in general. A new learning object in any language can be added to the course in the current version of XAPOS. Similarity allows choice between both educational example and learning objects similar to the actual one.

Addition of external web sources into the XAPOS navigation is a more distant future work. But we expect that our approach allows it. In this task there are more general technical problems and copyright protection of the external content must be fulfilled.

References

1. Aroyo, L., Dolog, P., Houben, G., Kravcik, M., Naeve, A., Nilsson, M., Wild, F.: Interoperability in Personalized Adaptive Learning. Educational Technology & Society 9(2), 4–18 (2006)
2. Barla, M., Bieliková, M.: Ordinary Web Pages as a Source for Metadata Acquisition for Open Corpus User Modeling. In: Proc. of IADIS WWW/Internet 2010, pp. 227–233. IADIS Press (2010)
3. Brusilovsky, P.: Methods and techniques of adaptive hypermedia. In: User Modeling and User-Adapted Interaction, pp. 87–129. Kluwer Academic Publishers (1996)
4. Brusilovsky, P.: Adaptive hypermedia. In: User Modelling and User Adapted Interaction, vol. 11(1), pp. 87–110 (2001)
5. De Bra, P., Calvi, L.: Aha! an open adaptive hypermedia architecture. The New Review of Hypermedia and Multimedia 1(4), 115–139 (1998)
6. De Bra, P., Houben, G.J., Wu, H.: AHAM: A Dexter-based Reference Model for Adaptive Hypermedia. In: Proceedings of the ACM Conference on Hypertext and Hypermedia, pp. 221–239 (1999)
7. De Bra, P., Aerts, A., Smits, D., Stash, N.: AHA! Meets AHAM. In: De Bra, P., Brusilovsky, P., Conejo, R. (eds.) AH 2002. LNCS, vol. 2347, pp. 388–391. Springer, Heidelberg (2002)
8. De Bra, P., Smits, D., Stash, N.: The Design of AHA! In: Proceedings of the seventeenth ACM Conference on Hypertext and Hypermedia, pp. 133–134 (2006)
9. Smits, D., De Bra, P.: GALE: A Highly Extensible Adaptive Hypermedia Engine. In: Proceedings of the Seventeenth ACM Conference on Hypertext and Hypermedia, pp. 63–72 (2006)
10. Dicheva, D., Dichev, C.: Helping Courseware Authors to Build Ontologies: the Case of TM4L. In: Proceeding of the 2007 conference on Artificial Intelligence in Education: Building Technology Rich Learning Contexts That Work, pp. 77–84 (2007)
11. Fortuna, B., Mladenič, D., Grobelnik, M.: Semi-automatic Construction of Topic Ontologies. In: Ackermann, M., Berendt, B., Grobelnik, M., Hotho, A., Mladenič, D., Semeraro, G., Spiliopoulou, M., Stumme, G., Svátek, V., van Someren, M. (eds.) EWMF 2005 and KDO 2005. LNCS (LNAI), vol. 4289, pp. 121–131. Springer, Heidelberg (2006)
12. Henze, N.: Personal Readers: Personalized Learning Object Readers for the Semantic Web. In: Proceeding of the 12th International Conference on Artificial Intelligence in Education (AIED 2005), pp. 15–22 (2005)
13. Holub, M., Bieliková, M.: An Inquiry into the Utilization of Behavior of Users in Personalized Web. Journal of Universal Computer Science 17(13), 1830–1853 (2011)
14. Jovanovic, J., Gašević, D., Devedžić, V.: TANGRAM for Personalized Learning Using the Semantic Web Technologies. Journal of Emerging Technologies in Web Intelligence 1(1), 6–21 (2009)

15. Kavcic, A.: The role of user models in adaptive hypermedia systems. In: 10th Mediterranean Electrotechnical Conference MEleCon 2000, pp. 119–122 (2000)
16. Kay, J.: The um toolkit for cooperative user modelling. In: User Modelling and User-Adapted Interaction, pp. 149–196 (1995)
17. Kay, J.: Stereotypes, Student Models and Scrutability. In: Gauthier, G., VanLehn, K., Frasson, C. (eds.) ITS 2000. LNCS, vol. 1839, pp. 19–30. Springer, Heidelberg (2000)
18. Kobsa, A.: User modeling: Recent work, prospects and hazards. In: Schneider-Hufschmidt, M., Kühme, T., Malinowski, U. (eds.) Adaptive user interfaces: Principles and practice, pp. 111–128 (1993)
19. Kompan, M., Bieliková, M.: Content-Based News Recommendation. In: Buccafurri, F., Semeraro, G. (eds.) EC-Web 2010. LNBIP, vol. 61, pp. 61–72. Springer, Heidelberg (2010)
20. Nebel, I., Smith, B., Paschke, R.: A user profiling component with the aid of user ontologies. In: Workshop Learning, Teaching, Knowledge, Adaptivity (2003)
21. Page, L., Brin, S., Motwani, R., Winograd, T.: The PageRank Citation Ranking: Bringing Order to the Web. Technical Report, Stanford InfoLab (1999)
22. Rich, E.: User modeling via stereotypes. In: Readings in Intelligent user Interfaces, pp. 329–342. Morgan Kaufmann Publishers Inc. (1998)
23. Salton, G., Buckley, C.: Term-weighting approaches in automatic text retrieval. Information Processing & Management 24(5), 513–523 (1988)
24. Šaloun, P., Velart, Z.: Concept Space Rating for Personalization of Learning Materials Based on Relations. In: Proc. of 4th International Workshop on Semantic Media Adaptation and Personalization, SMAP 2009, pp. 67–72 (2009)
25. Šaloun, P., Velart, Z., Nekula, J.: Navigation over Multilingual Content using one Concept Space: Controlled Experiment. In: Fifth International Workshop on Semantics Media Adaptation and Personalization SMAP 2010, pp. 1–6 (2010)
26. Šimko, M., Bieliková, M.: Automated Educational Course Metadata Generation Based on Semantics Discovery. In: Cress, U., Dimitrova, V., Specht, M. (eds.) EC-TEL 2009. LNCS, vol. 5794, pp. 99–105. Springer, Heidelberg (2009)
27. Šimko, M., Barla, M., Bieliková, M.: ALEF: A Framework for Adaptive Web-Based Learning 2.0. In: Reynolds, N., Turcsányi-Szabó, M. (eds.) KCKS 2010. IFIP AICT, vol. 324, pp. 367–378. Springer, Heidelberg (2010)
28. Sosnovsky, S., Brusilovsky, P., Lee, D.H., Zadorozhny, V., Zhou, X.: Re-assessing the Value of Adaptive Navigation Support in E-Learning Context. In: Nejdl, W., Kay, J., Pu, P., Herder, E. (eds.) AH 2008. LNCS, vol. 5149, pp. 193–203. Springer, Heidelberg (2008)
29. Tiňo, P., Polčicová, G.: Topographic organization of user preference patterns in collaborative filtering. Neural Network World 13(3), 311–324 (2003)
30. van Meteren, R., van Someren, M.: Using content-based filtering for recommendation. In: Potamias, G., Moustakis, V., van Someren, M. (eds.) ECML/MLNET Workshop on Machine Learning and the New Information Age, pp. 47–56 (2000)
31. Yudelson, M., Brusilovsky, P., Sosnovsky, S.: Acessing Interactive Example with Adaptive Navigation Support. In: Proceedings of the IEEE International Conference on Advanced Learning Technologies (ICALT 2004), pp. 842–843 (2004)
32. Yudelson, M., Gavrilova, T., Brusilovsky, P.: Towards User Modeling Meta-ontology. In: Ardissono, L., Brna, P., Mitrović, A. (eds.) UM 2005. LNCS (LNAI), vol. 3538, pp. 448–452. Springer, Heidelberg (2005)

Personalized and Adaptive Access to Services – The Semantic Web Services Approach

Marek Paralič, Peter Bednár, and Ján Paralič

Faculty of Electrical Engineering and Informatics of the Technical University of Košice, Letná 9, 04200 Košice, Slovakia
{Marek Paralic,Peter Bednar,Jan Paralic}@tuke.sk

Abstract. In this paper we present and compare experiences with two different approaches to the utilization of semantic technologies for the adaptive and personalized access to the services. By services we mean generally both web services accessible online and "traditional" services provided by the business or government organizations. We illustrate presented approaches based on our experiences from two projects, where applications in different areas, such as e-government, e-business and crisis management have been implemented. In these cases different approaches according to the used semantic services technology were applied.

Keywords: semantic service, ontology, personalized access, Semantic Business Process Management.

1 Introduction

Semantic services promise to automate tasks such as discovery, mediation, selection, composition, and invocation of services [1]. Each of mentioned tasks leads to the utilization of the services by users, who use them in different context, with different background, resources available and of course for different purposes.

In order to support automation of all tasks required to fulfil user needs, services have to be semantically described for different purposes. In this paper we will follow division of the semantic description to:

- *Information semantics* – specifies all domain specific ontologies (i.e. information model consisting of concepts and instances) that are necessary for service discovery and orchestration.
- *Functional and non-functional semantics* – enables and supports service discovery. Functional properties specify what service requires (what is the input) and what service provides (what is the output) or generally conditions that have to be met before or after the service invocation. Non-functional properties extend functional with any other aspects important for the consumer for service selection, such as policy, quality of service, etc.
- *Behavioural semantics* – enables and supports service orchestration and choreography. It specifies how the service provides functionality. It can specify external behaviour, i.e. protocol which consumer has to follow during invocation and/or internal behaviour, i.e. how service uses other services internally to achieve provided functionality.

I.E. Anagnostopoulos et al. (Eds.): Semantic Hyper/Multimedia Adaptation, SCI 418, pp. 231–247.
springerlink.com

We will describe how semantics on all of these levels formalized in WSMO [8], SAWSDL and OWL-S frameworks [5] was used in the context and user centric applications in the area of e-Government (Access-eGov solution) and crisis management (SEMCO-WS).

The paper is structured as follows. In chapter 2 semantic frameworks and ontologies for the semantic description of web services are described. In the chapter 3 the Access-eGov approach based on WSMO and in chapter 4 the SEMCO-WS approach based on OWL-S and the semantics of High Level Petri Nets are introduced. Chapter 5 contains related work description with Semantic Business Process Management in focus. In chapter 6 we shortly compare both approaches that utilise the semantic web services and we conclude our paper with lessons learned remarks.

2 Semantic Web Services

In this chapter, we describe standardized semantic frameworks and ontologies for the semantic description of web services.

2.1 WSMO

The WSMO [2, 8] is a framework and a conceptual model for describing semantic Web Services. The WSMO framework also provides a specification of WSML (Web Service Modeling Language), a formal language for WSMO ontologies, and the WSMX execution environment [3] as a reference implementation.

WSMO conceptual model consists of four major top-level elements: ontologies, Web Services, goals, and mediators:

Ontologies provide the formal semantics (terminology) used by other WSMO elements to describe the relevant aspects of the domains of discourse. WSMO specifies the following constituents as part of the description of ontology: concepts, relations, functions, axioms, and instances of concepts and relations, as well as non-functional properties, imported ontologies, and used mediators.

Web Services describe the computational entity providing access to services that provide some value in a domain. These descriptions comprise the capabilities, interfaces and internal workings of the Web Service. In particular, a Web Service description in WSMO consists of five sub-components - non-functional properties, capabilities, interfaces, imported ontologies, and used mediators:

- Non-functional properties are logical expressions that specify non-functional information. Recommended non-functional properties are accuracy, financial, network-related QoS (Quality of Service), performance, reliability, robustness, scalability, security, transactional, trust.
- Capability defines the Web Service by means of its functionality, namely in terms of preconditions, postconditions, assumptions and effects. A capability (therefore a Web Service) may be linked to certain goals, which are solved by the Web Service via mediators. Preconditions, assumptions, postconditions and effects are expressed through a set of axioms and a set of shared all-quantified variables.

- Interface describes how the functionality of the Web Service can be achieved (i.e. how the capability of a Web Service can be fulfilled). It describes the behavior of the service from the client's point of view (service choreography) and how the overall functionality of the service is achieved in terms of co-operation with the other services (service orchestration). Besides the choreography and orchestration, the interface is defined by importing ontologies, using mediators, and non-functional properties.
- A choreography description consists of the states represented by ontology, and the if-then rules that specify transitions between the states. The ontology that represents the states provides the vocabulary of the transition rules and contains the set of instances that change their values from one state to the other. The concepts of an ontology used for representing a state may have specified the grounding mechanism which binds service description to the concrete message specification (e.g. WSDL).
- Like for the choreography, an orchestration description consists of the states and guarded transitions. In addition, transition rules that have as postcondition the invocation of a mediator that links the orchestration with the choreography of a required Web Service can also appear in the orchestration.

Goals specify objectives that a client might have when consulting a Web Service, i.e. functionalities that a Web Service should provide from the users' perspective. In other words, Goals model the user's view in the Web Service usage process. Goals are described by a set of non-functional properties, imported ontologies, used mediators, the requested capability and the requested interface (see the Web Services description).

Mediators describe elements that aim to overcome structural, semantic or conceptual mismatches that appear between the different elements that build up a WSMO description. Mediator is the core concept to resolve incompatibilities on the data, process and protocol level, that is, to resolve mismatches between different used terminologies (the data level), how to communicate between Web Services (the protocol level) and on the level of combining Web Services (and goals) (the process level). Currently the specification covers four different types of mediators:

- *ggMediators* that link two goals that are in a relation. This link represents the refinement of the source goal into the target goal or state equivalence if both goals are substitutable. It allows a definition of sub-goal hierarchies and resolving mismatches between them.
- *ooMediators* that import the target ontology into the source ontology by resolving all the representation mismatches between the source and the target ontologies.
- *wgMediators* that link Web Services to goals (using choreography interface), meaning that the Web Service (totally or partially) fulfills the goal to which it is linked. wgMediators may explicitly state the difference between the two entities and map different vocabularies (through the use of ooMediators).
- *wwMediators* linking two or more Web Services for collaboration.

Each WSMO element has an attached set of annotations that are used in its definition. The general recommended annotation types are taken from the Dublin Core metadata standard, and they are for example the contributor, coverage, creator, date, description, format, etc.

2.2 SAWSDL

SAWSDL [4] defines a mechanism to associate semantic annotations with Web Services that are described using WSDL. In particular, SAWSDL defines URI reference mechanisms to the interface, operation, and message WSDL constructs to point to the semantic annotations defined in the externalized domain models. SAWSDL defines the following extension elements and attributes:

Annotating message types:

- *modelReference* extension attribute, which allows a creation of one-to-one associations of WSDL input and output type schema elements to the concepts in a semantic model;
- *schemaMapping* extension attribute for creation of many-to-many associations of WSDL input and output complex type schema elements to the concepts in a semantic model. It can point to a transformation (for example XSLT), from XML data to the external ontological data in RDF/OWL or in WSML;

Annotating operations:

- *precondition and effect* extension elements, which are used on the WSDL interface operations to specify conditions that must be held before and after respectively the operation is invoked. The conditions can be specified directly as an expression with format defined by the semantic language or by reference to the semantic model;
- *category* extension element, which provides a pointer to some taxonomy category, i.e. it models a service category on a WSDL interface element. The category element is intended to be used for taxonomy-based discovery.
- *action* extension element, used to annotate the operation element with a functional ontology concept.

2.3 OWL-S

OWL-S [5] is an OWL ontology for semantic description of Web Services. For annotated Web Services, it enables automatic Web Service discovery, invocation, composition, and interoperation. The structure of the OWL-S ontology provides three upper-level classes capturing essential types of knowledge about a service, namely a *ServiceProfile* (for service discovery), a *ServiceModel* (a process model for composition of services), and a *ServiceGrounding* (for interaction of services, including details about transport protocols - it associates profile and process concepts with the underlying service interfaces).

- The *service profile* tells "what the service does", in a way that is suitable for a service-seeking or matchmaking agent to determine whether the service meets its needs. The parameters of the service are expressed by functional and non-functional properties. The functional properties describe the inputs, outputs, preconditions and effects of the service (IOPEs). The non-functional properties describe the semi-structured information intended for human users for service discovery, e.g. service name, description, and parameters that can be used for further requirements on the service capabilities (e.g. security, measuring a quality of the service, geographical scope, etc.).
- The *service model* tells a client how to use the service, by detailing the semantic content of requests, the conditions under which particular outcomes will occur, and optionally also the step by step processes leading to those outcomes. The service is viewed as a process which defines the functional properties of the service (IOPEs), together with details of its constituent processes (in the case of composite services). The functional properties specified within the service model can be shared with the service profile.
- The *service grounding* specifies the details of how an agent can access a service, as e.g. communication protocol, message formats, and other service-specific details such as port numbers used in contacting the service. The grounding enables the execution of the Web Service by binding the abstract concepts of the service profile and process model to concrete messages and protocols. Although different message specifications are supported by OWL- S, the widely accepted WSDL is preferred.

OWL-S distinguishes between simple, or "atomic" services, and complex or "composite" services (processes). The atomic services can be directly invoked, have no sub-services, and are executed in a single step from the requester's point of view. The atomic services are used as elements of abstraction, no ongoing interaction between the user and the service is required. In contrast, composite services consist of multiple more primitive services, and may require an extended interaction or conversation between the requester and the set of services that are being utilized. The composite services define workflows for the sub-services, using control constructs such a sequence, split, if-then-else or iterate.

3 The Access-eGov Approach

The Access-eGov (AeG) solution[1] provides end users with a personalized (i.e. adapted to his/her personal data/situation) scenario of services corresponding to the given goal including detailed information on individual services in this scenario [6], [7]. The WSMO framework was extended by the top-level elements called Life Events (formal models of user's needs, consisting from multiple goals and services organised into generic scenario and expressed by orchestration construction defined as control-flow and data-flow sequences) and Services (generalisation of Web service concepts for enabling support of both electronic

[1] `http://www.accessegov.org/`

and traditional government services by means of a service profile, containing functional and non-functional properties, capabilities, and interfaces). The created system and supporting tools were tested for different real life event situations – marriage, setting up a new company and request for building permission.

3.1 Architecture and Functionality of the Access-eGov

The Access-eGov system architecture consists of four main functional modules (see Fig. 1):

- *AeG resource ontology* that persistently stores the life events and WSML representations of the goals, as well as supporting semantic information for the service annotation and the results of this process (i.e. service instances).
- *AeG core components* module that utilizes semantic information in life event and goal descriptions in order to provide the requested functionality to the user. It includes decomposition of a given life event or a goal into sub-goals, orchestration, composition, and mediation of the sub-goals within a workflow thread, semantic matching and discovery of the services for a given goal, as well as execution of the retrieved and resolved services.
- *Annotation tool* (AT) that supports transformation process from services to semantic services, whereby as services the traditional face-to-face services, electronic services as well as web services were took into consideration. The transformation process is managed by service templates and its result - required inputs and provided outputs, capability interfaces, and related workflow sequences, as well as non-functional properties - are stored in form of WSML representations in the resource ontology.
- *Personal Assistant client* (PAC) that as a web application enables the citizens to browse and navigate through the life event and corresponding sub-goals.

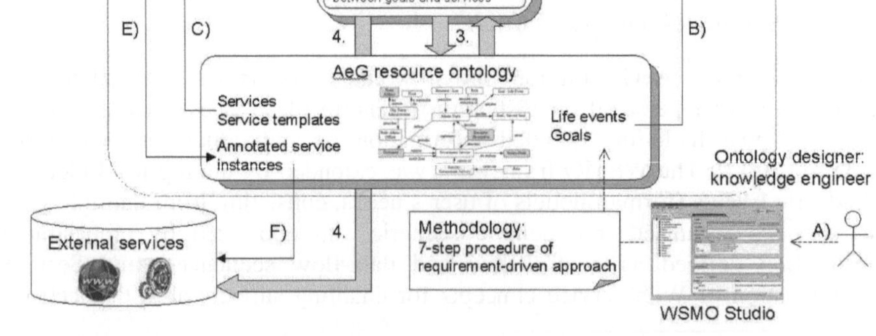

Fig. 1 Architecture and control flow within the Access-eGov platform [7]

Operations performed during the design time on the side of information provider (i.e. a public administration) are represented in Fig. 1 by thin arrows and are labeled with capital letters. For detailed description is the reader kindly asked to read the publication [7].

3.2 Personal Assistant Client – The User's View

PAC is a wizard-like application that enables to personalize and customize the thread of sub-goals by answering a set of customization questions, which can be defined in a process model of the semantic representation of corresponding life events and sub-goals.

At the beginning the user selects one of the predefined life events that are supported by corresponding public administration office. Each life event is a complex goal that can be achieved in different ways depending on personalized information given by the user during the process. Individual tasks to be fulfilled are displayed, whereby the starting tasks' list is usually incomplete and the user is guided to provide the necessary information to resolve open tasks into services or sub processes, which can offer the needed functionality. PAC supports the step-by-step like navigation by the personalised information gathering (see Fig. 2).

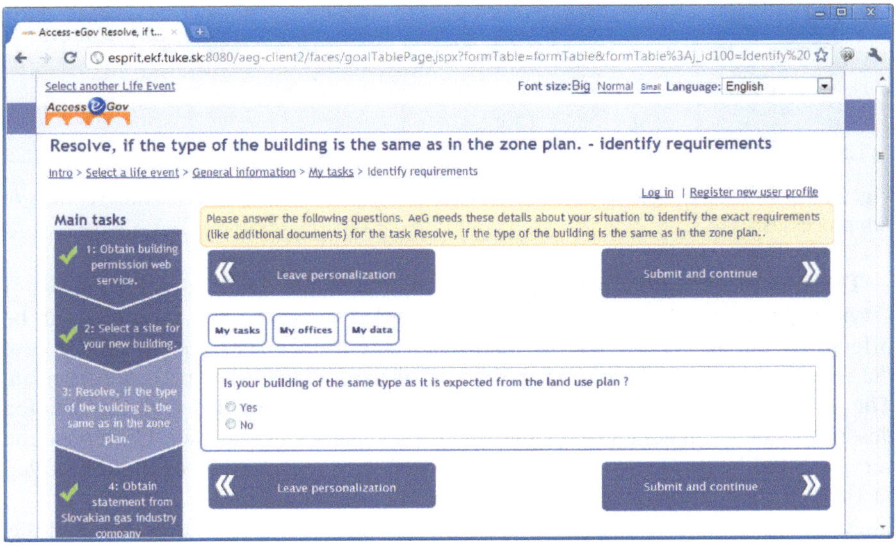

Fig. 2 PAC supports guided personalization process.

Personalization data can be changed anytime, where the user determines the order, in which the personalized questions are answered. The later approach is supported via "My data" tab available in PAC, that can be accessed anytime during the life event resolving process and changes are directly projected into the tasks' list (see Fig. 3).

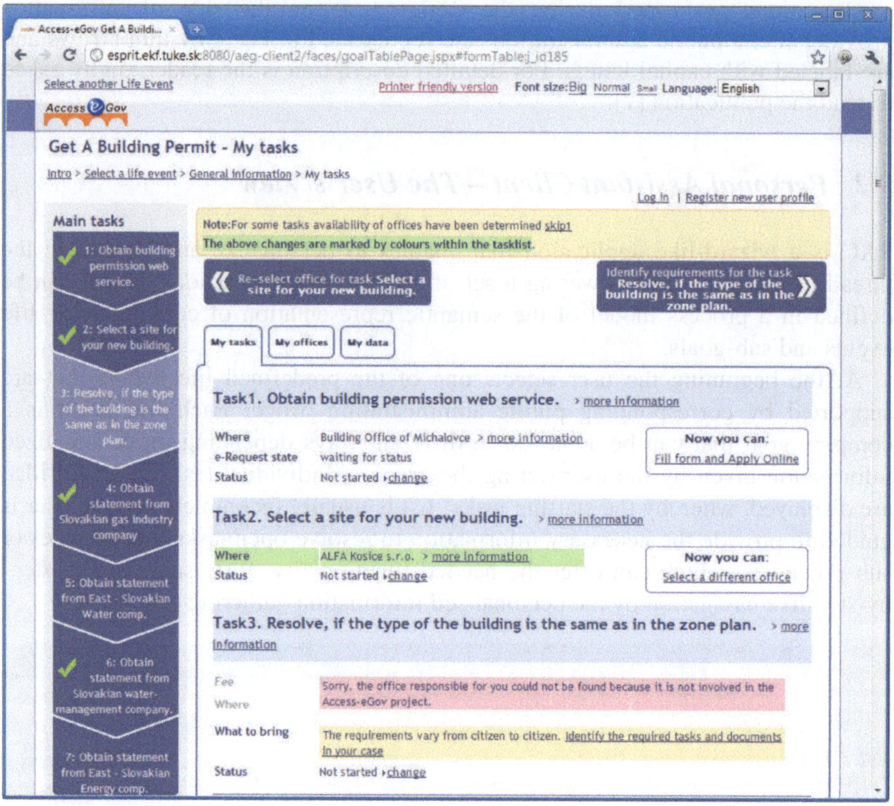

Fig. 3 "My tasks" tab – the personalized data can help to resolve open tasks into services or sub processes.

The goal of the whole personalization process of life event is to get the list of services together with exact requirements needed to enact them. Services can be offered by public offices also as traditional face-to-face services. In such a case the requirement could be e.g. specification of particular form type to be filled in. The service is than resolved in particular office (user is provided with the detailed description and contact data) responsible for the matter of this user. Otherwise the service could be an electronic service or a web service, so the form could be filled in via web application or sent by e-mail.

As far as all tasks are personalized, the result list of to-do tasks and subtasks contains exact information and services' descriptions, so the user has all information necessary to solve the requested life event in his/her personal life situation.

The illustrating figures contain data from the pilot application in Slovakia that was aiming to support citizens during the process of obtaining permits for building a house, including services related to the land-use planning and final approval proceedings. The "Get a Building Permit" life event was modeled to provide for citizens a personalized guidance through the whole process, often very complex and difficult to comprehend even for experienced laymen.

The practical deployment was designed to cover an extended area of Košice and Michalovce with around 350,000 inhabitants. All five branches of the main construction office of Košice Self-governing Region (KSR) were engaged in the testing. The process contained a construction supervision phase, modeled in the ontology. Service types as real estate agencies, project designers, and architects were included and respective services were annotated. Altogether, more than 200 annotated services were created. Example of electronic web service included into the life event solution that is automatically discovered and inserted into the services list was a web service for online checking and provision of the land-use plan for a given municipality. Service integration with the Slovak cadastre portal (www.katasterportal.sk) was also tested.

3.3 How the Personalization Is Achieved

Personalization of life events for the particular user case is handled as the part of the service or complex goal orchestration model. Orchestration is divided to the personalization part, where the service evaluates all information required in order to identify particular user case and service execution part where service invokes orchestrated services and provides outputs.

Orchestration model is based on Abstract State Machines (ASM). In ASM, information state is represented as the set of ontology instances (i.e. grounded facts). Information state is changed by the set of production rules in the form if-condition-then-action, where condition can be arbitrary logical expression. Actions can directly modify information state, i.e. add or delete instances or change their attribute values, or can be used for communication with service requester or with ASMs of orchestrated services. ASM can wait for data (receive communication action) or provide data (send communication action).

Information space is local to ASM, i.e. all changes to instances are visible to the ASM only and instances are not shared. However, it is possible to explicitly copy instances from information space of one ASM to another one using send/receive actions, which is the base mechanism for passing of inputs and receive outputs from the orchestrated service.

Important feature of ASMs is implicit parallelism. Conditions of all rules are evaluated in parallel and all rules are also fired in parallel with the exception that after the modification of the state, resulting ontology has to be consistent. If some kind of synchronization between activities is needed, it has to be implemented with the mutual conditions of rules.

Regarding the personalization, it is directly part of the orchestration ASM model of the service. It consists of the set of rules with receive action where the service requests information about the particular personal case. This information can be provided directly by citizen using the automatically generated form in PAC, reused from user profile, which stores common personal data about the citizen or it can be part of inputs requested by the service (i.e. from instances representing required documents).

Personalization rules can be mutually conditioned, i.e. depending on the answers to some personalization requests new requests can be generated. During the resolving of personalization requests, system will at first check if the required type of the instance is stored in the user profile and if not it will collect all uncovered types of instances and generate a form in PAC, which will collect the necessary information from the user.

Note that personalization block of rules is only formally divided from the execution part of the service so as it was noted before, all rules - both personalization as well as orchestration, are evaluated and executed in parallel, so there is no limitation that the service is fully customized before any orchestrated service is invoked. This also allows that condition of some personalization rules depends on information provided by the orchestrated services.

From the point of view of the whole life event process, AeG personalization is always local for service or goal, i.e. since the life event is represented as complex goal, this goal has assigned orchestration model which at first identifies general information about the user case and according to this information decomposes life event to sub-goals. Sub-goals can have recursively assigned orchestration models, which subsequently ask more specific information about the user case and decompose sub-goals to sub-sub-goals etc.

4 The SEMCO-WS Approach

Other experience comes from the Semco-WS project [9], where not only services were annotated, but also the data processed during the composite service execution. In this way, the workflow enactment engine is able to use the ontology and stored knowledge in construction and execution of workflows of web services. The principal feature of the module is its ability to reuse also already existing data in an automated manner, not requiring the user to enter the data into an already constructed workflow. Distributed workflows use the semantics of High Level Petri Nets to describe the logic, data and state, whereby the transitions are semantically annotated by OWL-S profiles, which semantically describe the activity of transforming input data to the output as well as the preconditions and effects that apply before and after transformation. In this way each token produced in the process of workflow execution can be annotated with the corresponding transition profile.

4.1 Architecture and Functionality of the SEMCO-WS

The SEMCO-WS system is built upon the K-Wf Grid[2] solution [10], whereby the knowledge store and the web service orchestration modules were further extended. The knowledge store is decentralized (each of the modules of SEMCO-WS interacting with the ontology has its own knowledge store) and orchestration considers the data gathered from the user during the personalization process.

---

[2] `http://www.kwfgrid.eu/`

Basic architecture of the SEMCO-WS system is depicted in Fig. 4 and consists of the following main modules:

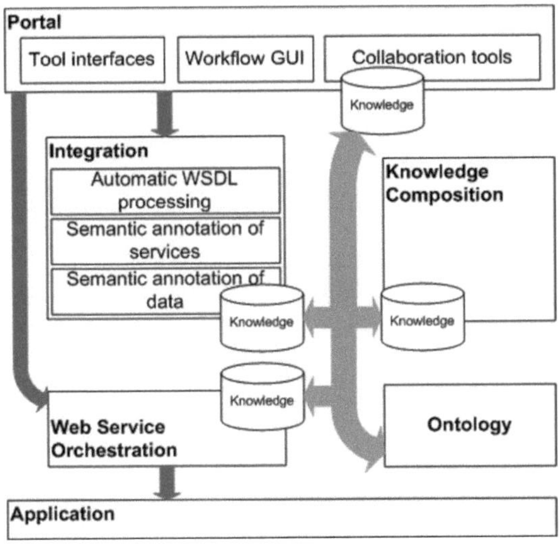

Fig. 4 Architecture of the SEMCO-WS system [9]

- *Knowledge Composition* that is responsible for processing of monitoring data and learning from past workflows,
- *Integration* that offers the means for semantic annotation of grid and web services and enables creation of semantic web services based on OWL-S representation.
- *Web Service Orchestration* that combines the advantages of automatic workflow composition with the possibility of reuse already computed or simply existing (personalized) data during the phase of creating the executable workflow from an abstract one [9, 11].
- *Portal* that covers the workflow editor for manual and automatic composition of workflows of semantic web services, whereby collaborative work of multiple users, as well as the context recognition based on textual notes is supported [12]. Also the execution of resulting workflow is supported.

For detailed description of the system architecture is the reader kindly asked to read the publication [9].

4.2 The User View of the SEMCO-WS

User interface depicted on Fig. 5 contains graphical tool for defining and execution of workflow of web services in order to achieve requested goal defined as an ontology instance [11]. The upper toolbox contains buttons that enable the editing of the workflow. In the left part of the window is panel for setting up the parameters of utilized web services and in the middle is the Multi-User Workflow Editor (MUWE). In the right side is the context-sensitive help that displays information gathered either from the monitoring process or from the direct user input from similar situation.

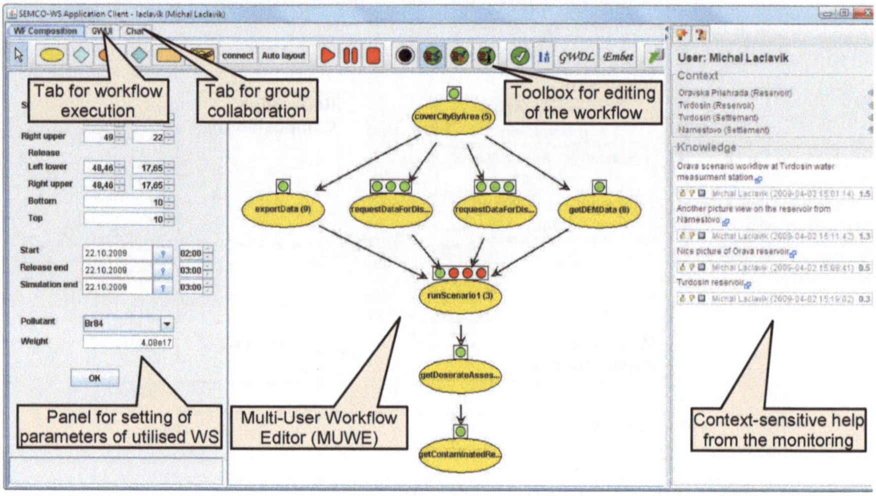

Fig. 5 SEMCO-WS user interface

Besides these tools the user interface contains also modified Grid Workflow User Editor (GWUI tab) for the workflow execution and a communication tool for group collaboration (Chat tab).

The user interaction begins with the specification of the goal, whereby the method of automatic workflow composition takes the information from goal description and uses it for creation of the workflow. The orchestration process takes into consideration not only input and output parameters of the composing web services, but also the pre- and post-conditions defined as firs order logic formulas. After setting up the input data, it is possible to run the executable workflow, whereby the data produced by web services are annotated and stored in the knowledge store for later use in case that the web service should be called under the same conditions.

System was tested in the domain of crisis management and proved the expectation of creating the workflow in a more effective way, whereby also the scalability aspect was tested on own test set and third party test sets [11]. Created tool was tested also in context of the Web Services Challenge 2009 competition focused on semantic composition of web service chains, whereby it achieved first place in the Architecture category and third in the Performance category[3].

---

[3] `http://ws-challenge.georgetown.edu/wsc09/downloads/WSCResult2009.pdf`

4.3 How the Personalization Is Achieved

In the SEMCO-WS system we tried to move our focus from functional and non-functional semantics to the context of service utilization in a way that considers the service selection including the data instances, which will be used. As far as it is already known during the orchestration phase, which goal should be achieved, not only input/output and pre-/post-conditions from semantic description of services are taken into consideration, but also QoS parameters from the monitoring module (as non-functional properties), as well as the existing data. Existing data given by the user during the personalization phase (at the beginning of the workflow construction), are during the orchestration phase evolved and semantically annotated (i.e. the metadata are added).

Underlying representation of the evolving workflow as High Level Petri Net supports the idea of semantically enriched data, as far as we think in context of tokens moving between places via transitions. During the first phase of workflow construction an abstract workflow (not executable) is created, whereby backtracking from final activity (that produces requested result) to the initial activities is used. Transitions are connected with input places providing the input data that are consumed and to the output places as targets for the produced data. Data produced by the transitions are semantically annotated and occurs in form of OWL instances that will be stored in places. For transitions can be automatically created OWL-S profile, which semantically describes the activity of transforming input data to the output as well as the preconditions and effects that apply before and after transformation. These profiles are used not only for searching the web service candidates for the activity execution during the second phase of workflow construction, but also for annotation of tokens that are processed by the transition [13]. Producing of annotated tokens in execution phase in end effect also influences the process of workflow refinement. Before searching for the service candidates, which are able to realize the transition, the repository of existing annotated data tokens is requested. If the matching process is successful, the workflow is changed by adding the found data and calling a service is not necessary anymore.

However as already mentioned in the beginning of this section, the annotated data in the form of tokens are not only used for workflow execution improvement, but also during the orchestration phase and serves thus for the personalization purposes. The backtracking process searches for such a sequence of web services' call, where the requested resulting goal can be produced from available input data. Therefore each transition (that represents web service call) has an additional version that transforms metadata describing the input to the metadata describing the output based on a functional dependence of the output from the input. During the transformation of the abstract workflow to the executable one, we have available a workflow, where all output places are annotated by metadata instances describing the user's request, based on which the web services will be inserted. So during the orchestration phase the available personal input data from the user are enriched with semantic annotations and additional transformation rules are used for producing the annotated data as potential result of transition – i.e. service execution. In this way personalized data are available also by matching decisions, which would be not possible to make just having the input data. Practical tests were made in crisis management domain, where the target goal was the identification of a pollution assessment and prediction of the consequences in case of radiological emergence. For example the necessary information about accuracy and covered geographical region of the metrological model was derived as metadata created by the Dispersion service [14].

5 Related Work

Semantic web services are dealing with two main aspects related to the personalization: service choreography, i.e. how the user can interact with one particular service and service orchestration, i.e. how many services can be composed in order to achieve complex goal. The vision of semantic business process modelling, formulated in [15] and further elaborated, aims at achieving a higher degree of automation in discovery and mediation of co-operating services [16]. The use of semantic technologies, namely Semantic Web services and underlying ontologies in the process modelling, service configuration, execution, monitoring, and analysis is envisioned as a method that can overcome the heterogeneity and incompatibility problems towards the semantically interoperable services. It may also help to reduce the human intervention throughout the life cycle of business process modelling [17].

Semantically enhanced business process modelling and workflow management is in focus of research and standardisation organisations such as the Object Management Group (OMG, http://www.omg.org), W3C consortium (http://www.w3c.org), OASIS group (http://www.oasis-open.org), or Workflow Management Coalition (WfMC, http://www.wfmc.org). In the European context, particular solutions were provided as outcomes of several FP6 and FP7 research projects, mostly integrated in the FInES cluster initiative [19]. Some of the projects are as follows:

- STASIS (FP6-034980, http://www.stasis-project.net) provides an infrastructure for semantic mapping of services by means of a distributed peer-to-peer repository and shareable ontology structure [20];
- SUPER (FP6-026850, http://www.ip-super.org) provides a framework for semantic business process management, including generic formal languages, process models, and shareable ontology resources [21];
- SPIKE (FP7-217098, http://www.spike-project.eu) provides an environment for service-based and process-oriented collaboration in virtual business alliances [22];
- COIN (FP7-216256, http://www.coin-ip.eu) targets the long-lasting enterprise collaboration, networking, and interoperability by integrating services and business processes in a generic service platform [23];
- NisB (FP7-256955, http://www.nisb-project.eu) is aiming at a provision of user-centric tools for hierarchical interoperability of enterprises, by means of designing and applying various business model archetypes and principles of dynamic business ecosystems [24].

All mentioned research efforts try to overcome the gap between the business view of the Business Process Management (BPM) and the IT view by using semantic technologies. These difficulties, often referred to as the Business-IT gap, are caused partly by the lack of understanding of the business needs by IT experts on the one hand and of technical details unknown to the business experts on the other hand [17]. It corresponds to the vision of Semantic Business Process Management (SBPM) [18] attempts to improve the level of automation in process modelling, configuration, execution, monitoring, and analysis by using ontologies and Semantic Web Services technologies. In this way the human intervention throughout the life cycle of business process modelling can be reduced.

6 Conclusion

In this paper we described two different ways, how the semantic web services technologies can support the personalization aspects. Our experiences come from two successful projects applied in two different application domains - Access-eGov in e-Government and SEMCO-WS in crisis management domain. If we compare both approaches, we can focus on the scope, where the personalization is applied and on the possibility of data reuse:

Local vs. global personalization - AeG model for personalization is local. Typical for AeG processes is hierarchical structure of goals with the life event as a root complex goal. Personalization is performed incrementally in each sub-goal or service execution. As opposite, SEMCO-WS approach is more global in our opinion, and this is determined by the global planning method where at first complete abstract workflow is created, then each transition is resolved to service and only after all transitions are resolved, workflow is executed. Both approaches have some advantages and disadvantages, for example, local personalization is more modular. On the other side global personalization can take into the consideration global structure of the process. Note that technically, SEMCO-WS approach can be also used for the hierarchical decomposition, but then each sub-goal has to be explicitly modeled bottom-up as a complex service before the orchestration workflow is planned.

Data reusing – Both approaches support reusing of existing available data during the process of personalization. The way how this is implemented is very similar; since the state of the process is locally represented for one executed orchestration model in both approaches (state is stored in tokens for one Petri net in case of SEMCO-WS solution or as instances for one ASM in case of AeG). Data reusing mechanism is based on the global external storage, where already known information is stored during the process execution. This is implemented as a token repository in SEMCO-WS or as a user profile in AeG. Difference is that in AeG approach, data are used dynamically during process execution when in SEMCO-WS data are used during the workflow resolution (i.e. after the planning before the process is executed).

Personalization part of the orchestration model in the Access-eGov system causes behavior that resembles discussion with a clerk, who step by step asks the questions to identify user's case in order to find the solution of his/her life event situation. In SEMCO-WS semantic web services providers have to define additional transformation rules for metadata processing that enable (after applying service specific rules) utilization of inserted personalized data in modified form.

In our opinion both examples proved that semantic web services can support personalization process as a substantial part of the technology and have potential for further development.

References

1. de Bruijn, J., Kerrigan, M., Keller, U., Lausen, H., Scicluna, J.: Modeling Semantic Web Services. Springer, Heidelberg (2008)
2. Roman, D., Keller, U., Lausen, H., De Bruijn, J., Lara, R., Stollberg, M., Polleres, A., Feier, C., Bussler, C., Fensel, D., et al.: Web Service Modeling Ontology. Applied Ontology 1(1), 77–106 (2005)
3. Haselwanter, T., Kotinurmi, P., Moran, M., Vitvar, T., Zaremba, M.: WSMX: A Semantic Service Oriented Middleware for B2B Integration. In: Dan, A., Lamersdorf, W. (eds.) ICSOC 2006. LNCS, vol. 4294, pp. 477–483. Springer, Heidelberg (2006)
4. Akkiraju, R., Farrell, J., Miller, J., Nagarajan, M., Schmidt, M., Sheth, A., Verma, K.: Web Service Semantics - WSDLS, Tech. rep., LSDIS Lab (2005), http://lsdis.cs.uga.edu/projects/meteors/wsdl-s/
5. Martin, D., et al.: OWL-S: Semantic Markup for Web Services. Member submission, W3C (2004), http://www.w3.org/Submission/OWL-S/
6. Bednár, P., Furdik, K., Paralic, M., Sabol, T., Skokan, M.: Semantic integration of government services - the access-eGov approach. In: eChallenges e-2008 Conference, pp. 1-8. IOS Press, Amsterdam (2008)
7. Furdik, K., Klischewski, R., Paralic, M., Sabol, T., Skokan, M.: E-government service integration and provision using semantic technologies. In: Electronic Government. Schriftenreihe, Informatic, vol. 30, pp. 273–280. Trauner Verlag, Linz (2009)
8. Web Service Modeling Ontology, http://www.wsmo.org/
9. Habala, O., Paralič, M., Rozinajová, V., Bartalos, P.: Semantically-Aided Data-Aware Service Workflow Composition. In: Nielsen, M., Kučera, A., Miltersen, P.B., Palamidessi, C., Tůma, P., Valencia, F. (eds.) SOFSEM 2009. LNCS, vol. 5404, pp. 317–328. Springer, Heidelberg (2009)
10. Balogh, Z., Gatial, E., Laclavik, M., Maliska, M., Hluchý, L.: Knowledge-Based Runtime Prediction of Stateful Web Services for Optimal Workflow Construction. In: Wyrzykowski, R., Dongarra, J., Meyer, N., Waśniewski, J. (eds.) PPAM 2005. LNCS, vol. 3911, pp. 599–607. Springer, Heidelberg (2006)
11. Bartalos, P., Bielikova, M.: Semantic Web Service Composition Framework Based on Parallel Processing, E-Commerce Technology. In: IEEE International Conference on Commerce and Enterprise Computing, pp. 495–498 (2009)
12. Rozinajova, V., Kasan, M., Navrat, P.: Towards More Effective Support of Service Composition: Utilizing AI-planning, Semantics and User's Assistance. In: International Conference on Next Generation Web Services Practices, pp. 50–55 (2009)
13. Habala, O., Simo, B., Gatial, E., Hluchy, L.: Automatic Data Reuse in Grid Workflow Composition. In: Bubak, M., van Albada, G.D., Dongarra, J., Sloot, P.M.A. (eds.) ICCS 2008, Part I. LNCS, vol. 5101, pp. 194–202. Springer, Heidelberg (2008)

14. Bartalos, P., Gazak, M.: Semantic web services based crisis information system exploiting automated workflow composition. In: GCCP 2009: Proc. of the 2009 Int. Workshop on Grid Computing for Complex Problems, SAS, pp. 44–51 (2009)
15. Hepp, M., Leymann, F., Domingue, J., Wahler, A., Fensel, D.: Semantic Business Process Management: A Vision Towards Using Semantic Web Services for Business Process Management. In: Proceedings of the IEEE International Conference on e-Business Engineering (ICEBE 2005), Beijing, China, IEEE Press, pp. 535–540 (2005)
16. Pedrinaci, C., Brelage, C., van Lessen, T., Domingue, J., Karastoyanova, D., Leymann, F.: Semantic Business Process Management: Scaling up the Management of Business Processes. In: Proceedings of the 2nd IEEE International Conference on Semantic Computing (ICSC 2008), Santa Clara, CA, pp. 546–553. IEEE Press (2008)
17. Karastoyanova, D., van Lessen, T., Leymann, F., Ma, Z., Nitzsche, J., Wetzstein, B., Bhiri, S., Hauswirth, M., Zaremba, M.: A Reference Architecture for Semantic Business Process Management Systems. In: Multikonferenz Wirtschaftsinformatik 2008, pp. 1727–1738. GITO-Verlag, Berlin (2008)
18. Hepp, M., Leymann, F., Domingue, J., Wahler, A., Fensel, D.: Semantic Business Process Management: A Vision Towards Using Semantic Web Services for Business Process Management. In: Proceedings of the IEEE ICEBE 2005, Beijing, China, pp. 535–540 (2005)
19. Missikoff, M., Drissi, S., Giesecke, R., Grilo, A., Li, M., Mehandjiev, N., Werth, D.: Future Internet Enterprise Systems (FInES). Research Roadmap. Final report, ver. 3.0, European Communities (June 2010)
20. Beneventano, D., Dahlem, N., El Haoum, S., Hahn, A., Montanari, D., Reinelt, M.: Ontology-driven Semantic Mapping. In: Enterprise Interoperability III, pp. 329–341. Springer, London (2008)
21. Frankowski, J., Kupidura, H., Rubach, P., Szczekocka, E.: Business Process Management for Convergent Services Provisioning Using the SUPER Platform. In: International Conference on Intelligence in service delivery Networks (ICIN), Bordeaux, France (2008)
22. Broser, C., Fritsch, C., Gmelch, O., Pernul, G., Schillinger, R.: SPIKE - A Collaboration Platform for Short-Term Virtual Business Alliances. In: Proceedings of the Workshop on Technological trends in Enterprise Systems for SMEs and large enterprises: Heading towards the Future Internet, London, UK during DIGIBIZ 2009, Springer, Heidelberg (2009)
23. Huber, S., Carrez, C., Suttner, H.: Development of Innovative Services Enhancing Interoperability in Cross-Organizational Business Processes. In: van Sinderen, M., Johnson, P. (eds.) IWEI 2011, Part II . LNBIP, vol. 76, pp. 75–88. Springer, Heidelberg (2011)
24. Otto, B., Reichert, A., Sagmon, M.: The Network is the Business (2011), Project White Paper `http://www.nisb-project.eu/uploads/media/NisB_Project_White_Paper.pdf`

Chapter 3: Social and Context-Aware Adaptation

Context awareness and adaptation are very closely related terms both in computer science and real-life situations. In addition, context awareness is connected to location awareness, where location defines spatial constraints, especially in terms of nomadic users. Context awareness is also related with ubiquitous or pervasive computing applications. The first article presents a prototype system which combines user profile preferences and context information as an assistive application to a nomadic user. Towards this direction and as the number of users in a certain society/community increases, the problem of similarity computation among users for adaptation recommendation provision becomes a challenging task. The identification of the most influential users in social communities stands as a very interesting area of research, especially for targeted advertising. Given that recommender systems can be also considered as online social networks, the second article enhance the first approach and tries to build further the idea of influential users for generic social network environments. Finally, this chapter covers some context-aware adaptation issues for client-server web applications in terms of privacy policies and anonymity.

Article 3.1

Title: "With a Little Help from My Friends": Context Aware Help and Guidance for Using the Social Network
Authors: Nasim Mahmud, Kris Luyten, and Karin Coninx

In context of increasing mobile computing, this contribution provides a discussion on help systems that glean answers from human. The authors concentrate on social issues (e.g. availability and interruptibility) along with the issues of relevance and reliability for seeking the most appropriate person to ask for any information or help. They also present the software prototype of a help system called *Ubiquitous Help System*, which combines a user profile and preferences and context information for assistance to a nomadic user to find people who can support him in a large scale ubiquitous computing environment. Web to Peer is used as the communication framework. User profile is set by the user, the FOAF data are used as well. Central component of presented approach is context-aware search for human guidance where the context is defined as internal (such as user preferences) and external (such as location).

Article 3.2

Title: Influential Users in Social Networks
Authors: Dimitrios Vogiatzis

An online social network is conceived as a web-like structure of interacting nodes. The links might be formally established, such as those set explicitly in facebook, twitter and other similar social networks; or they might be implicit such as those formed by email exchanges, postings in fora, co-authorship of documents etc. By virtue of the content created, and knowledge possessed by each user, as well as by the user interactions, an enormous nexus of information is created. Social networks have always been present in the form of groups of acquaintances, co-workers, alumnae, etc. and they were the subject of study of various disciplines. A relatively recent change is the growth of online social networks, in terms of users' participation and diversity. Making sense out of this information can be a formidable challenge. In this article various aspects of influential users in an online social network (SN) are examined and the term "influential" is defined in this context by focusing on its purpose or usefulness.

Article 3.3

Title: A Client-Side Privacy Framework for Web Personalization
Authors: C. Kolias, V. Kolias, G. Kambourakis, and E. Kayafas

Kolias et al. elaborate issue of privacy in adaptive web, which is more and more important as personalized web applications together with social networks are more and more in use. The authors discuss privacy policies, approaches to anonymization and pseudonymity employed when data stored on server side, and pros and cons when private data is stored on the client side. They propose an abstract architecture that enables users to fine-tune their privacy level according to trust they put on different applications. The user profiles are maintained on the client side. Presented approach differentiates itself in the profile manipulation and privacy negotiation process, where both the client and the service provider cooperate by exchanging a series of messages for a personalization action to take place.

“With a Little Help from My Friends”: Context Aware Help and Guidance Using the Social Network

Nasim Mahmud, Kris Luyten, and Karin Coninx

Hasselt University - tUL - IBBT,
Expertise Centre for Digital Media,
Wetenschapspark 2, B-3590 Diepenbeek, Belgium
{nasim.mahmud,kris.luyten,karin.coninx}@uhasselt.be

Abstract. People are more mobile than ever before. People predominantly need information or help that they can not anticipate and plan for before they begin their journey. There is an abundance of information around us. But the users seek precise and fine–grained information that is not provided by the computing environment but required to carry out their tasks. Especially nomadic users have little choice than to call upon help from other people present in their vicinity. There are social issues (e.g. availability and interruptibility) along with the issues of relevance and reliability to find the most appropriate person to ask for any information or help. Identifying this person is still a challenge today. In this chapter we provide a thorough and grounded discussion of the state of the art of help systems that glean answers from human. We also present the software prototype *Ubiquitous–Help–System (UHS)* that has been developed to demonstrate how using a social network in combination with the user profile and preferences, can assist a nomadic user. The system assists the nomadic user to find people who can support him in a large scale ubiquitous computing environment.

1 Introduction

In our everyday life we spend a remarkable amount of time searching for some information. The process of finding proper information (e.g. a book, journal, and website) and being able to identify and extract the relevant information requires time and effort [11]. While looking for information we often seek help from the people we know. If we have a specific question from a field in which a friend or colleague is an expert, it is more efficient to consult her or him directly [11, 28, 29]. But, in real–life, most of the questions we have and the type of information we look for, are so diverged that our everyday social network may not be sufficiently large enough to come up with an answer.

I.E. Anagnostopoulos et al. (Eds.): Semantic Hyper/Multimedia Adaptation, SCI 418, pp. 251–269.
springerlink.com

Fortunately, there are ways to broadcast our queries to the people who are not necessarily directly acquainted with us. For example, a mailing list or a discussion forum can be a mechanism to broadcast queries and get answers for our questions. Another form of source for relatively practical information is wikis. A mailing list can be used to receive direct responses to a question, albeit a mailing list is usually limited within a small community sharing a particular background. An online forum can offer a greater scope where the users can pose questions and get answers from other knowledgeable users. It can also serve as a repository of old questions and answers. Since the questions and answers are usually public, forums act as a discussion board as well as a source of knowledge gathering. Wikis also serve as knowledge repository, the only difference from the traditional web is that the wikis are usually the users generated repository of collective intelligence. Broadcasting messages is good for technical questions that can be answered in one step with a single answer or with multiple plausible answers [14]. This works efficiently, mostly for technical questions where a question is well defined, and have a straight answer.

There are abundance of information everywhere around us. But finding the right piece of required information in time is a challenging task. Furthermore, it is more difficult for persons on the go to find the relevant and reliable piece of information. Mobile people have less time to verify an information before accepting it. Context awareness can help the users to extract right information relatively easily. With the increased amount of available information, context awareness and social awareness can serve as a filtering mechanism.

Awareness of context and activity provides ubiquitous computing environment with the ability to adapt its services in order to best meet its users' expectations by extracting the users' need from the context. Here context is defined as any information that can be used to characterize the situation of an entity whereas an entity is a person, place or object that is considered relevant to the interaction between a user and an application, including the user and applications themselves [2].

Users activity and context are interwoven characteristics that give meaning to any natural form of communication or interaction. Despite the fact these are interwoven characteristics, both context awareness [1, 35] and activity awareness [18, 30] are often considered separated from each other.

With the advancement of modern mobile technology, computing devices are becoming ubiquitous, surrounding us virtually all the time. These devices give computational power and communication capabilities that we can exploit in our daily life activity. For example, Web searching has become the preferred way of looking for information to support various tasks (e.g looking for a nearby pizzeria, checking the train schedule, etc.). Web search is still the most dominant form of searching and accessing information.

But people are more frequently connected when mobile than ever before. Mobile people are exposed to newer situations where they need more contextual information (e.g., a tourist may need help and direction from his current location). They seek for more and more fine–grained information from the context. This leads to information relevance not only being determined by the content but also by extrinsic properties, related to users' current context. Popular search engines are updating their search techniques to cope with the abundance of information and to search and retrieve the correct result. Mainstream search engines (such as Bing[1] and Google[2]) have applied social search technique to retrieve and rank search result using social network information. With the increasing amount of available information, users still struggle with finding the right information that they can rely on. The web search provides way to find web pages that have the information one is looking for. There are new search engines that attempt to find people who have the answer. For example, a relatively new social search engine, Aardvark [3] finds people that have the information one is looking for. Most of these approaches assume that the user is searching the Web on a stationary computing device. And those approaches do not take into account the user's current context, leaving very little information filtering, and context aware help and support for nomadic users.

Moreover, in many situations we prefer asking a person who we can relate ourselves to, by means of some social relations. This gives us comfort, certainty and also ensures immediate feedback. When people need information or help, still one effective and yet natural way people react is to ask an expert in the domain for help. People usually turn to other people for help that they can rely on in order to get quick information or recommendation [12, 17, 13].

When we visit a new place, sometimes we are perplexed by the amount of available information in the surroundings. In this kind of situations as travelers we, instead of trying to extract the required information from the environment, often prefer to ask someone nearby for some precise information we need. For example, instead of following the sign, or checking the 'departure board' found in the train station, sometimes we prefer to ask someone in the neighborhood. We expect to get reliable information from someone who is familiar with the place (e.g., we prefer asking the guy in uniform, assuming that he is a reliable source of information). This is not only because we feel comfortable while we have access to authentic information; sometimes it is rather hard to get fine grained and reliable information from any other available sources. But there is also the interpersonal communication barrier while seeking for information or help from a person we do not have any common ground with. In the paper [26] we have described a new way of selecting the most appropriate person to get help from by using the social network of the user in combination with the users' context, goals and activities.

We argue that exploiting social network, user's personal preference, presence of other users in the vicinity and their preferences enable a new era of interaction that helps the user to accomplish tasks more efficiently. It allows

---

[1] Bing `http://www.bing.com` - Last accessed: November 22, 2011 07:00 CET
[2] Google `http://www.google.com` - Last accessed: November 22, 2011 07:00 CET
[3] Aardvark, `http://www.vark.com/` - Last accessed: November 22, 2011 19:00 CET

to select a suitable person in the vicinity as a source of fine-grained information in a large-scale ubiquitous computing environment. In our ongoing work we demonstrate how a person within the neighborhood (both in terms of location as well as social network) can be invited using *UHS* to provide help.

2 State of the Art

The era of context aware applications began with the introduction of the Active Badge Location System[35], an infrared technology based tracking system capable of tracking its users current location to forward incoming telephone calls closer to the user. From then, most of the context-aware systems focus on the external context [19, 31]. External context or physical context is basically referred to the class of context where context data is collected via physical sensors. It includes context data of physical environment e.g. location, distance, time, temperature, air pressure, lighting levels, surrounding users etc [16, 31].

In order to provide personalized services according to the user preferences, focus on internal context is necessary. Internal context [16] describes a state of the user. It comprises of task and emotional state of user, the cognitive domains, such as information retrieval, decision making, situation monitoring, and so on. Models have been proposed to capture the internal context, e.g. Schmidt et al. [33] propose hierarchical structure of context to codify both internal and external elements of the context. At the top of the hierarchy, human factors includes user, social environment and task. Physical environment includes conditions, infrastructure and location.

Helping nomadic people for providing context aware information has been proposed e.g. the Cyberguide [1, 24] mobile assistant or tour guide that is aware of the location and the orientation of its user and provides information about the surrounding space.

A conceptual model has been proposed [22], showing how social awareness through instant messaging can help mobile learners by providing on-demand sharing of knowledge. In this model social awareness has been described as a *mental concept* where users become aware of social networking by accessing to the same social network or resources regardless of their location and learning context. Thus the learners remain aware of available tutors, lecture and knowledgeable peers should they encounter a learning problem for which they need to consult someone.

Asking questions and seeking for help from other people is one of the most common ways for people to solve real–life problems in a social environment. Systems that support people sharing knowledge or expertise has been a research topic for at least 15 years [37]. So far this field of research has been largely studied within organizational settings or online communities [3, 4]. These have been explored in a series of CSCW studies [7]. Systems that utilize social networks have also been studied [15, 21]. These systems attempt to leverage social networks within an organization or community to help find and locate suitable person who can help. However, while many systems have been built, there is still a lack of exploration in context aware help system that takes both user's context and social whereabouts into account.

Community based question answering site has recently become popular, where a member can post a question online and another member may reply to the question. In this kind of question answering site, usually the user receives reward points for answering questions. Popular sites, such as Yahoo! Answers [4] offers free service whereas Google Answers (closed since December 2006) used to impose fees for asking questions. Use of similar question answering site is one of the popular alternatives for searching answers for those subjective questions that usually search engines (such as Bing, Google, Yahoo![5]) cannot help with. Search engines can not help to find useful solution when the problem is hard to define.

Answer Garden [5, 6] is one of the earliest cooperative work tool, providing an expert locating system that is designed to help improve 'organizational memory' by providing database for commonly and frequently asked question. Whereas, new questions are automatically routed to appropriate experts. Once a new question is answered by an expert, the new question and its answer can be inserted into a database.

Answer Garden aims to eliminate the need for answering many simple questions by experts which has already been answered previously. Within the system, the question is checked against a database. If a user is unable to find an answer or find the answer incomplete, the user may ask his question through the system and the system then routes the question to an expert in the domain. And then the expert answers the user through electronic mail. If it was answered previously, the experts no longer have to answer the same questions over and over again. Instead, they can concentrate on more interesting problems. The system utilizes a branching network of diagnostic questions that helps users to find the answers they want. Answer Garden is not the first one of its kind, but it showed how relatively simple combination of well-known concepts can provide a platform for a new kind of cooperative work application in form of a question answering system. Answer Garden system is designed for stationary computer users and does not utilize any kind of context information.

CityFlocks [11] is a newer and context aware mobile system that enables nomadic visitors or new residents in a city to acquire knowledge about the city from the local residents. It also allows the users to share their experiences with the local residents and other users by digitally annotating, commenting and rating any artifacts in the city. It specifically aims to lower the existing access barrier for information. This system is inspired by the fact that asking

[4] Yahoo! Answers, `http://answers.yahoo.com/` - Last accessed: November 22, 2011 19:00 CET

[5] Yahoo! `http://www.yahoo.com` - Last accessed: November 22, 2011 07:00 CET

other people for advice or 'doing what other people do' is a seemingly popular form for navigating in new environments. It allows the user to get direct and indirect social navigation and information about a specific place of interest. A user can see other users' comments and rating about the respective places that are stored in the system. As opposed to this indirect communication, the system provides users with the option to establish a direct communication (such as, voice or short text message) with the residents of the city, who can provide them with direct advice. However the system does not consider the contexts of the information seeker and information provider.

VizWiz [10] is one of the most recent question answering system. It allows a blind person to recruit remote sighted workers to help them with visual problems. The sighted persons recruited by the blind person to answer the question are provided by workers on Amazon Mechanical Turk [6] . The system provides intelligent mechanism to recruit multiple workers before they are needed. It is a picture based mobile question and answering system for blind persons. It allows posing a question together with a picture of an object of interest, to an answer provider who is physically apart from the scene. The answer provider is a sighted person who can identify the object or text in the picture and can provide her reply with voice. Currently available automatic recognition technologies, such as image or character recognition are error prone and cannot provide reliable translation particularly when the picture is not so clear or the text or label is on some uneven surface, such as textual description or label of any canned food.

VizWiz system is one of the most interesting examples where an effective or accurate answer can only come from another human. This system is a partially mobile system, meaning that half of the system rely on traditional stationary computer systems. The recruited persons who are going to provide answer of any question asked by the user, typically are replying remotely from a stationary computer. VizWiz does not take into account the context of the user asking the question, nor the context of the sighted person helping or answering the question. Moreover, part of the system relies on the responses from the workers from Amazon. That means there is no common ground or social relationship between the user asking the question and the person replying; making the answer an issue of reliability and comfort.

Currently most of the existing approaches exploiting context awareness focus on the external context and there has been little work that addresses internal context. Though this has resulted in some useful context aware and activity aware applications and systems, most of these approaches don't consider more personal and social issues. Existing approaches also do not consider the unpredictable type and amount of human's need for fine-grained or too customized queries in mind that can be only addressed by other human being. The Ubiquitous–Help–System addresses this limitation by using social network, user preference, and presence of other user in the vicinity.

[6] Amazon Mechanical Turk, `http://www.mturk.com/` - Last accessed: November 22, 2011 01:30 CET

3 Our Approach: The Ubiquitous–Help–System(UHS)

3.1 Typical Usage Scenario

This section presents a typical usage scenario where UHS can help a user to find context aware help and guidance.

Fig. 1 A day in Linda's life. Linda has an appointment with someone in the university, which she has never visited before.

Linda is a regular commuter; she uses train to go to her office. She regularly passes by Brussels by train. It is somewhere in the middle on her way to the work from home. She has visited Brussels many times, but still this city is big enough for her to get lost! Today she needs to visit a university in Brussels. She has an appointment with Susan who works for the university. Linda has a rough idea about how to go to the university campus. But she wants to know some simple but practical information about the university; where exactly is it, how to get there, which transport is cheaper, faster or safer? She wants to know which direction she should follow after she has reached the nearest station (see figure 1, second from the top-left). Linda knows that she would be able to go to the university campus somehow but she also wants to know which path she should follow to get to the venue.

Linda was wondering if there is someone on the train who knows about that university (see figure 1, third from the bottom-left); Linda would feel rather lucky if she could find someone on the same train who is also going to the university! And in the office hours it is not a coincident to find someone with a similar or same destination. But she was wondering how!

This scenario illustrates the basic requirements that a social and context aware help system should be able to complement:

1. People need in-situ information or help that they cannot anticipate and plan for beforehand
2. Some practical information can only come from relevant persons
3. When people have a query, several other related queries usually follow
4. It is difficult to find and identify someone from the crowd who is available and reliable, and who can provide suitable help
5. Finding a help provider without interrupting her or him is a cumbersome task

3.2 The System Description

We have developed runtime software prototype the *Ubiquitous–Help–System (UHS)* that exploits two typical properties of end-users searching for help: *locality* and *common ground*. Locality implies that users often seek help and guidance related to the area they are located in or are heading to. For example, enquiring about the exact location of the baby food in a particular super market is less likely to appear in mind while someone is away from the super market. Similarly, people often want or need to know fine-grained information about destination station while they are on board towards the destination.

Asking a stranger for help is often difficult because people are usually hesitant to approach people they are not acquainted with. The second property, *common ground*, indicates users feel more comfortable receiving help and guidance of other users they have something in common with. UHS uses a combination of the social network of a user and the context of use of all users in the social network to seek for the most appropriate help and guidance.

Techniques originating from the semantic web initiative as proposed by Tim Berners-Lee [23] are used to connect different people with each other. A participant is identified by a Friend-Of-A-Friend (FOAF)[7] profile. This profile contains information about the social relations of the user, next to traditional information that identifies the user (such as name, address, hobbies, school or work environment,...). Figure 2 gives an example of a FOAF profile that is used by the UHS system. This small example shows that FOAF is a Resource Description Framework (RDF) [9] vocabulary for describing people and social

[7] Brickley D., Miller L. FOAF specification versions 0.1, `http://xmlns.com/foaf/0.1/` - Last Accessed: November 22, 2011 02:30 CET

networks. This allows us to easily query relations between different FOAF files and build a graph of related FOAF profiles (e.g. figure 3).

```
<foaf:Person rdf:ID="me">
<foaf:name>Noah Jonson</foaf:name>
<foaf:title>Mr.</foaf:title>
<foaf:givenname>Noah</foaf:givenname>
<foaf:family_name>Jonson</foaf:family_name>
<foaf:mbox_sha1sum>617e3e9d9ac9239dbd5dc26c411e20fd6a393d83</foaf:mbox_sha1sum>
<foaf:homepage rdf:resource="http://research.example.com/~noah/"/>

<foaf:knows>
<foaf:Person>
<foaf:name>Emma Jonson</foaf:name>
<foaf:mbox_sha1sum>b513b6f7a7718967523b69fb38133bd21a5a42a0</foaf:mbox_sha1sum>
</foaf:Person>
</foaf:knows>
```

Fig. 2 A social user profile using the FOAF scheme

UHS uses the process enlisted below to obtain help and guidance. The different steps of this process are:

1. A user has a question and seeks for guidance. The user can access the UHS using her mobile device and input the question.
2. The UHS connects with the UHS server and sends the question together with the context data and user profile to the server.
3. Upon receiving enquiry from any UHS client through the server, the receiving clients run a matching algorithm (see algorithm 1).
4. The algorithm takes input from two information sources: local information such as profile and preference information from its local device, which is basically information about the owner of the UHS, and information received from the enquiring UHS client which is the information about a user in the neighborhood who is seeking help.
5. Then the algorithm tries to find a match; upon finding any match the UHS client informs its user so that she can decide whether she wants to address the enquiry by answering using the available communication channel (e.g. the chat system included in UHS, instant messaging or even just talking with each other).

Figure 3 depicts this process where a nomadic user (black circle in the figure 3(a)) needs assistance and is seeking for guidance. She uses her UHS client (figure 3(b)) to connect to the UHS server. By exploiting FOAF the client sends her query to peers (gray circles in figure 3(a)) who are present in the vicinity and also share some common ground with the user. A peer (the gray circle with an X mark, in figure 3(a)) is found as a match and is ready to respond. It notifies its user about the query. The user responded and communication between the nomadic user, who needs some help and the relevant person who is willing to help begins.

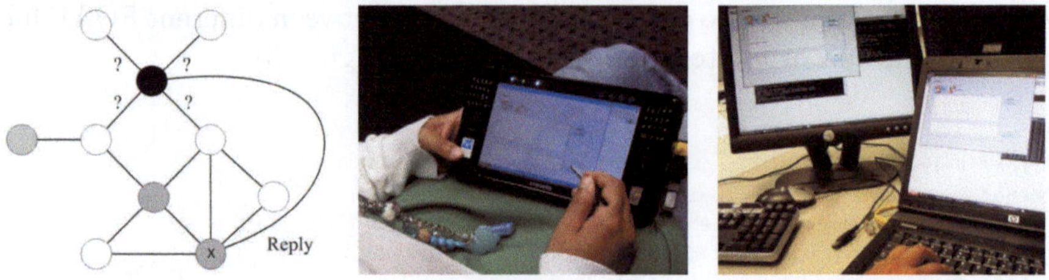

(a) (b) (c)

Fig. 3 The Ubiquitous–Help–System in action. 3(a) Visualizes the social network, a nomadic user (the black circle in the graph) needs assistance and is seeking for guidance. 3(b) Shows a mobile user querying for assistance and 3(c) Shows another user receiving the request.

Algorithm 1. UHS matching algorithm

```
Input (selfProfile, selfPreferene, rcvProfile, rcvPreference)
while queryReceived = true do
   Parse(selfProfile)
   Parse(selfPreferene)
   Parse(rcvUserProfile)
   Parse(rcvUserPreference)
   if isFOAF(selfProfile, rcvUserProfile) = true then
      if selfPreference.status! = busy then
         FindMatch(selfPreference, rcvPreference)
      end if
   end if
   if matchFound = true then
      NotifyUser(listWords,matchedRatio)
   end if
end while

FindMatch(selfPreference,rcvPreference)
selfKeywords ← Extract(selfProfile)
rcvKeywords ← Extract(rcvProfile)
matchFound = false
for each selfKeyword do
   for each rcvKeywords do
      if selfKeyword = rcvKeyword then
         matchFound = true
         countMatch ← countMatch + 1
         listWords ← addToList(selfKeyWord)
      end if
   end for
end for
matchedRatio ← countMatch/selfKeyWords * 100
return listWords, matchedRatio
```

In our approach, user profile (see figure 2) is stored on the UHS client device. User profile is described using FOAF, a descriptive vocabulary expressed using the Resource Description Framework (RDF) [9]. Each UHS client can process user profile and find friend of a friend in the neighborhood. This feature made it possible to avoid use of a centralized server, which is commonly used in social network systems where the server is responsible to determine relationship among the users.

The next section will provide more details on the architecture and the behavior of the separate components of the system.

4 UHS Architecture

The UHS prototype runs on mobile computing devices such as PDAs and laptops. It keeps track of the user's profile, preference, presence of other users in the vicinity, preferences of other users, current location and time. Figure 4 shows the basic structure of a client and communication with other clients in the neighborhood.

The client controller is responsible for maintaining all communication and computation. An HTTP/XML–based communication framework (see section 4.1) is used to facilitate a UHS client to communicate with other UHS clients. The UHS client can send and receive plain text for chatting and

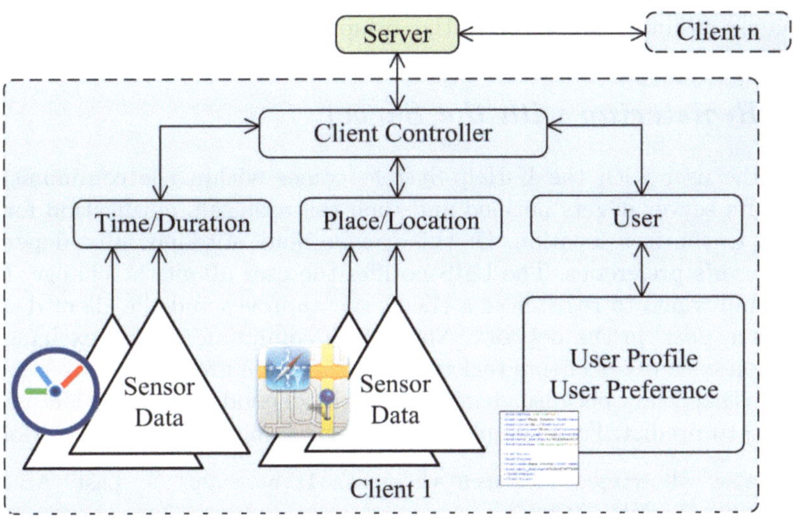

Fig. 4 The UHS client structure. *Client 1* is communicating with other clients, such as *Client n*. In order to assist the user, the client controller utilizes information from the user's profile, preference, current location and time.

also send and receive attached files such as a regular email. When a user needs help and asks a query, the controller sends the query to the other clients via the server. The controllers of the receiving clients initiate individual processing for a profile and preference matching. If it finds a match, depending on the users preference (such as interruptibility state) it notifies its user. The controller takes the user profile and preference (see figure 4.3) to compute the match together with the shared public profiles and preferences received from other users. The system is designed to utilize off–the–shelf "sensors" such as the time from the system, location from the location provider (such as GPS and Ubisense). The system takes a mixed-initiative approach, meaning that the user can always override the decision that the system perceived from the context.

4.1 Web to Peer

We have used web to peer (W2P) [34][8] as the communication framework. W2P is a middleware framework that enables easy communication among the peers in a network. W2P server is designed as a web application, which is deployable in any servlet container such as Tomcat[9] or Jetty[10].

In the communication framework, a W2P client announces itself as a peer in the network. As a client declares itself a peer it gets incoming and outgoing message channels. Then it can communicate with other clients registered in the same group via the W2P server. The server acts as a message gateway for all the clients registered in a group. A client in the vicinity of a server, are assumed registered to the same group e.g. all the clients inside the train wagon are assumed registered to the group called train-server.

4.2 Registering with the Server

When the user with the U-Help-System comes within the communication range of a server it gets notified and then the user gets notification for registering to the new location. Or this can be done automatically, depending on the user's preference. The UHS notifies the user about the change. If she agrees and wants to register as a client, she approves and the client declares itself as a 'peer' in the network. Now it can communicate and exchange information with other clients registered on the same group on the server.

This is necessary because vicinity or nearby depends on several factors that are hard to predict. For example, a person standing on the train platform is

[8] Available at `http://research.edm.uhasselt.be/~w2p/` - Last Accessed: November 22, 2011 02:30 CET

[9] Apache Tomcat, `http://tomcat.apache.org/` - Last Accessed: November 22, 2011 02:30 CET

[10] Jetty Web Server, `http://www.mortbay.org/jetty/` - Last Accessed: November 22, 2011 02:30 CET

nearby to the person sitting inside the train. But as soon as the train leaves the station this relation is no more valid. In such case, if a client is within the communication range of a server inside the train wagon, the user can effectively decline to registering to the 'train-server' and remain registered as peer in the 'station-server.'

4.3 Selecting Profile and Preference

The system can automatically select or suggest a profile, e.g. a profile for school or home. The user can edit her own preferences. Initially a set of criteria is defined, for example one criterion is hobby and there is list of hobbies that a user can select from the list. Alternatively, the user can also insert any other property that is supported by the FOAF specification, so our algorithm can process this information.

For each profile selected there are different kinds of preferences e.g., a profile called outdoor-profile may have a preference which might allow the user's friend to chat with her, but on the other hand another preference might not allow to do so.

The user preferences are specified by a XML description file that can be associated with (FOAF) profile. There is a predefined set of preferences to make it simpler, at the same time a user can also create custom preferences.

4.4 Querying

Users can initiate a query and broadcast it over the network; all the clients present in the vicinity and are registered to the same group will receive the query. The client will process the query to match with its user profile and preferences. RDF [9] query language SPARQL [32] is used to allow efficient queries over the distributed knowledge base scattered in different client devices within the vicinity. A matching algorithm (see algorithm 1) matches the profile and preference and suggests its user whether she should respond to the query or not. For example, a query for searching someone who can provide information regarding a local university should be matched with a profile containing the university name and with key words e.g. student or employee. At the same time the client program will also look at the preference for the interruptibility state. The client does not immediately respond to the matched query, rather it asks its user that someone needs help and she probably can respond (see figure 5).

4.5 Selecting Context

Context is not only location information. There are different notions of context which is not only location, rather location, time etc [33]. Presence of

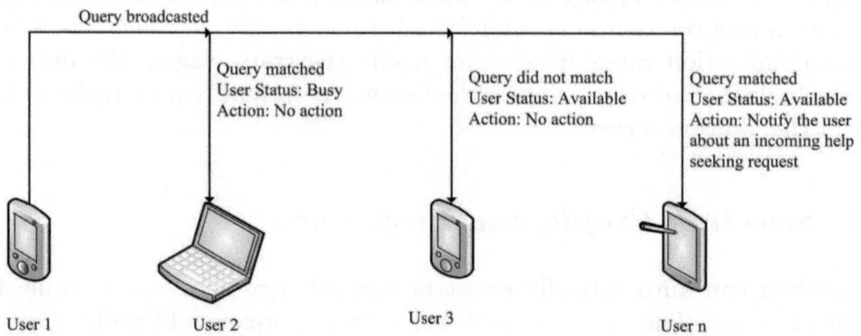

Fig. 5 Steps involved in a query. *User 1* has initiated a query that matched both with the profile and preferences of the *User n*. The system then notifies the user to respond to the query.

other users in the vicinity and the users preferences also play an important role in the interaction. Selecting a context would be done by exploiting location sensors, time and preference. In our current prototype, the location information is manually specified by the user. Eventually this limitation will be replaced by a location sensing module that exploits the combination of positioning systems, such as GPS, Wi-Fi and UbiSense[11] [8] technology. Once the system has its location information, it can query the database and can suggest or select a profile depending on the users preference. A user can allow the system to select a profile automatically or she can control this herself by selecting a profile from the suggested set of profiles. For example, the system will select a profile such as 'on the train,' while the user is on the train and she allows the system to select a profile automatically depending on the available context information. Otherwise, depending on the available context information the system may suggest:

- On the train
- On the way to the university
- On the way to an interview

And the user will be able to select one from the list.

The UHS was successfully used in a simulated networked environment of vehicular network using realistic data to observe efficiency of information delivery using informed decision on data routing where UHS provided the awareness, reported in [25, 36].

Furthermore, based on the UHS framework, we developed Ubiquitous Help System for Person with Dementia, a context and social aware mobile system to assist person with dementia, reported in [27]. The system works in a semi-automatic approach that provides navigational assistance and context aware

[11] UbiSense, `http://www.ubisense.net/` - Last accessed: November 22, 2011 01:30 CET

help to people with dementia. It provides help and enables awareness between the person with dementia and his caregiver. The system stores the type of social relationship for each contact and the current availability status of each person. When necessary this system dynamically finds a potential caregiver who can assist remotely based on the task that the person with dementia wants to complete. Furthermore, it provides help only when it is necessary, without unnecessarily patronizing its user.

5 Discussion and Future Work

Q&A systems based online social networks have recently gained a growing interest (such as ChaCha [12], Quora [13]). The most popular general purpose social networking site, Facebook has recently introduced 'Facebook Questions' [14] a social Q&A service which also gained researchers' interest. None of these solutions guarantee the timeliness of response. There is also a growing interest for crowdsourcing based Q&A systems. Crowdsourcing based Q&A systems cannot guarantee the reliability and comfort, since the workers at the crowdsourcing center do not share a common ground or do not have a social relation with the user asking for help. This solution is also unable provide in–situ help where physical presence of the 'help provider' is expected. How people ask question or provide answer to a question, widely varies depending on number of factors (e.g., their cognitive patterns, social orientation and culture [20]) that a crowdsourcing based Q&A system cannot address. A context aware, mobile and social aware help system can address this gap by providing information or help from reliable sources (e.g., from friends) in the context, in a direct and interactive way.

The UHS presented in our earlier work, reported in [26], shows how social awareness and context awareness together can be utilized to help nomadic people in finding reliable and fine–grained information comfortably. In large public spaces such as train stations, airports or train wagons, typically people are nomadic and need some situational information. Very often those queries are too specific. Moreover, the type of people varies widely, depending on different factors (e.g. age, profession, skill level, etc) that represent the inner context of a user. So, it is a bigger research challenge to tackle users and their activities in such open public places. However tracking a nomadic user is not only a privacy issue, but also a matter of viability. Therefore, helping the user by rapidly tracking her/his context, interest, need, preference, skill or ability, and providing appropriate information help is difficult. UHS took a semi-automatic approach to handle this. It uses algorithms to approximate, and at the same time asks the user to validate its choice.

[12] ChaCha, `http://www.chacha.com/` - Last accessed: November 22, 2011 01:30 CET

[13] Quora, `http://www.quora.com/` - Last accessed: November 22, 2011 01:30 CET

[14] Facebook Questions, `http://www.facebook.com/blog.php?post=411795942130`

The web is not situation-aware, and real–life scenario is yet too dynamic to be incorporated and made available for being searched. Moreover, web does not take the context of use into account. Therefore, a peer to peer interaction is necessary for many reasons. Not only for certainty, comfort and immediate feedback but also to accommodate people with different range of abilities. This is another reason why we often prefer to ask another human being compared to web based information searching.

This view of context and context-awareness differs from many of the previous approaches discussed in this chapter. In this work our focus was on creating context-aware applications that integrates awareness of the users' social network as a central component to enable context–aware search for *human* guidance. It takes into account both the internal context such as preference as well as external context such as location.

Acknowledgement. Part of the research at EDM is funded the Flemish Government. Funding for this research was also provided by the Research Foundation – Flanders (F.W.O. Vlaanderen, project CoLaSUE, number G.0439.08N).

References

1. Abowd, G.D., Atkeson, C.G., Hong, J., Long, S., Kooper, R., Pinkerton, M.: Cyberguide: a mobile context-aware tour guide. Wirel. Netw. 3(5), 421–433 (1997), `http://dx.doi.org/10.1023/A:1019194325861`
2. Abowd, G.D., Dey, A.K., Brown, P.J., Davies, N., Smith, M., Steggles, P.: Towards a Better Understanding of Context and Context-Awareness. In: Gellersen, H.-W. (ed.) HUC 1999. LNCS, vol. 1707, pp. 304–307. Springer, Heidelberg (1999)
3. Ackerman, M.S.: Augmenting organizational memory: a field study of answer garden. ACM Trans. Inf. Syst. 16, 203–224 (1998), doi: `http://doi.acm.org/10.1145/290159.290160`
4. Ackerman, M.S.: The intellectual challenge of cscw: the gap between social requirements and technical feasibility. Hum.-Comput. Interact. 15, 179–203 (2000), doi: `http://dx.doi.org/10.1207/S15327051HCI1523_5`
5. Ackerman, M.S., Malone, T.W.: Answer garden: a tool for growing organizational memory. In: Proceedings of the ACM SIGOIS and IEEE CS TC-OA Conference on Office Information Systems, COCS 1990, Cambridge, Massachusetts, United States, pp. 31–39. ACM, New York (1990), doi: `http://doi.acm.org/10.1145/91474.91485`
6. Ackerman, M.S., McDonald, D.W.: Answer garden 2: merging organizational memory with collaborative help. In: Proceedings of the 1996 ACM Conference on Computer Supported Cooperative Work, CSCW 1996, pp. 97–105. ACM, New York (1996), doi: `http://doi.acm.org/10.1145/240080.240203`
7. Ackerman, M.S., Wulf, V., Pipek, V.: Sharing Expertise: Beyond Knowledge Management. MIT Press, Cambridge (2002)
8. Aksenov, P., Luyten, K., Coninx, K.: Coping with variability of location sensing in large-scale ubicomp environments. In: ICUMT, pp. 1–5 (2009)

9. Beckett, D., McBride, B.: Rdf/xml syntax specification (revised). Tech. rep., W3C (2004), http://www.w3.org/TR/rdf-syntax-grammar/.
10. Bigham, J.P., Jayant, C., Ji, H., Little, G., Miller, A., Miller, R.C., Tatarowicz, A., White, B., White, S., Yeh, T.: Vizwiz: nearly real-time answers to visual questions. In: Proceedings of the 2010 International Cross Disciplinary Conference on Web Accessibility (W4A 2010), pp. 1–24. ACM, New York (2010), doi: http://doi.acm.org/10.1145/1805986.1806020
11. Bilandzic, M., Foth, M., De Luca, A.: Cityflocks: designing social navigation for urban mobile information systems. In: Proceedings of the 7th ACM Conference on Designing Interactive Systems, DIS 2008, Cape Town, South Africa, pp. 174–183. ACM Press, New York (2008), doi: http://doi.acm.org/10.1145/1394445.1394464
12. Nardi, B. A., Whittaker, S., Schwarz, H.: It's not what you know it's who you know. First Monday 5(5) (May1, 2000), http://firstmonday.org/htbin/cgiwrap/bin/ojs/index.php/fm/article/view/741/650 (last accessed: April 7, 2011)
13. Borgatti, S.P., Cross, R.: A relational view of information seeking and learning in social networks. Manage. Sci. 49, 432–445 (2003), doi: http://portal.acm.org/citation.cfm?id=970411.970435
14. Ehrlich, K., Shami, N.S.: Searching for expertise. In: Proceeding of the Twenty-sixth Annual SIGCHI Conference on Human Factors in Computing Systems, CHI 2008, pp. 1093–1096. ACM, New York (2008), doi: http://doi.acm.org/10.1145/1357054.1357224
15. Foner, L.N.: Yenta: A multi-agent, referral-based matchmaking system. In: Proceedings of the First International Conference on Autonomous Agents (Agents 1997), pp. 301–307. ACM Press (1997)
16. Gwizdka, J.: What's in the context? In: Computer Human Interaction CHI 2000 (2000)
17. Hertzum, M., Pejtersen, A.M.: The information-seeking practices of engineers: searching for documents as well as for people. Inf. Process. Manage. 36, 761–778 (2000), http://portal.acm.org/citation.cfm?id=350075.350093, doi:10.1016/S0306-4573(00)00011-X
18. Hodges, M.R., Pollack, M.E.: An 'Object-Use Fingerprint': The Use of Electronic Sensors for Human Identification. In: Krumm, J., Abowd, G.D., Seneviratne, A., Strang, T. (eds.) UbiComp 2007. LNCS, vol. 4717, pp. 289–303. Springer, Heidelberg (2007)
19. Hong, J., Suh, E., Kim, S.: Context-aware systems: A literature review and classification. Expert Systems with Applications (2008), http://dx.doi.org/10.1016/j.eswa.2008.10.071, doi:10.1016/j.eswa.2008.10.071
20. Yang, J., Morris, M., Teevan, J., Adamic, L., Ackerman, M.: Culture matters: A survey study of social qa behavior. In: International AAAI Conference on Weblogs and Social Media, 2011 (2011)
21. Kautz, H., Selman, B., Shah, M.: The hidden web. AI Magazine 18, 27–36 (1997)
22. Kekwaletswe, R.M., Ngambi, D.: Ubiquitous social presence: Context-awareness in a mobile learning environment. In: International Conference on Sensor Networks, Ubiquitous, and Trustworthy Computing, vol. 2, pp. 90–95 (2006), doi: http://doi.ieeecomputersociety.org/10.1109/SUTC.2006.142

23. Lee, T.B.: Weaving the Web. Harper, San Francisco (1999)
24. Long, S., Kooper, R., Abowd, G.D., Atkeson, C.G.: Rapid prototyping of mobile context-aware applications: the cyberguide case study. In: MobiCom 1996: Proceedings of the 2nd Annual International Conference on Mobile Computing and Networking, pp. 97–107. ACM, New York (1996), doi:http://doi.acm.org/10.1145/236387.236412
25. Mahmud, N., Aksenov, P., Yasar, A.-U.-H., Preuveneers, D., Luyten, K., Coninx, K., Berbers, Y.: Geo-Social Interaction: Context-Aware Help in Large Scale Public Spaces. In: de Ruyter, B., Wichert, R., Keyson, D.V., Markopoulos, P., Streitz, N., Divitini, M., Georgantas, N., Mana Gomez, A. (eds.) AmI 2010. LNCS, vol. 6439, pp. 107–116. Springer, Heidelberg (2010)
26. Mahmud, N., Luyten, K., Coninx, K.: Context aware help and guidance for large-scale public spaces. In: Proceedings of the 2009 Fourth International Workshop on Semantic Media Adaptation and Personalization, SMAP 2009, pp. 105–110. IEEE Computer Society, Washington, DC, USA (2009), doi:http://dx.doi.org/10.1109/SMAP.2009.18
27. Mahmud, N., Vogt, J., Luyten, K., Slegers, K., Van Den Bergh, J., Coninx, K.: Dazed and confused considered normal: an approach to create interactive systems for people with dementia. In: Proceedings of the Third International Conference on Human-Centred Software Engineering, HCSGE 2010, pp. 119–134. Springer, Heidelberg (2010), http://portal.acm.org/citation.cfm?id=1939212.1939225
28. McDonald, D.W., Ackerman, M.S.: Just talk to me: a field study of expertise location. In: Proceedings of the 1998 ACM Conference on Computer Supported Cooperative Work, CSCW 1998, pp. 315–324. ACM, New York (1998), doi:http://doi.acm.org/10.1145/289444.289506
29. McDonald, D.W., Ackerman, M.S.: Expertise recommender: a flexible recommendation system and architecture. In: Proceedings of the 2000 ACM Conference on Computer Supported Cooperative Work, CSCW 2000, pp. 231–240. ACM, New York (2000), doi:http://doi.acm.org/10.1145/358916.358994
30. Philipose, M., Fishkin, K.P., Perkowitz, M., Patterson, D.J., Fox, D., Kautz, H., Hahnel, D.: Inferring activities from interactions with objects. IEEE Pervasive Computing 3(4), 50–57 (2004), doi:http://dx.doi.org/10.1109/MPRV.2004.7
31. Prekop, P., Burnett, M.: Activities, context and ubiquitous computing. Computer Communications 26(11), 1168–1176 (2003), doi: http://dx.doi.org/10.1016/S0140-3664(02)00251-7, doi: 10.1016/S0140-3664(02)00251-7
32. Prud'hommeaux, E., Seaborne, A.: SPARQL query language for RDF. Recommentation, W3C (2008), http://www.w3.org/TR/rdf-sparql-query/
33. Schmidt, A., Beigl, M., Gellersen, H.W.: There is more to context than location. Computers and Graphics 23, 893–901 (1998)
34. Vanderhulst, G., Luyten, K., Coninx, K.: Middleware for ubiquitous service-oriented spaces on the web. In: AINAW 2007: Proceedings of the 21st International Conference on Advanced Information Networking and Applications Workshops, pp. 1001–1006. IEEE Computer Society, Washington, DC, USA (2007), doi:http://dx.doi.org/10.1109/AINAW.2007.248
35. Want, R., Hopper, A., Falcão, V., Gibbons, J.: The active badge location system. ACM Trans. Inf. Syst. 10(1), 91–102 (1992), doi: http://doi.acm.org/10.1145/128756.128759

36. Yasar, A.U.H., Mahmud, N., Preuveneers, D., Luyten, K., Coninx, K., Berbers, Y.: Where people and cars meet: social interactions to improve information sharing in large scale vehicular networks. In: Proceedings of the, ACM Symposium on Applied Computing, SAC 2010, pp. 1188–1194. ACM, New York (2010), doi:http://doi.acm.org/10.1145/1774088.1774339
37. Zhang, J.: Understanding and augmenting expertise networks. Ph.D. thesis, The University of Michigan (2008), http://hdl.handle.net/2027.42/58450

Influential Users in Social Networks

Dimitrios Vogiatzis

Institute of Informatics and Telecommunications,
NCSR "D" and The American College of Greece
e-mail: dimitrv@iit.demokritos.gr

Abstract. A study of the influential users in online social networks is the focus of this work. Social networks expand both in terms of membership and diversity. User driven content creation is growing, and yet this information potential remains largely untapped. Future search engines focusing in social networks should take into account both the content and the structural properties of the nodes. Whereas a social network bears a superficial similarity to the Web, it is different in the sense that it connects primarily individuals rather than pages of content. Not all individuals are equally important for any given task, therefore the influential ones should be detected, in that vein we review facets of influence in social networks.

1 Introduction

An online social network is conceived as a web-like structure of interacting nodes. The links might be formally established, such as those set explicitly in facebook, twitter and other similar social networks; or they might be implicit such as those formed by email exchanges, postings in fora, co-authorship of documents etc. By virtue of the content created, and knowledge possessed by each user, as well as by the user interactions, an enormous nexus of information is created. Social networks have always been present in the form of groups of acquaintances, co-workers, alumnae, etc. and they were the subject of study of various disciplines. A relatively recent change is the growth of online social networks, in terms of users' participation and diversity. Making sense out of this information can be a formidable challenge. In the current study, we are interested in various aspects of influential users in an online social network (SN). Next, we will try to define the term influential by focusing on its purpose or usefulness.

An influential user might be an expert user, that is someone that possesses extensive knowledge on a topic or range of topics. For instance, in a SN that focuses on movies, the experts will be more knowledgeable about movies, actors, directors etc. than the rest of users. Thus, if movie is to be recommended to a new user, the knowledge of the experts might be influential.

I.E. Anagnostopoulos et al. (Eds.): Semantic Hyper/Multimedia Adaptation, SCI 418, pp. 271–295.
springerlink.com

In a different setting, we could imagine a company promoting a product or service in a SN with the purpose of maximising product penetration, while minimising the promotion cost. The most influential users would be the prime target, because they could spread the furthest the news of this product and possibly purchases of the product of service would reach the maximum possible level. Thus users of high network value are sought.

A corporate or an academic establishment contains the so called tacit knowledge, that is knowledge, skills, abilities that have not been explicitly recorded in databases, and yet they exist in the minds of people that are employed in such an environment. A SN would connect those people, but some computational procedures must be in place for detection of foremost experts in a given topic.

Thus the point of the current paper is, to report on work in influence on online social network from different perspectives. Notably, the focus is on the location of influential users in a social networks (i.e. their structural properties). Search procedures for locating influential users, calculation of influence based on the content generated, and dynamic phenomena are also reported.

The rest of paper is organised as follows: In Sect. 2 areas of research related to influence are reported. Then, in Sect. 3 we refer to the characteristics of influence as well as to methods of deriving it. In Sect. 4, we refer to existing systems or approaches that implement the methods mentioned in the previous section. Next, in Sect. 5 we refer to counter arguments against the usefulness of influential users, which will helps us frame the limits of expertise. Conclusions are drawn in Sect. 6.

2 Related Work

A concept that is related to that of influence is trust. In a social network that involves promotion of products or services, trust is of paramount importance. For instance, many models have been developed to ensure reliable transactions in online auctions. An overview of trust models can be found in [26]. Whereas, it cannot not be denied that trust and influence are to certain degree intermingled, they are not however identical. Thus, it would be reasonable to assume that an information source, say a user in an electronic community, that is trustworthy is eventually more influential. However, it can be claimed that a deeper understanding of the two concepts involves their separate study. In the study of influence generation, and spread of it, we consider different models from those of the trust, the models are more "macroscopic". In the study of trust two dominant models, between interacting parts, have been developed: the cognitive and the game theoretic. According the former, there are underlying beliefs in the interacting agents, and their behaviour depends on those beliefs. The game theoretic model, considers probabilities, tries to optimise objective functions and takes into account past history. We do not adopt such models when dealing with influence. Finally, we would place reputation as a concept related to trust, a social version of trust perhaps, which is also something out of scope in this study.

Web search engines create indexes of documents at very efficient ways. Until recently the focus of search engines was on documents and not on people. This has started to change with a plethora of search engines that search for people in social networks and elsewhere, such as: PIPL, [1] Snitch.name, [2] and Folowen, [3] to name a few. Some of the engines have gone a little further, for instance NNDB Mapper [4] builds a graph of connections for people that are contained in the NNDB database. Whereas, Microsoft Academic Search [5] builds a similar graph of connections in the domain of scientific publications, and also provides two widely accepted measures that are indicators of influence or importance in scientific publications, i.e. G-Index, and H-Index.

An earlier review of expert finding systems has been undertaken in [29]. The current review expands on that in several respects. First, we consider the broader issue of influence as well as we refer to models of diffusion of influence. Second, we examine the issue of influence in the context of social networks, where graph structural properties affect search for experts. Finally, we consider dynamic properties of social networks and their consequences for influential users.

3 Facets of Influence in Social Networks

The simplest approach to implement and yet the hardest to maintain is the explicit, manual recording of influential users in the sense of experts. Thus, in a corporate environment there are could be indices of employees along with their expertise expressed as a series of keywords. It is up to the employee to submit a list of terms that describes his expertise (see also [7]). However, there are limitations to this approach that are related to the reliability of creating such a list, as well as keeping the list updated. Moreover, it presupposes a central registry to enforce the maintenance of this list which is not available in informal electronic communities.

In the current work, we consider automated processes for the discovery of influential users. The substrate that connects users is an online social network, that is represented as a graph where nodes correspond to users and links represent interactions or relations. Influence is a property that refers to the *network value* of a node, and depends on its connections, or on its location in the graph. Moreover, it can also depend on the knowledge that the node represents or is related to.

We address the following issues: what leads the influential users to be productive? how can we search for influential users? what is the computational complexity of search? is there also a social cost? what are the structural properties of influential users? how can the influence be associated with the content created?

Influence in social networks can be analysed from different perspectives; the point of the current section is to detect parameters or facets of influence that will

[1] `http://www.pipl.com/`

[2] `http://snitch.name`

[3] `http://folowen.com/`

[4] `mapper.nnedb.com`

[5] `http://academic.research.microsoft.com/`

allow a systematic classification of existing systems. The following have been identified: *search for influential users, structural properties of influential users, content based computation of influence, dynamic properties of social networks*, and *mechanisms for diffusion of influence*. The parameters of influence are analysed in the following subsections. Finally, the facets of influence are summarised in Table 2.

3.1 Search for Influential Users in a Social Network

Locating influential users usually requires a search in the SN graph. In that, a query is issued from the seeker, and is forwarded to neighbouring users, until a node is reached, whose expertise is relevant to the query. A method for neighbour selection is of major importance in a social network, since the edges between nodes (users) are meaningful, and they form the structure of the network. Consequently neighbour selection during the search can influence the accuracy in expert discovery.

Some of the strategies that have been suggested are the *breadth first search*, the *random walk* search, the *best connected* search, the *weak tie* search, the *strong tie*, the *hamming distance* search, and *information scent* search (see Sect. 4.8). Other search strategies that have been used include: the *PageRank*, *HITS* and *z-score* (see Sect. 4.3).

The performance of the search strategies depends on the structural properties of the social network. Moreover, the aforementioned search strategies can be evaluated in terms of success rate in finding an expert user, who is able to answer the query. The performance, should take also into account the length of the path from the seeker to the expert, the shorter the better. The determination of whether someone is expert in the query topic can be based on the content generated by the expert.

The nature of the search in a SN may also involve human labour. For instance, each query reaching a node might involve some human processing in order to answer it. Thus the social cost of query processing should be taken into account, which can be analysed as: the number of candidate experts involved in the query processing, or how often a candidate expert is consulted (see Sect. 4.8).

In another case, influential users might be sought because they are the ones to be targeted in a direct marketing case. The objective would be to minimise marketing cost and to maximise spread of influence in the sense of conversion of neighbours, and neighbours of neighbours, of the influential users to a new product or service. Given the linear-threshold model, or the cascade model of influence diffusion, it is NP-hard to optimally determine the most influential users. However, a greedy hill-climbing strategy can be used to find a solution to this problem that is at most 63% of the optimal (see Sects. 4.5, 4.8).

3.2 Structural Properties of Influential Users

Given that users are represented as nodes in a graph, we are interested in the structural properties of nodes. The connectivity (or degree centrality) of a node , might refer to the number of incident edges, the direction of edges (i.e. the in-degree or the out-degree), as well as to the weight of the edges.

It is not however the case that the most influential are necessarily the most connected ones. Connections might denote popularity because of expertise, but there are cases where the users of a SN might be stratified in a hierarchy. Interaction might be more probable between neighbouring levels of the hierarchy, and improbable between distant layers. Thus a user at the top of the hierarchy, might be the overall expert, and might allocate his time among users that are slightly less expert than he is (see Sect. 4.3). Moreover, the status of a node in an organisation also confers influence (see Sect. 4.7).

Another important property is value of k-shell decomposition k_s. The k-shell decomposition proves to be a most relevant measure of the influence of a single node. High values of k_s refer to nodes at the core of the SN. On the other hand degree centralility characterises nodes that can be found at the periphery of a social network (see Sect. 4.4).

3.3 Content Based Computation of Influence

The content created by the users in a social network can be used as a measure of their influence or expertise. As a matter of fact, we will be using the two terms interchangeably.

First, let us assume that the users rate the items they interact with, and that each item can be decomposed into a set of features. Influential or expert users can be discovered for *specific feature values*. There is not a dichotomy between influential and non influential users, but rather grades of influence. Each item, will possibly have different experts for its different feature values. The *level of influence* of a user will be determined by the depth or the extend of his interaction with that feature [8]. In the estimation of the influence domain specific knowledge will be taken into account (see Sect. 4.9).

Content based detection of influential users has also been studied in [24], which stipulates that influential (or trustworthy) users are the ones that can predict accurately the desires of other users, when asked to do so in a recommender system. In particular, the influence of a user concerns the predictions made over all items: *user profile level trust*, or the predictions for a specific item: *item level trust*, (see Sect. 4.9).

Another similar measure of expertise is the so called *contribution*, which essentially expresses the proximity of the predictions of the expert against the desires of the users [8], it can be derived from the following linear model: $f : E \times C \to P$, where P is the prediction of a user's preference, E is the expertise of the said expert (which can be domain dependent), and C is the sought contribution [8] (see Sect. 4.9). Moreover, a related approach has been followed in [6] where candidate experts are sought and combined with similar users in a hybrid expert and collaborative filtering system for movie recommendation.

Assuming that the items for which expertise is sought cannot be reduced into a series of features, and assuming that ratings of items by the users are not available, then a different approach is needed. For instance, in an academic or corporate environment documents of various sorts are generated. In this context, a differentiation can be made between *expert profiling*, that is building a model of what the expert user knows about all topics, and *expert finding*, that is discovery of an expert in a given topic [3]. The seeker of expertise issues a query of terms, that constitutes a topic, and he eventually obtains the experts that are related to the topic. Notably, three text data techniques have been developed to produce a ranking of the users based on their expertise: $p(ca|q)$, where ca is a candidate expert and q is the query. The first technique, models the knowledge of the expert from the documents he is associated with. The second one, finds the documents relevant to a topic and then discovers the relevant experts. The third one, directly discovers the candidates relevant to a query. In the aforementioned models there is the need to associate users with documents. This can be achieved by considering the author of a document, or the names mentioned within the document.

Existing organisational ontologies can be exploited in the process of expert discovery by enhancing the models mentioned in the previous paragraph. First, documents which are "close" enough in the ontology are deemed similar and this information can be used to increase the expertise level of an expert. Second, the position of a user in the organisational hierarchy, can also be used to enhance the above models especially in the context of sparse data (see Sect. 4.7).

3.4 Dynamic Properties of a Network of Influential Users

A social network is not expected to be static, rather it evolves in many and different ways. Nodes can be added or deleted; edges also can be also be removed, added or rewired. The evolution of a social network is closely related to its structure. In particular, the degree of the nodes can play an important role. Thus, according to the Barabasi Model there a *preferential attachment* of novel nodes to high degree nodes [25]. The above affects the structural properties of the network and consequently it will affect influence measures that depend on the graph properties of the network, as well related search procedures for influential users. Additionally, there are models that describe the deletion of nodes, edges, as well as models that introduce an aging parameter for edges (see Sect. 4.12).

Moreover, there are studies that relate the productivity of user nodes to there structural properties. Thus, the more productive a user node is the more attention he receives from other users in the network, which could be represented as links from the network population to the producing user. In addition, decreased attention leads to decreased productivity [13] (see Sect. 4.10).

Further studies have shown that not all links are the same. There are links from followers and links from friends to an influential user, where friends are the ones with which the user has interacted with. Differentiating between the kind of links is important, because an increase in the number of followers leads to an increase in the number of posts produced by the influential user up to a certain point. On the other hand, an increase in the number of friends increases the number of posts, which does not get saturated [14] (see Sect. 4.11).

3.5 *Diffusion of Influence*

The spread of news, ideas, epidemics etc., from one or multiple points of origin can be simulated with diffusion models. Such models have their origin in physics, sociology, epidemiology and elsewhere. In this study we are interested in models that have been designed, implemented, and evaluated in online social networks, and in particular model that operate at discrete time steps.

The *linear threshold model* stipulates that an inactive node v will be activated provided the weighted sum of its neighbours exceeds a certain threshold: $\sum_w b_{v,w} \geq \theta_v$, where w represents a neighbour of v, and θ_v is the threshold of v (see Sect. 4.5). In real networks, the threshold is not given, and an important issue is to estimate it. In that vein, an approach based on active learning with a weighted sampling is proposed [5]. This method has been evaluated on some data sets from the Pajek site [6].

The *independent cascade model* assumes the existence of an initial set of active nodes. A given active node v can activate each of its neighbours w with a probability of $p_{v,w}$ and there is a single chance of activation (see Sect. 4.5).

One node is considered infected, and all other nodes are in the susceptible state in *susceptibility interference recovery (SIR)* model. A node will infect its neighbours with probability β, and subsequently the node will enter in the recovery state, in which it can no longer infect. This model is almost identical to the independent cascade model (see Sect. 4.4).

In the *susceptibility interference susceptibility (SIS)* model, the probability of a node to infect each of its neighbours is β. But the inflecting node will remain in the susceptibility mode with probability λ (see Sect. 4.4).

Moreover the form of diffusion or the information cascade as it is also know, is investigated in [22]. The authors discovered the most frequent information cascades at an unnamed data set that included purchases of four product categories, where also users could recommend each other products. They discovered that most frequent cascades are small, their frequency depends on the product type, and they approximately follow a heavy-tailed distribution.

4 Existing Influence Calculation Systems

Various approaches of SN analysis are presented that incorporate the facets of influence that were noted and analysed in the previous section. It should be mentioned that the presentation is meant to be representative rather than exhaustive. Moreover, Table 1 displays the actual SNs that were investigated and the issue that was under study.

[6] `http://vlado.fmf.uni-lj.si/pub/networks/data/`

4.1 Referral Web

One of the earliest and influential attempts to experts discovery in a social network is the Referral Web project [15], [16]. The point was to discover a path form the seeker of information to an expert who may be able and is willing to respond. The path will help the seeker to establish some level of trust to the expert, and the expert through the path will see its association to the seeker and this will affect his willingness to respond.

The social network in question, is an implicit one that can be constructed by analysing web documents to find relations between people. Once a user (the seeker) enters the Referral Web system, a search engine, which may not part of the system, will search for documents created by the said user, or documents in which the said user is mentioned. In those documents other users are mentioned, and a link can be created from the seeker to those users provided their names are proximal. At a later stage, some of the links from the seeker to the other user can be pruned by setting a threshold on the Jackard coefficient on all pairs of nodes. As more users join the system, the network is expanded.

Two important issues are modeled: the further, in number of hops, the expert is from the seeker A, the less probable is to have accurate information: $p(A,d) = A^{ad}$, where d is the number of hops and α is a scaling factor. The second issue refers to the responsiveness of the expert, the closer he is the more likely to respond: $p(R,d) = R^{\beta d}$, where β is a scaling factor. Naturally, the A, R parameters have to be set to values relevant to the said social network.

4.2 Generative Language Models

Supposing that a graph represents the interactions of a group of people, ranking the involved people based on their expertise with graph based measures is addressed in [9]. It is a multi step process that has been applied in exchanged emails at IBM as well as to synthetic data. In the first step, email were classifying according to topic, based on some keywords. In the second step, undirected edges were drawn between the nodes that have exchanged email. The third step, involved providing direction to the edges formed during the previous step. An edge points from of node of lesser expertise to a node of higher expertise. Obtaining those edges involved using a trained a classifier to tell the direction of the edge. Finally, on the directed graph which represents the relative expertise of pairs of people, 4 different graph based algorithms are applied to compute the ranking of all users. These are: the *affinity, successor, PageRank, positional power function* and *HITS*. Affinity refers to the number of times a node has served as direct expert to other nodes, in other words a path of length one connects the expert from the one receiving advice. Successor is the same as affinity, but the said paths can be of any length. PageRank judges a page by number of links that point to it from other pages, the more the better. But the pages providing the links are not considered of equal importance, but rather the same criterion of number of links is applied to them. The adapted HITS assigns two scores to a node, the hub score and the authority score. The authority is presented as a criterion of expertise. The positional power function is defined as,

$$r_i = \sum_{j \in S_i} \frac{1}{n}(r_j + 1) \tag{1}$$

where S_i denotes the successors of node i, and r_j are their ranks.

An evaluation was performed for step three (determination of edge directions), on an email data set from IBM, that involved some classifiers, including an ideal classifier (i.e. correctly identifying the direction of edge in 100% of the cases) and a classifier based on maximum entropy (ME). For the ideal classifier the PageRank algorithm provided the best overall performance, whereas in the case of ME the all four algorithms had similar performance.

4.3 Bow Tie Structure

Graph based algorithms are evaluated at a Java forum to rate experts, in addition the behaviour of the users of the network is modeled in order to gain a deeper understanding of the working of the network and by extension on the performance of algorithms [31]. A graph has been formed to represent the postings (i.e. answers and replies) in a Java forum where nodes represent users, and edges are drawn from the seeker making to the users who have replied to it.

The *bow tie structure* is a good model of this forum; this structure has three parts: the *in*, *core* and *out*. In the Java forum, *in* refers to users that mostly post questions, in the *core* there are users that post questions or replies to other users of the *core*, and in *out* there are users that help core users. The first remark stemming out of this structure is that most users are located in the *in* part. A similar bow tie analysis of the Web suggests that users are located in equal analogy in all three parts of the said structure. A second remark is that a few users answer many questions, and many users answer a few questions. A third remark is that the high in-degree nodes tend to answer questions from whoever asks them, thus no assortative behaviour is observed. Finally, low in-degree nodes do not to reply to high in-degree nodes.

The graph based ranking algorithms that have been compared are the Z-score measure, Expertise Rank and HITS authority. The gold standard for comparison has been set by human labour, which sets users into 5 levels. All the above algorithms are closely correlated with that of the human rating, but the best results stem from the Z-score which is simpler than PageRank and HITS. To explain the above results two models of the users's behaviour have been developed.

First model. The probability p_A of a user u with expertise level $L(u)$ of asking a question is modeled as,

$$p_A(u) = \frac{[L(u)+1]^{-1}}{\sum_v [L(v)+1]^{-1}} \tag{2}$$

The probability of user u to reply to a question posed by user a is modeled as,

$$p_R(u,a) = \frac{e^{L(u)-L(a)}}{\sum_v e^{L(v)-L(a)}} \tag{3}$$

It has been shown that (3) is a good model of the real Java forum data. That is data produced by this equation (i.e. synthetic data) have similar structural properties to the real Java forum data (i.e. similar bow tie structure). And the performance of ranking algorithms on the synthetic data is similar to that of the real data, i.e. Z-score is better than PageRank or HITS.

Second model. Takes into account that the experts being under time constraints, might answer the questions that make best use of their expertise. This is interpreted that expert users tend to answer users' questions that are slightly less expert than they are. This is captured by the following equation,

$$p_H(u,a) = \frac{e^{L(a)-L(u)}}{\sum_v e^{L(a)-L(v)}}, \quad when\, L(u) > L(a) \tag{4}$$

In this case PageRank has the best performance, and that is because the links do not point to the highest experts, but rather to users with higher expertise. That is because expertise of the lower level experts is propagated to higher level experts. On the other hand HITS is worse than PageRank because novices are helped by medium level experts, who have a low hub score, and they confer a low authority score to the ones helping them.

4.4 K-Shell and SIS

Graph theoretic measures are considered in [18] [19] to detect influential nodes. The influence of node i with degree k, and k_s value in k-shell decomposition, according to the susceptibility interference recovery (SIR) model is,

$$M(k_s,k) = \sum_{i \in Y(k_s,k)} \frac{M_i}{N(k_s,k)} \tag{5}$$

where M_i, is the size of the affected population, $N(k_s,k)$ is the number of nodes with values k_s and k, and $Y(k_s,k)$ is the set of all nodes with values k_s and k, $M(k_s,k)$ represents the average size of the population that is affected.

It has been experimentally shown that the value of $M(k_s,k)$ varies greatly with k_s, on the other hand the value of k is rather inconsequential. Similar results have been obtained from the study of $M(k_s,C_B)$, where C_B denotes the in-betweenness centrality of a node. Thus the in-betweenness centrality or the degree of a node are not good predictors of the influence of a node, whereas the value k_s is indeed a good predictor. The is that high values of k, or C_B characterise nodes that are mostly at the periphery of a network; on the other hand the k_s is a very good indicator of the centrality of node and thus of its influence.

The next study concerned the spread of influence once not a single node is considered, but rather multiple nodes act as initiators of influence. The experiments suggest that choosing the nodes with the highest k_s is worse than choosing nodes with high k. The reason for that is the nodes of high k_s tend to occur next to each other. The remedy suggested, is to select nodes of high k_s that are not directly linked to each other.

In the Susceptibility Infection Susceptibility (SIS) model, a node can pass from susceptibility to infection, and then back to susceptibility without ending up in an immune state. The SIS model describes systems that reach a dynamic equilibrium state, in which the number of infectious individuals that reach the susceptible state equals the individuals that pass from the susceptible to the infectious state. The diffusion of the influence of node i, denoted as $\rho_i(t)$ is the probability that node i is infected at time t. In the final dynamic state $\rho_i(t \to \infty)$ becomes independent of time t. Thus, the asymptotic probability that nodes of k_s, k values are infected is given by,

$$\rho(k_s, k) = \sum_{i \in Y(k_s,k)} \frac{\rho_i(t \to \infty)}{N(k_s, k)} \tag{6}$$

where only the nodes of high k_s are important in the spread of influence or infection.

4.5 Computational Complexity of Influence Diffusion

One of the studies in influence diffusion approached the problem of what is the initial set of users that should be targeted in a hypothetical marketing action in order to maximise adoption of a product (see section 4.6). Since the discovery of the optimal solution is an NP-hard problem, the authors came up with an approximate greedy algorithm.

Next, the computational complexity of the greedy algorithm was estimated in [17], under the linear threshold and the cascade models. The first result was that the optimal solution for influence optimisation can be approximated to within a factor of $1 - 1/e - \varepsilon$, where e is the basis of natural algorithms and ε is any positive real number.

The authors were interested in an experimental evaluation of the greedy algorithm against three algorithms that perform, high degree, distance centrality and random spread of influence for the linear threshold and the weighted cascade models. In the high degree approach, the highest degree nodes are targeted first. In the distance centrality, the nodes that have short paths to all other nodes are targeted, finally in the random approach nodes are randomly selected. The social network in question is the co-authorship network in high energy physics and in theoretical computer science. The results show that the greedy method outperformed all other methods including the one with high degree nodes, showing that high degree nodes should not necessarily be the prime target for direct marketing.

4.6 Spread of Influence

If information of a product or service was to reach maximum spread in a social network, which is the initial group of users that should be target, given that each user can influence its neighbours and targeting a single user has a certain cost? Alternatively, this can be stated as how to take advantage of the "word of mouth" with minimum cost and a maximum reach? In [10] this problem has been modeled as follows,

$$P(X_i|\mathbf{X}^k,\mathbf{Y},\mathbf{M}) = \sum_{C(\mathbf{N}_i^u)} P(X_i,\mathbf{N}_i^u|\mathbf{X}^k,\mathbf{Y},\mathbf{M}) \tag{7}$$

where X_i is a boolean variable denoting whether user i purchased the service or product that is been promoted, and this depends on the features of the product $\mathbf{Y}$, on the subject of users $\mathbf{X}^k$ for who it is known whether they purchased the product and on the market actions taken denoting by $\mathbf{M}$, where $\mathbf{M}$ is a boolean vector representing whether a marketing or promotion action has been carried out for each user. $\mathbf{N}_i$ represents the neighbours of user i, and $\mathbf{N}_i^u$ are the neighbours for who it is unknown whether they purchased the said product, and $C(\mathbf{N}_i^u)$ represents all possible purchase decisions of the neighbours of i. Eventually, the above formula involves the calculation of various probabilities, including the: $P(X_i|\mathbf{N}_i)$. We concentrate on that because it represents the influence of the neighbours to user i. The calculation of the above quantity is application specific, and the authors provide a example of calculation given a collaborative filtering system that involves the movie lens data set. Moreover, it is important to notice that in 7 the summations are exponential in the number of neighbours of i, thus approximations must be used.

Next, the *expected lift in profit* from marketing to customer i, that is without taking into account the influence of his neighbours is defined as follows,

$$\begin{aligned} ELP_i(\mathbf{X}^k,\mathbf{Y},\mathbf{M}) = r_1 P[X_i = 1|\mathbf{X}^k,\mathbf{Y},f_i^1(\mathbf{M})] - \\ r_o P[X_i = 1|\mathbf{X}^k,\mathbf{Y},f_i^0(\mathbf{M})] - c \end{aligned} \tag{8}$$

where $f_i^1(\mathbf{M})(f_i^0(\mathbf{M}))$ denotes that the said product has been marketed (not marketed) to user i expressed as $M_i = 1, (M_i = 0)$, c is the cost of marketing, r_i (r_0) is the revenue if marketing is (not) performed

The lift in profit for the whole network after performing a marketing action M is defined as,

$$\begin{aligned} ELP(\mathbf{X}^k,\mathbf{Y},\mathbf{M}) = \sum_{i=1}^{n} r_i P(X_i = 1|\mathbf{X}^k,\mathbf{Y},\mathbf{M}) - \\ r_0 \sum_{i=1}^{n} P(X_i = 1|\mathbf{X}^k,\mathbf{Y},\mathbf{M_0}) - |\mathbf{M}|c \end{aligned} \tag{9}$$

where $\mathbf{M_0}$ is a vector of zeros, denoting that no marketing action has been performed.

The issue is to find a configuration of $\mathbf{M}$ that maximises the *ELP*, this would require trying $2^{|\mathbf{M}|}$ different assignments, which is intractable. Instead some heuristics have been proposed:

Greedy search: Start with $\mathbf{M_0}$. Set $M_1 = 1$, if the ELP raises, keep it, otherwise discard it and proceed with M_2 up until $\mathbf{M}_{|\mathbf{M}|}$. Then start from M_1 again, the process stops when there are no more changes in ELP.

4.7 Generative Language Models

Next, we examine an approach for the discovery of influential users based solely on the content generated by them or related to them. Thus no information about an explicit or implicit social network is taken into account. A document set from the University of Tilburgh [7] is considered, which is bilingual and includes documents of various types such as homepages, publications and course descriptions. In the current work the influential users are defined to be the experts of a domain [3].

It is assumed that a seeker of experts will issue a query to an expert finding engine. Expertise retrieval can be reduced to two closely related process: the *expert finding* and the *expert profiling*. The former means to discover users that are knowledgeable about a specific topic, whereas the latter refers to finding the topics a user in knowledgeable about. An approach is to relate users with topics, assuming that the users author documents that contain the said topics. Three different generative language models are employed, and at later stage they are enhanced. The expert finding is reduced to estimating the probability $p(ca|q)$, where ca is a candidate expert and q is a query posed to the system by a user,

$$p(ca|q) = \frac{p(q|ca)p(ca)}{p(q)} \tag{10}$$

Expert profiling involves calculating the following,

$$profile(ca) = [s(ca,k_1), s(ca,k_2), \ldots, s(ca,k_n)] \tag{11}$$

where k_i represents a topic; $s(ca,k_i)$ represents ca's expertise on topic k_i and is computed as $p(k_i|ca)$.

Three models have been developed to discover candidate experts ca given a topic set by the query q terms.

Model-1. Models the knowledge of and expert from the documents he is associated with; thus a *candidate language model* θ_{ca} will be inferred for query q, given that t are the terms of the query, $n(t,q)$ is the frequency of term t in query q,

$$p(q|\theta_{ca}) = \prod_{t \in q} p(t|\theta_{ca})^{n(t,q)} \tag{12}$$

[7] `http://ilk.uvt.nl/uvt-expert-collection/`

Model-2. Discover the documents relevant to a topic and then discovers the relevant experts. The probability $p(q|d)$ is estimated with a *document language model* θ_d

$$p(q|ca) = \sum_d p(q|d)p(d|ca) \tag{13}$$

$$p(q|\theta_d) = \prod_{t \in q} p(t|\theta_d)^{n(t,q)} \tag{14}$$

Model 3. Directly discovers the candidates relevant to a query, that is it estimates $p(ca|q)$. This is described by the following equation, where θ_k is a query model, θ_{ca} is candidate expert model, and KL is the Kullback-Leibler divergence,

$$KL(\theta_k||\theta_{ca}) = \sum_t p(t|\theta_k) log \frac{p(t|\theta_k)}{p(t|\theta_{ca})} \tag{15}$$

$p(t|\theta_k)$ is estimated with the following equation, $p(t|\theta_{ca})$ is estimated as in 12.

$$p(t|\theta_k) \approx p(t|q) = \frac{p(t,q_1,\ldots,q_m)}{p(q_1,\ldots,q_m)} \tag{16}$$

In the above models an estimate of probability that a candidate expert *ca* has produced document *d*, $p(d|ca)$ is required. In the current work $p(d|ca) \in \{1,0\}$, depending on whether *ca* has authored a document.

Furthermore, information about the relatedness of documents stemming from a corporate *ontology* can be integrated as well as information about the *authority* of a person in the corporation. Existing organisational ontologies can be exploited in the process of expert discovery, by enhancing the models mentioned in the previous paragraphs. First, documents which are "close" enough in the ontology are similar and this information can be used to increase the expertise level of an expert. Thus,

$$p\prime(q|ca) = \lambda p(q|ca) + (1-\lambda) \sum_{q'} p(q|q\prime)p(q\prime|ca) \tag{17}$$

where $p\prime(q|ca)$ is a new estimate of $p(q|ca)$, given $p(q|q\prime)$, which in turn can be estimated by the length of the path of the queries q, q' in a topic hierarchy. The position of a user in the organisational hierarchy, can also be used to enhance the above models especially in the context of sparse data,

$$p\prime(q|ca) = (1 - \sum_{ou \in OU(ca)} \lambda_{ou}) \times p(q|ca) + \sum_{ou \in OU(ca)} \lambda_{ou} \times p(q|ou) \tag{18}$$

where $OU(ca)$ is the set of corporate organisation units into which candidate expert belongs to *ca*, $p(q|ou)$ expresses the probability of q belonging to *ou*, and λ_{ou} denotes the importance of organisational unit *ou* This can be estimated by replacing *ca* with *ou* in any of the three models.

4.8 Search for Influential Users in a Social Network

Locating an influential user who is knowledgeable about a topic, i.e. an expert, may involve a graph search initiated by a seeker node up to goal node. The content that is generated or related to a node, determines its expertise [30].

A social network by the Enron emails has been studied. [8] The nodes correspond to humans, and the edges are drawn from the sender to the receiver of an email. The seeker node issues a query made of terms which traverses the SN. The match between a query and a node can be measured with the TF/IDF criterion, and once the match is above a certain threshold, it is considered that an expert has been found, and the expert can be questioned. The aim of this work, i.e. to search for an expert, is similar to the Referral Web project [15], but the implementation differs.

Six techniques for graph search have been proposed to guide the search from a seeker node to a goal node, and they are applied recursively. They are the: Breadth first search (BFS), random walk search (RWS), weak ties search (WTS), strong ties search (STS), hamming distance search (HDS), information scent search (ISS). In BFS all neighbours of the seeker nodes receive the query. It is guaranteed to find the expert node, albeit at huge computational cost. In RWS only one neighbour of the seeker node, selected randomly, receives the query. The neighbour with the highest out-degree connectivity is selected in BCS. In WTS (STS) the neighbour who has received the most (least) messages from the seeker node is selected. HDS selects a neighbour who has the most uncommon friends with the seeker node. According to ISS the neighbour with the highest match with the query is selected.

In evaluating search methods the following computational criteria can be considered: the success rate in locating an expert and the length of the search path. The authors have also considered the possibility of users contributing in the search process. Thus as the search progresses, according to one the aforementioned techniques, the nodes that are getting involved might entail the labour of the corresponding humans. For instance, a human might have to read the query and decide whether he can answer it. Thus there is an additional amount of social cost that stems from the human processing of the queries. In particular, the number of people involved, the frequency of involvement of each person, and the depth of the query chain form part of the social cost. For example, a shallow path from a seeker to a goal node, makes it more probable for an expert to respond to a seeker.

The results show the out-degree based strategies, such as BCS and HDS are better than the other methods in terms of both computational and social criteria.

4.9 Influential Users for Well Structured Items

A number of approaches have been suggested for detecting the most influential users among the ones having interacted with well structured items.

The definition of an *influential user* can be reduced into two components: the *expertise* and the *contribution*, according to [8]. The former is related to the content

[8] `http://www.cs.cmu.edu/~enron/`

of items evaluated by a user and the latter refers to the influence of a user to the rest of the users. Influential users are sought in the domain of movies at the movie lens and the IMDB data sets. However, the ideas developed are not domain dependent, and can subsequently be extended to other domains also. The data items in this case concern the features of movies, that include: actors, genres, directors, etc. In particular influential users are sought for specific feature values, that is for specific actors, genres and directors etc. The point of this work was to enhance a collaborative filtering system with influential users. A similar approach has been followed in [2], which aimed at "importing" influential users from an external source, in order to enrich an existing system.

The expertise, is related to the items a user has interacted with, as well as to the depth of interaction. For example, user's u expertise regarding movie director d is defined as follows,

$$expertise_{ud} = 2\frac{movies\_directed(S_u)}{movies\_directed_d} + \frac{directing\_period(S_u)}{directing\_period_d} + \frac{genres\_directed(S_u)}{genres\_directed_d} \tag{19}$$

where S_u represents the movies seen by user u.

The contribution of a user to the community, is defined as the accuracy of this user's predictions of the preferences of all community members. The way expertise (E) and contribution (C) interact in order to predict the rating P of a feature item has been defined as follows, $f : E \times C \to P$. To compute C, a single layer linear perceptron is trained with the delta rule, to minimise the difference between the preferences of all users of the community to that of the expert, for a specific feature value.

The contribution of a user has also been considered in [24], where it is called trust. It comes into two types. The first type is the *profile trust* defined as follows,

$$trust^P(u) = \frac{CorrectSet(u)}{RecSet(u)} \tag{20}$$

$$\tag{21}$$

where user u has recommended $RecSet(u)$ items, out of which $CorrectSet(u)$ have been adopted.

The second type is more refined, because a user's opinion is formed at an item level rather than over all items as in the previous equation,

$$trust^I(u,i) = \frac{(c_k, i_k) \in CorrectSet(u) : i_k = i}{(c_k, i_k) \in RecSet(u) : i_k = i} \tag{22}$$

where i is a item, c_k is the rating of this item by the user.

An empirical study of the evolution of influential users was undertaken in [8]. The movie lens data set was investigated in order to derive the most influential users in term of expertise, which is domain specific; and their influence, which roughly refers to how closely they predict other users' preferences. The study, involved the evolution of influence in terms of the addition of new users. Results have shown that most of the influential users appear early in the life of this data set. Moreover, the evolution of the social network produces users of increased influence in some domains (e.g. movie genres and keywords) but not in others (e.g. movie directors and actors).

4.10 Productivity and Influential Users

In some social networks users post content and they acquire followers. An issue to investigate is whether the number of followers increases their productivity.

An investigation has been carried out in YouTube [13]. Let n_t represent the videos uploaded at time t, v_t the average number of views of the n_t videos, then a linear regression model was build for contributors who were active for more $T > 10$ time periods,

$$n_{t+1} \sim a \cdot log_{10} v_t + \beta \tag{23}$$

The time periods at which a contributor is active were split into two groups, the ones at which the attention received was higher than average (a good period denoted by G) and the rest into which the attention was lower than average (denoted by B). The second test the authors performed, was to examined whether it is more likely for a contributor to be more productive after a good than after a bad period. The following equations denote the average productivity following a good or bad period,

$$n^G = \frac{1}{(T-1)/2} \sum_{s \in G} n_{s+1} \tag{24}$$

$$n^B = \frac{1}{(T-1)/2} \sum_{s \in B} n_{s+1} \tag{25}$$

Then $\Delta = n^G - n^B$ expresses the change of a contributor's productivity at different attention levels. It was experimentally attested that $\Delta > 0$ meaning that users contribute more after receiving more attention. Thus on average a user is more productive after a good than after a bad period. Also, there is difference in the Δ value based on the number of weeks a user is active; thus for users being active for a few time periods bad periods influence heavily their productivity, but this is much less severe for users that are active for more time periods.

Next, the authors tried to establish a causal relation between the attention and productivity, with the Granger test of causality [12]. Let $\overline{v}_t$ represent the average views of all videos by followers, and $\overline{n}_t$ the average of all videos uploads by contributors at time period t. The Granger test was performed on the hypothesis that $\overline{v}_t$ Grander causes $\overline{n}_t$, with a p-value of 0.01, where as the reverse hypothesis gave a p-value of 0.61. Which essentially states that the number of views increases productivity.

4.11 Link Types

Not all links between users in a social network have equal importance. That was the issue of an investigation in Twitter [14]. Notably the authors tried to classify the followers of a user into simple followers and friends. Subsequently they studied the effects of the two groups on the productivity of the user in terms of posts. Twitter allows a user to post direct and indirect updates, where a direct one aims at a single user and an indirect updates aims at the whole community. Based on this, a user's follower will fall into the friend category if he has posted at least two direct messages to the user.

An increase in the number of followers and in the number of friends leads to increase in the user's productivity as measured by the number of posts. However, the productivity saturates for the number of followers, but not for the number of friends. Thus the more friends a user has, the more productive he becomes. Another observation is that the proportion of friends versus the followers is very small for most users. Moreover, the number of friends increases as the number of followers increase, but the number of friends saturates.

4.12 Evolution of Social Networks and Consequences for Influence

The evolution of a social network refers to dynamic phenomena, such as the addition and removal of nodes, or edges. The evolution is of paramount importance in the evolution of influence, where graph theoretic measures are considered that take into account structural properties of nodes, such as the degree. A discussion of the main models can be found in [25], where the emphasis is on networks of indistinguishable nodes.

The most cited model is perhaps the Barabási-Albert model (BA), it contains two phases,

- *Growth*: Starting with a small number m_0 of nodes, a new node is added, with $m \leq m_0$ edges that link the new node to m different nodes already in the system.
- *Preferential attachment*: The probability of connecting the new node to node i of degree k is defined as,

$$p(k_i) = \frac{k_i}{\sum_j k_j} \tag{26}$$

If the degree of every new node at its introduction at time step t_i is $k_i(t_i) = m$ then,

$$k_i(t) = m\left(\frac{t}{t_i}\right)^{\beta}, \quad \beta = 0.5 \tag{27}$$

Asymptotically ($t \to \infty$), the degree distribution follows a power law,

$$P(k) \sim 2m^{1/\beta} k^{-\gamma}, \quad with\ \gamma = \frac{1}{\beta} + 1 = 3 \tag{28}$$

In many real networks there is a preferential attachment of the new nodes, i.e. $p(k) \sim k^a$. For instance, in Internet at domain level, and at a citation network $a \simeq 1$, thus the *linear preferential attachment* (see 26) describes the dynamics of such a network.

Non-linear preferential attachment

In case $a \neq 1$, the *preferential attachment* is *non-linear*. In particular, for $a > 1$ a single node connects to almost every node. For $a < 1$, the degree distribution is a stretched exponential. For instance in the neuroscience co-authorship and the actor collaboration network $a = 0.8 \pm 1$. Non-linear preferential attachment has been investigated in [21].

Adding and removing edges and nodes

An extension of the BA model is to consider adding edges, rewiring edges, as well as adding new nodes. Let m be the number of new edges that are added according to linear preferential attachment, m edges are rewired with probability q, and a new node is added with probability $1 - p - q$. The result is that a power law is present, with a saturation at small values of k. This model is a good approximation of an actor collaboration network [1].

In another approach new edges are added between existing nodes and existing edges can be removed. Let us assume that c new edges are added at every time step, which connect two unconnected nodes i and j with a probability proportional to the product of their degrees, then,

$$\beta = \frac{1+2c}{2(1+c)} \tag{29}$$

$$\gamma = 2 + \frac{1}{1+2c} \tag{30}$$

in equations 28, 27. When the network expands $c > 0$, and when the network shrinks $c < 0$. The above model is experimentally tested with a co-authorship network [11].

Preferential attachment and heterogeneous graphs

The previous approaches consider all nodes to be identical, which consequently make-up a homogeneous network. In a heterogeneous network graph nodes would be different, and the attachment of new node to an existing node would depend upon their similarity or *affinity*, as well as to the degree of the old node [27]. In particular, each node can be found in a state $x_i \in \mathbb{R}$, let $\sigma : \mathbb{R}^2 \to \mathbb{R}$ be a function that maps the states of two nodes to their affinity.

A new node u_a with m links is added to the network at each time step, each link is attached to an existing node k_i with the following probability,

$$p(k_i) = \frac{\pi(u_i)}{\sum_j \pi(u_j)}, \qquad \pi(u_i) = k_i \cdot \sigma(x_i, x_a) \tag{31}$$

Let $w(x) = \langle \sigma(x,y) \rangle_y$, which expresses the average propensity of each node being in state x to attract a link from a new node, based solely on its state. It was discovered that such heterogeneous networks exhibit a power law behaviour $k^{-\gamma(w)}$, with $\gamma(w) = 1 + 2/w$. Thus a heterogeneous network with preferential attachment exhibits a richer behaviour than the homogenous counterpart.

Table 1 Social networks investigated

Issue Investigated	Data set	Reference
k-shell in influence diffusion	LiveJournal.com Email of UCL CNI IMDB adult	Sect. 4.4, [19]
Search for experts	Enron Email	Sect. 4.8, [30]
Search for experts	IBM Email	Sect. 4.2, [9]
Chain of reference to an expert	Public Web documents	Sect. 4.1, [16] [15]
Link Types	Twitter	Sect. 4.11, [14]
Search for experts	Java Forum	Sect. 4.3, [31]
Expert Finding	Movie Lens	Sect. 4.6, 4.9, [8], [10]
Expert Finding	IMDB	Sect. 4.9, [8]
Productivity	YouTube	Sect. 4.10, [13]
Expert finding or profiling with unigram language models	Uviv. of Tilburg Docs	Sect. 4.7, [3]
Computational complexity of maximising influence diffusion	Co-authorship graphs in high energy physics & in theoretical computer science	Sect. 4.5, [17]
Evolution of social networks	Neuroscience co-authorship Citation Actor collaboration	Sect. 4.12, [25]
Active Learning of diffusion model parameters (linear threshold)	Pajek: Geom Pajek: Netsci Pajek: Zewail Pajek: Lederberg	Sect. 3.5, [5]

Table 2 Facets of Influence

Structural Properties	Search for influence	Diffusion of Influence	Content based Calculation of Influence	Cost of search	Dynamic Properties
Degree	Breadth first search	Linear threshold	Unigram language models	Computational	Barabasi Albert model
K-shell decomposition	Weak ties	Cascade	Features & ratings	Social	Productivity & Attention
In-betweeness centrality	Strong ties	SIR	TF/IDF		
	Hamming distance	SIS			
	Best connected search				
	Information scent				
	Greedy hill-climbing				
	Page rank				
	Hits				
	Z-score				

5 Counter Arguments

Up to now we have been looking for influential or experts user in a social network, i.e the focus was on an elite group. It seems there is a counter argument to that called crowdsourcing. Which roughly refers a groups of people solving collectively a given problem [4]. The quality of the solution produced by crowdsourcing is not the average of the solutions of the individual members, but rather an aggregation of all solutions. Web technologies provide the necessary substrate for a crowd to act collectively.

Crowdsourcing has been found particularly pertinent in problems that are cognitive, that is problems that involve factual questions, where the answer concerns the present or the future. For example, who will be world soccer champion? Coordination of a course of actions and cooperation problems can also be addressed with crowdsourcing. In addition, three qualities of a crowd are considered as important for successful crowdsourcing [28]. The first, is that the individuals should be diverse, they should independent and the whole process of crowdsourcing should be decentralised.

An investigation was carried in Wikipedia out to discover what are the characteristics of contributors [20]. The issue to investigate was whether the elite or the common users (i.e. crowds) dominate the number of edits and the changes in content. The elite users can form part of the administrators group, a group whose members are peer selected to have special privileges, such as blocking a page. Also, users could be considered as the elite because of their high productivity. It was discovered that while the participation of the elite group was instrumental in the early days of wikipedia, later on there is a shift in the productivity from the elite to the common users' groups. This is not due to decline of the individuals productivity of the elite group, but rather to the masses that join wikipedia in order to contribute. Notably, decline of the elite users is accompanied by an increase in the percentage of edits made by users with less than 100 edits. A similar initial rise and decline in the influence of elite users was attested at the popular social bookmarking site del.icio.us

6 Conclusions and Future Work

We have reviewed ways of determining the influence of individual users in online social networks.

Determining influence based solely on content generated by a user or content with which the user has interacted with, depends on the content's form. Thus the content can be a set of well structured items, where each item can be reduced to a set of features and relevant ratings can be assigned to whole items or to features of items. Alternatively, the items can be documents where the determination of the user's level of expertise is difficult. In that case, determination of the user's profile is more feasible, where the profile would contain document topics. Moreover, the status of a user in a corporate environment can be used in determining his influence.

Various structural characteristics of nodes/users can be used as a criterion for determining influence. First, the number of incident edges to a node seems to be weak criterion for determining influence, high degree nodes correspond to hubs, located at the periphery of a network. On the other hand the k_s (k-shell decomposition) value is stronger indicator of the influence of a node because such nodes form the core of the network. Thus if an information item is to reach the maximum extent, it should start spreading from a high k_s node. However, if multiple nodes are selected as initial points, then high degree nodes might lead to more extended spread of information than a selection of high k_s nodes. That is because high k_s nodes are concentrated at the core of the network, close to one another. These conclusions are based on the SIS and SIR spread models. Another study that employed a greedy algorithm to determine a set of very influential users, but not necessarily the most influential ones, has confirmed that targeting high degree nodes in a direct marketing strategy is very bad under the linear threshold or the cascaded model of influence diffusion. Instead a different greedy search strategy has been proposed with feasible computational complexity and good results regarding the discovery of influential nodes. Finally, another argument against the coincidence of high degree with high influence is from a Java forum, where an expert might choose to help users that are slightly less expert than he is.

A large scale study was undertaken in four social network sites: Flickr, YouTube, LiveJournal and Orkut [23], which is related to the search of influential users. The degree distribution of nodes follows a power law. In-degree is similar to out-degree, this is notably different from the web where hubs are different from authorities, which in turn might explain the relative inefficiency of the HITS algorithm in seeking influential users in the systems we reviewed. A small network diameter was observed, which is caused by the presence of a power law. Moreover, the average path length seems to very small, because of a large and densely connected core. Notably, a core seems to be present in all networks surrounded by small clusters of low degree nodes. Low degree nodes at the periphery seem to be opposite to a previously stated arguments against seeking high degree nodes.

We would see several directions for future research in influence at social networks. One path are hybrid systems for influential user discovery that will consider the graph properties along with the content generated by the users. Thus combinations of hamming distance search or best connected search should be combined with generative language models. Furthermore, paths from seeker of the information to the influential or expert user can be used as a form of explanation in a collaborative filtering system, a chain of references that would substantiate product recommendations to a user. Finally, the vulnerability of social network to malicious attacks that aim to promote or demote ideas or products can also be investigated relative to the structural properties of the network. In that vein the role of the core of the network or the nodes' connectivity should be studied. Moreover, all the above issues should be considered in the light of networks that evolve by growing or shrinking.

References

1. Albert, R., Barabási, A.L.: Topology of complex networks: Local events and universality. Physical Review Letters 85(24) (2000)
2. Amatriain, X., Lathia, N., Pujol, J.M., Kwak, H., Oliver, N.: The wisdom of the few: a collaborative filtering approach based on expert opinions from the web. In: Proceedings of the 32nd International ACM SIGIR Conference on Research and Development in Information Retrieval, SIGIR, pp. 532–539. ACM, New York (2009)
3. Balog, K., Bogers, T., Azzopardi, L., de Rijke, M., van den Bosch, A.: Broad expertise retrieval in sparse data environments. In: Proceedings of the 30th Annual International ACM SIGIR Conference on Research and Development in Information Retrieval, SIGIR, pp. 551–558. ACM, New York (2007)
4. Bradham, D.C.: Crowdsourcing as a Model for Problem Solving: An Introduction and Cases. Covergence: The International Journal of Research into New Media Technologies 14, 75–90 (2008)
5. Cao, T., Wu, X., Hu, T.X., Wang, S.: Active Learning of Model Parameters for Influence Maximization. In: Gunopulos, D., Hofmann, T., Malerba, D., Vazirgiannis, M. (eds.) ECML PKDD 2011. LNCS, vol. 6911, pp. 280–295. Springer, Heidelberg (2011)
6. Cho, J., Kwon, K., Park, Y.: Collaborative filtering using dual information sources. IEEE Intelligent Systems 22, 30–38 (2007)
7. Davenport, T., Prusak, L.: Working Knowledge. Harvard Business Press (1998)
8. Dima, M., Vogiatzis, D., Paliouras, G.: Expert based prediction of user preferences. In: 5th International Workshop on Semantic Media Adaptation and Personalization, SMAP (2010)
9. Dom, B., Eiron, I., Cozzi, A., Zhang, Y.: Graph-based ranking algorithms for e-mail expertise analysis. In: Proceedings of the 8th ACM SIGMOD Workshop on Research Issues in Data Mining and Knowledge Discovery, DMKD, pp. 42–48. ACM, New York (2003)
10. Domingos, P., Richardson, M.: Mining the network value of customers. In: Proceedings of the Seventh ACM SIGKDD International Conference on Knowledge Discovery and Data Mining, KDD 2001, pp. 57–66. ACM, New York (2001)
11. Dorogotsev, S., Mendes, J.F.F.: Scaling behaviour of developing and decaying networks. Europhys. Lett. 52(33) (2000)
12. Granger, C.: Investigating causal relations by econometric models and cross-spectral methods. Econometrica 37(3), 424–438 (1969)
13. Huberman, B.A., Romero, D.M., Wu, F.: Crowdsourcing, attention and productivity. Journal of Information Science 35(6), 758–765 (2009)
14. Hubermand, B.A., Romero, D.M., Wu, F.: Social networks that matter: Twitter under the microscope. arXiv:0812.1045v1 (2008)
15. Kautz, H., Selman, B., Shah, M.: Referral web: combining social networks and collaborative filtering. Commun. ACM 40, 63–65 (1997)
16. Kautz, H., Selman, B., Shah, M.: The Hidden Web. AI Magazine 18(2), 27–36 (1997)
17. Kempe, D., Kleinberg, J., Tardos, E.: Maximizing the spread of influence through a social network. In: Proceedings of the Ninth ACM SIGKDD International Conference on Knowledge Discovery and Data Mining, pp. 137–146. ACM, New York (2003)
18. Kitsak, M., Gallos, L.K., Havlin, S., Liljeros, F., Muchnik, L., Stanley, H.E., Makse, H.A.: Identifying influential spreaders in complex networks. Nature Physics 6, 888–903 (2010)
19. Kitsak, M., Gallos, L.K., Havlin, S., Liljeros, F., Muchnik, L., Stanley, H.E., Makse, H.A.: Identifying influential spreaders in complex networks. arXiv:1001.5285 (2010)

20. Kittur, A., Chi, E., Pendelton, B.A., Suh, B., Mytkowicz, T.: Power of the Few vs. Wisdom of the Crowd: Wikipedia and the Rise of the Bourgeoisie. In: 25th Annual ACM Conference on Human Factors in Computing Systems, CHI (2007)
21. Krapivsky, P., Redner, S., Leyvraz, F.: Connectivity of Growing Random Networks. Physical Review Letters 85(21) (2000)
22. Leskovec, J., Singh, A., Kleinberg, J.M.: Patterns of influence in a recommedation network. In: Proceedings of the 10th Pacific-Asia Conference on Knowledge Discovery and Data Mining, PKDD (2006)
23. Mislove, A., Marcon, M., Gummadi, K.P., Druschel, P., Bhattacharjee, B.: Measurement and analysis of online social networks. In: Proceedings of the 7th ACM SIGCOMM Conference on Internet Measurement, IMC, pp. 29–42. ACM, New York (2007)
24. O'Donovan, J., Smyth, B.: Trust in recommender systems. In: Proceedings of the 10th International Conference on Intelligent User Interfaces, IUI 2005, pp. 167–174. ACM, New York (2005)
25. Réka, A., László, B.A.: Statistical mechanics of complex networks. Rev. Mod. Phys. 74(1), 47–97 (2002)
26. Sabater, J., Sierra, C.: Review on computational trust and reputation models. Artif. Intell. Rev. 24, 33–60 (2005)
27. Santiago, A., Benito, R.: An extended formalism for preferential attachment in heteregenous complex networks. EPL 82 (5) (2008)
28. Surowiecki, J.: The wisdom of crowds. Anchow (2005)
29. Yimam, D., Kobsa, A.: Expert finding systems for organisations: Problem and domain analysis and the demoir approach. Journal of Organizational Computing and Electronic Commerce 13(1), 1–24 (2003)
30. Zhang, J., Ackerman, M.S.: Searching for expertise in social networks: a simulation of potential strategies. In: Proceedings of the 2005 International ACM SIGGROUP Conference on Supporting Group Work, pp. 71–80. ACM, New York (2005)
31. Zhang, J., Ackerman, M.S., Adamic, L.: Expertise networks in online communities: structure and algorithms. In: Proceedings of the 16th International Conference on World Wide Web, WWW, pp. 221–230. ACM, New York (2007)

A Client-Side Privacy Framework for Web Personalization

C. Kolias, V. Kolias, G. Kambourakis, and E. Kayafas

University of the Aegean, #National Technical University of Athens
{1kkolias,3gkamb}@aegean.gr, vkolias@medialab.ntua.gr,
kayafas@cs.ntua.gr

Abstract. Personalization of web applications is the complex process of dynamically rendering the application responsive to the unique needs of individual users. Nevertheless, the information required for achieving the personalization procedures is usually gathered and stored beyond the user's control. This is a situation that raises serious privacy concerns to the end-users and may drive them to reject the application. For example, when browsing an adaptive e-commerce website, users are not aware which behavior will be monitored and logged, how it will be processed, how long it will be stored, and with whom it will be shared in the long run, thus they may hesitate to visit the website. In this chapter after an introduction to the state of the art in privacy preserving personalized web applications we present an abstract architecture that enables users to fine-tune their privacy level (and in result their personalization experience) according to the trust they put on different applications. Since the data is stored on the client side, this approach by definition enhances user privacy.

Keywords: Adaptive Web, privacy, client-side personalization.

1 Introduction

As contemporary websites become increasingly complex and informationally massive, the need for rapid access to information has become greater than ever before. "Personalized" or "user-adaptive" applications and websites emerged to feel the void and this is the main reason behind their great success. User adaptive websites is a special category of adaptive hypermedia applications that can be accessed from the web and possess the ability to adjust their content, behavior and structure to the user's own preferences. As a result both parties benefit from this situation: the users locate the information in which they are really interested easily, spend less time in searching, and feel more comfortable with a website that appears to be custom-tailored to their own preferences. On the other hand, the website achieves its goals without the need of creating multiple versions of the same web page for each distinct user.

I.E. Anagnostopoulos et al. (Eds.): Semantic Hyper/Multimedia Adaptation, SCI 418, pp. 297–316.
springerlink.com

Nevertheless, the method of the adaptation process is not trivial and it usually involves the monitoring, logging, and analysing of user's behaviour for long periods of time (and for many interactions with the website) before it becomes really efficient. Of course, this practice raises serious privacy concerns since users are usually reluctant to reveal personal or private information through the Internet and to put their trust on a e-commerce organization to store and administer this kind of data securely. But this mistrust may have catastrophic results for example, on the e-commerce realm.

In this chapter we initially explore the privacy risks that escort adaptive hypermedia web applications and present an overview of the proposed mechanisms that attempt to provide personalized content with respect to user privacy. Next, we propose a high level architecture that enables users to take control over the profiles the adaptive websites would normally construct on their behalf. By doing so, users are able to declare their own privacy preferences by specifying which information they consider sensitive. Under these circumstances, user profiles can be shared across different websites according to the privacy settings specified by the user.

The rest of the chapter is organized as follows: Next section discusses the applications and the characteristics of traditional adaptive hypermedia applications and attempts to stress out the security risks involved. Section 3 presents various mechanisms that were proposed by the research community in order to enhance the privacy of the user in personalized web. These approaches are also organized into large categories having similar characteristics. Section 4 includes a discussion about the strengths and weaknesses of each category of the approaches found in the literature. The proposed framework is described in Section 5, while Section 6 elaborates on the advantages and disadvantages of our solution. This chapter is an extended version of the work presented in the 3rd International Workshop on Semantic Media Adaptation and Personalization (SMAP 2008) [1].

2 Adaptive Hypermedia and User Privacy

Adaptive hypermedia applications made their first steps in the early 1990s and started becoming increasingly popular after 1996 with the emergence of the World Wide Web (WWW). According to the definition given in [2] adaptive hypermedia applications are considered as hypermedia systems which adapt their content, structure and/or presentation of the networked hypermedia objects, to each individual user's characteristics, usage behavior and/or usage environment. As stated in [3], adaptive hypermedia can be utilized within the context of various spheres of action, including educational hypermedia applications, on-line information systems, on-line help systems, information retrieval hypermedia, institutional hypermedia, and systems for managing personalized views in information spaces. Today, most research efforts and industrial implementations mainly focus on on-line advertising systems, location aware social applications, e-commerce websites and e-learning fields which have benefited by personalization to a great extent, e.g., by improving the quality of service provided and consequently the user experience. For example, a personalized educational website will automatically adapt its methodologies and content according to the knowledge level of the students, their general interests and their progress, in order to make the learning curve smoother. Contrariwise, a conventional e-learning application will try to find a "golden path" of teaching method for all students and will present the same information to them without discrimination. Similarly, an adaptive e-commerce application will significantly assist their visitors in the shopping process by recommending products that possibly meet their preferences, for example automatically redirecting them to product categories that reflect their likings.

Generally, adaptation can be distinguished into two types: (a) adaptive presentation, and (b) adaptive navigation support. The former refers to actions such as modifying textual and multimedia content of the website, while the latter to actions such as link hiding, sorting, annotation, direct guidance and hypertext map adaptation [2]. Adaptation in the context of the Web may also include: (a) personalized item description whose complexity is geared towards the presumed level of user expertise, (b) tailored presentations that take into account the preferences of each user regarding the item presentation and media types, (c) various recommendations that are based on recognized interests and goals of the user and data or information from various sensors like GPS and accelerometer most likely seen in mobile devices.

Adaptive websites must be distinguished from adaptable websites which generally follow a less complex concept and are easier in their implementation. More specifically, in contrast to adaptable websites where the user has manual control over its appearance and structure, in adaptive websites the personalization process is usually adjusted automatically by the website. This however happens not in an opt-in or voluntary basis but without the user being aware of the internal site mechanics. Simply put, the user is not in control of how the site behaves. This transparent to the web page visitor process is based on dynamically built user models which represent users' behavior to the system. These models are constantly and dynamically constructed by acquiring specific user data every time a user visits such an adaptive website. According to [2] user data that constitute models are classified into the following categories:

- **User data**: information about user's preferences and other personal information (for example, name, gender, and similar).
- **Usage data**: information about the user's behavior on various web pages. For example, what pages the user visited and how long the user remained connected with each one of them.
- **Environment data**: information about the user's network location characteristics revealed by the client machine, e.g., IP address, memory, screen resolution, browser type.

User modeling is an active research field of Artificial Intelligence (AI). Many approaches exist for creating accurate user models but the common denominator of all is that their accuracy is analogous to the amount of user data gathered. As a result, adaptive websites require a large amount of information in order to better personalize their content. However, gathering personal information, some of which may be sensitive, raises serious privacy concerns.

Recent studies prove that up to 87% of Internet users are highly concerned about their digital privacy and up to 41% would abandon a site that requires registration information [4]. Also 37% of the users do not see enough value in personalization and are not willing to take privacy risks associated with such features [5]. Due to their unique characteristics and requirements an adaptive website may pose additional security risks which may discourage users from interacting with it. In most of the cases the user information harvest is conducted in a concealed manner on behalf of the adaptive site, usually for fear of disturbing and consequently discouraging the users from their normal interaction. This basically, means that the users are unaware of when their actions are being monitored. Most importantly the users are not in control of the type of data that are collected.

Moreover, the user has to trust the website to store sensitive personal information in a secure manner. This means that appropriate procedures must be followed within the organization so that random mistakes that reveal private user data never occur and that websites are capable of bouncing malicious entities that attempt to gain access to that information. This of course is something that a common and sometimes even privacy-savvy user cannot evaluate and assess and that is why this kind of trust between web companies and users is hard to be built.

Another point of concern for the user is that since data gathered from the website lay outside the user's control, there is no guarantee that this data will not be shared with third parties. In most of the cases the users do not feel comfortable knowing that their private data travel (possibly unprotected) through the Internet and all this information may end up to a company they do not know or do not trust.

Adaptive websites utilize algorithms for creating user models which predict user likings and future user behavior. This process of analyzing user behavior in order to extract conclusion about a user's personality usually is one additional reason for user discomfort towards this kind of websites.

Finally, users may object to the idea of this categorization of humans into groups that bear specific characteristics. For many this is viewed as a kind a social, political or economical discrimination.

3 Related Work

In this section we will analyze some approaches that have been followed in order to enhance the user's privacy in the personalized web. We have organized the approaches based on the privacy preservation mechanism and the basic categories are: a) approaches that make use of privacy policies, b) approaches that are based on anonymization or pseudonymous access to the websites, c) approaches that maintain the data on the client side.

3.1 Privacy Policies

A simplistic solution followed by many websites is to commit to a predefined privacy policy, which would be publicly accessible in the site and contain an analytical description of the purpose collecting user data and the type of data itself. The users could then inspect these policies and evaluate them to decide if the website corresponds to their privacy requirements or not.

For this reason the Worldwide Web Consortium (W3C) [6] recommended A P3P Preference Exchange Language (APPEL) [7] and the Platform for Privacy Preferences (P3P) [8] standards. P3P is a language based on XML, which provides a standardized way for the service providers (running adaptive sites) to express their privacy policies. APPEL is a similar language which enables users to express privacy preferences. Consequently, tools like the AT&T Privacy Bird [9] have been created which evaluate if a website's privacy policy is in accordance with the user's privacy preferences. If not, then the utility advices the user to leave the site.

Authors in [10] were motivated by the fact that P3P tools do not provide a negotiation mechanism, they proposed a framework which supports a type of privacy/personalization bargaining. The system named PriveAd is founded upon P3P and APPEL and extends them in order to enable negotiation possibilities. Users are able to explicitly define preferences statements. A Policy Evaluation Unit, installed on the client side which is the core component of the infrastructure and its main duties is to compare the incoming privacy policy proposal sent by the server with the preferences of the client. The negotiation Strategy component is responsible for calculating counterproposals. The server then commits to the decided policy.

In [11] the author discusses how negotiation techniques can overcome the drawbacks of static privacy policies and that it is possible to implement dynamic policies using existing technologies. The negotiation process presented in this work is modeled as a Bayesian game. The approach is implemented as an extension to P3P to allow software-supported negotiations in e-commerce, and it is integrated in the Mozilla browser.

Another solution for websites is to bear a privacy seal, i.e., a special mark awarded by an independent organization, verifying that the website respects user privacy policies (at least up to some extent). Such is the TRUSTe [12] privacy seal.

3.2 Anonymization and Pseudonymity

For the majority of applications that offer personalized services, the data needed for the personalization are typically stored centrally, to a database beyond the user's control. It is possible for service providers to respect their privacy policies or not and monitor and collect more data than agreed. In any case, the users may not always trust a service even to store any of their personal information. Even in the case where the service provider is honest, these centralized databases may attract the attention of malicious entities and lure serious attackers. The most common technical approach adopted in such cases is pseudonymous or anonymous interaction with the provider of the personalized services. According to this technique the provider allows the user to interact with the service by using a pseudonym (alias) rather than the real name. In the anonymization approaches the identity of the user is simply concealed or masqueraded. Thus, the personal information the user discloses during the interaction with the website cannot be directly associated with the user's physical entity.

In some approaches, instead of a direct communication with the website, intermediate services are used like the Lucent Personalized Web Assistant (LPWA) [13] and the Janus Personalized Web Anonymizer [14]. The former is based in the concept of pseudonymous aliases to achieve Web browsing in a personalized and at the same time secure manner. The system generates aliases that consist of an alias username, an alias password and an alias e-mail address for each web site the user visits. The information required by the user is a single user ID (e-mail address) and a secret (i.e., a password). The system is comprised by three main components namely (a) persona generator, (b) browsing proxy and (c) e-mail forwarder. The persona generator generates a unique, static persona for each web site during the first visit to the site. The browsing proxy redirects the connection on the TCP level and strips private information from HTTP headers. The e-mail forwarder forwards to the user the e-mails addressed to the corresponding persona. In this way the Web server is able to register, sign-in and recognize a returning user. The Web server can personalize the service accordingly, but is not able to associate that user with the real identity of a physical entity. The Janus Personalized Web Anonymizer follows a much similar approach.

The authors in [15] propose an infrastructure that guarantees pseudonymous communication between a client and a number of servers. The proposed system utilizes Knowledge Query and Manipulation Language (KQML) [16] and is based on the existence of a mix network between the user's applications and user modeling servers. This architecture achieves anonymization by concealing both the identity of the users and the location (that is the IP address) of the user modeling servers in the network.

Masks [17] is a distributed, consent-based privacy architecture that provides anonymity on a group level to the users. The system comprises of two main components: a Masks server, which acts as an intermediate between users and Web sites, and a privacy and security agent (PSA), which acts as an intermediate between users and the Masks server and is installed on the client's browser. The Masks server organizes users into groups of similar interests and distributes masks, which is temporary identifier assigned to each user to declare the interest in a given topic or specific site.

A widespread approach is based on the K-anonymity [18] principle. K-anonymity has been widely used in the context of Location Based Services (LBS). According to K-anonymity a request is considered private, if is indistinguishable from K-1 other requests where K is a variable usually defined by the user. In this way the probability of identifying the user cannot exceed 1/K. In such schemes a trusted server which acts like an anonymizer is employed. The user sends the query containing its private location to the anonymizer instead of the service. The anonymizer modifies the query to contain a larger region which contains that user plus K-1 other users and forwards the query to the service. The service then replies to the anonymizer with a number of elements (such as points of interest) which may include redundant information. The anonymizer then filters the redundant information and forwards the result to the user. The work in [19] describes a full scale implementation of the above scheme which allows the user to define the K variable thus setting different privacy for different levels of trust.

The MobiPriv system presented in [20] is based on the observation that most K-anonymous techniques degrade their performance or fail to serve some clients requests if K-1 users are not present in the area close to the requesting user during the time the query was sent. For example, the anonymization engine may (a) have to wait until K-1 other requests are made in the region of the request, (b) expand the region in search for K-1 other requests or (c) continue to split a larger region until the region contains at most K-1 other users. MobiPriv handles the above situations by generating realistic and diverse fake users (dummies) instead of discarding the query. The proposals in [21], [22] describe similar approaches with the use of some dummy information.

The K-anonymity approach is widely used in Location Based Services (LBS). Traditional K-anonymity may prove inappropriate for more complex personalization services such as those requiring other personal information rather than just the user location. The reason is that K-anonymity ignores the possibility that profile information may serve to further identify an individual. For example, if K users have totally different profiles, then an adversary can easily distinguish each requesting user even if they share the same location stigma.

The proposed anonymization scheme in [23] generalizes both location and profiles in a way controlled by the user so that all users in the selected region would seem to share both the same profiles and location signatures. To achieve this, location data is generalized so that the selected region includes at least additional K– 1 users with identical profiles to ones of the user. Thus, even if a malicious entity gains somehow access to the request information, it cannot distinguish the real requester from other K–1 users as they have the same probability of submitting a query.

The authors in [24] propose a concept much similar to the k-anonymity although it is modified to suit personalized web services context. Instead of transmitting a query which contains location information this model assumes that the query contains the triplet <d, q, t> where d is the user personal information, q is the query to service, and t is the time that the query was placed. Their system assumes the existence of an intermediate server with which the user first registers. In this context this server generalizes the personal information d of the user and forwards the new triplet <d', q, t> to the service. For example if a user's age is 25 the triplet will contain the record age:20-30. The authors introduce the notion (k,w)-online-anonymity. A query <d', q, t> in the server logs is said to have (k,w)-online-anonymity if there are at least k queries $<d'_i, q_i, t_i>$ in the query log such that each $<d'_i, q_i, t_i>$ has $d'_i = d'$, and $|t-t_i| \leq w$. The query log is said to have (k,w)-online-anonymity if all queries in the log have (k,w)-online-anonymity. The system groups the users with similar characteristics according to both their personal information and query time so that the records in the server logs provide (k,w)-online anonymity.

The work in [25] proposes a new anonymity method namely, Personalized-Granular_k-anonymity. This method takes into consideration that in a set of personal information, some attributes may have different privacy importance depending on their values. For example, a patient's status may not be sensitive information if the patient is running on fever due to common cold but it might be if it is due to a non treatable disease.

3.3 User Data on Client Side

The very little control over personal information when it resides on a remote location has lead to approaches that keep this information on the client side. Such systems are PrivAd [10] and Adnostic[26] for personalized advertizing. These systems preserve user privacy by performing all relevant behavior tracking processes on the client. For achieving this, the client must initially download all possible advertisements from the advertiser's servers, and then locally filter and choose the appropriate advertisement to display. Although the two systems are quite different in concept, the basic approach for enhancing user privacy is the same: keep sensitive information local to the user.

The authors in [27] are motivated by the fact that the browser "knows more" about its user than a content provider. Indeed, browsing in a website is just a small fraction of the total browsing time of the user. Moreover, the browser itself is a much safer place for storing personal information of the user. The proposed system, namely Reprive, conducts classification based on Naive Bayess classifier to categorize the user's browsing history. Based on the user's habits, a user interest profile is built and maintained inside the browser. RePrive controls the dissemination of this information to providers by explicitly requesting permission by the user to do so. The transfer of such data is done using a custom protocol on top of HTTP. RePrive also supports high level policies to automate the process of requesting permission.

In [28] the authors note that traditional personalization approaches (i.e., gathering and analyzing data on server side) possess the critical limitation of being unaware of the of user's every day local activity, such as schedules, favorite websites and personal emails. In this way, the personalization results may be poor. The proposed system, namely, CRESDUP is an agent based implementation that collects and mines the private data of user at the client side. The efficiency of the system is tested against personalized RSS news feeds scenarios. The main advantage of the system is that the resulting user model (a) is updated much more frequently, (b) supports explicit user feedback and (c) the personalization evolves data that the user would normally hesitate to share and the service provider would be impossible to obtain, such as basic personal information, recently visited URLs, self-made documents, calendar and schedules, favorite websites. As with all client-side approaches this system preserves privacy but on the downside, the server has to send all possible information and this is later filtered on the client by a special component of the system. This might be a viable solution in syndicate feeds scenarios where feeds themselves are small in size but it is unlikely to be applicable in e-commerce sites for example.

In [29] UCAIR, a client-side agent targeting personalized web search, is presented. UCAIR is implemented as a web browser plug-in which is able to perform reordering of the Google search engine results. This system does not require any manual actions by the user to rate the results. It is able to implicitly conclude on the quality of the search context by monitoring the user's search queries, history information and user behavior on the pages. Since UCAIR's personalization process is entirely conducted on the client side the user does not need to worry about any privacy implications.

Finally, in [29] the authors follow a much different approach for achieving privacy in personalized web applications. They introduce a system for the creation and enforcement of privacy preserving browser cookie policies through explicit user interaction. Browser cookies are used extensively in personalization procedures but browsers lack of effective cookie management. Thus, there are ways for a mistrusted and potentially dangerous web application, to access personal information of the user by exploiting cookies installed through interaction with trusted applications. The proposed approach tries to fit between the two opposites of common browsing, i.e. full cookie disclosure and full cookie conceal, by automatically determining the value of a cookie to enable a cost-to-benefit analysis to the user. It is implemented as a Firefox extension and it is fault tolerant.

4 Discussion

Each of the categories presented above has its own shortcomings and inefficiencies. For example, the problem with the privacy policies is that sometimes the text of the privacy policy document can be quite lengthy, generic or fuzzy, and may contain technical terms. As a proven fact users are not willing to invest time for actions that are not immediately relevant with their purpose, and in many cases, they cannot fully understand. In such situations the user cannot continue interacting with the site if privacy compromises are not made. The application will simply not scale down its services to meet the user's privacy demands. Moreover, it is clear that non expert users may find it difficult to write and read complex policies. An additional drawback is that when negotiation protocols are employed they may require several rounds to pick a commonly agreed policy. On the other hand, privacy seals are known to be problematical too. Firstly, there are different types of privacy seals and most users are not familiar with their meaning and do not know what level of privacy is verified and how this is exactly done. Also, non expert users are not able to distinguish between a genuine seal and a forged one. What is more, some seals are not awarded by trustworthy organizations.

As far as the anonymization approaches, they seem to behave efficiently in recommendation systems but they appear to be ineffective when the real name of the user is required (e.g., when a bank transaction needs to take place). On the other hand, it is possible to override effective pseudonymity by employing techniques of social engineering, while at the same time the risk of associating a pseudonym with a real name after monitoring and sophisticated analysis of data exchanged is never eliminated. Moreover, these tools do not cope well with Javascript applications, Java applets and Flash objects. Finally, there seems to be an additional communication overhead associated with each request/response.

The drawbacks that escort client-side based personalization are enumerated as follows: (a) the personalization service cannot classify users with common characteristics into groups and treat them accordingly, (b) the client machine gathers the attention of attackers and therefore the user must take additional actions to protect it more effectively, (c) personalization may not be possible if the user attempts to receive it from different client machines, (d) transmission of personal information might happen more frequently from the client to the site, (e) data analysis for the extraction of an effective user model can be a computationally intense process which might not be possible to achieve when the client has limited amount of resources (such as a smartphone or PDA), (f) these type of approaches are usually application specific and do not promote the collaboration and exchange of user models among different websites needed for faster and more complete user adaptation. So far a method for preserving privacy in personalized applications that acts as a panacea does not exist. A general conclusion is that when a user requires personalization he must be willing to make some privacy sacrifices.

5 The Proposed Architecture

In this section the architecture of the proposed approach is described in detail. The reader may notice that this architecture was based in the conclusions discussed in the previous paragraph. It resembles some client-side personalization approaches in the sense that user profiles are maintained on the client, although our approach differentiates itself in the profile manipulation and privacy negotiation process. Both the client and the service provider must cooperate by exchanging a series of messages, for a personalization action to take place. The fact that personal user information is stored on the client side exclusively, grants the user greater security. Additionally, it provides the user with a trade-off mechanism between privacy and personalization. In other words, it makes it possible for the user to sacrifice a better personalization experience to enjoy increased privacy. The architecture presented in this section is by no means a complete implemented system. It exists merely to provide the blueprint for such applications to be developed.

5.1 Architecture Components Description

Our model assumes that the following entities participate throughout the lifetime of a typical personalization action:

- The client: the entity that requests personalized services and hosts the user model.
- The user: modeling agent – the program responsible for constructing the user model on the client.
- The adaptive service provider: the application on the service provider that offers personalized content.
- The user profiles: XML documents that contain information about the user preferences and habits.
- The user data request document: XML document that describes the necessary information required for a personalization action to be initiated in the adaptive service provider's side.
- The privacy preferences document: XML document that contains the rules for the type of the data that are allowed to be sent from the client to the adaptive service provider.
- The user data response document: XML document that contains the personal information extracted from the user's profile.

The rest of this section presents a more comprehensive description of these entities as well as their specific role in the architecture. An overview of the architecture components is depicted in Figure 3.

5.1.1 Client

The client is the entity that makes requests for personalized content from adaptive websites. It is also the physical location where the user model – containing the user's personal information - resides. It is to be noted that at all times the client communicates with the adaptive sites through the modeling agent which is also installed at the client side. During a typical personalization lifecycle, the client will be requested to make decisions about which personal information is to be shared, and provide information when necessary. The idea is to keep this interaction to a minimum, especially after a full user model has been constructed and plenty of rules exist in the privacy preferences document.

5.1.2 User Modeling Agent

The user modeling agent is the core the proposed system. It is a client side service, which is primarily monitoring the user behavior in order to construct a user model. Describing the user agent's implementation as well as how the user models are created is out of scope of this chapter. It is worth mentioning that classic machine learning algorithms would normally be utilized for such tasks. Secondarily, the agent aims to assist the users in the personalization experience by acting as an intermediate between them and the service, thus automating any negotiation process.

From a high level perspective, the agent is responsible for presenting information to the user in human readable form. It analyzes user behavior aiming to construct the user model per se or requests the user to explicitly define profile parameters whenever necessary. It is also responsible for constructing, organizing and storing user data such as privacy preferences in documents. Finally, it produces the according responses by evaluating user's privacy preferences. Figure 1 presents the main components of the user modeling agent.

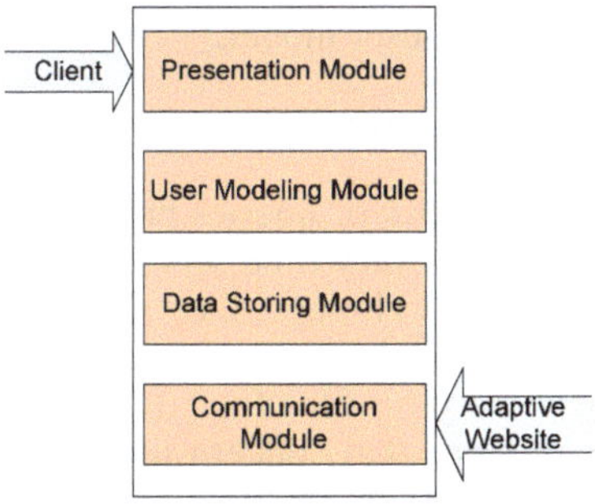

Fig. 1 User Modeling Agent main components

5.1.3 Service Provider

With the term service provider we refer to the adaptive website which the client attempts to visit and more accurately the component that carries out personalization activities. Unlike most modern practices the proposed architecture assumes that the service provider will never store any user data. Such behavior be achieved for example by having the client reject any cookies that originate from a specific website for example.

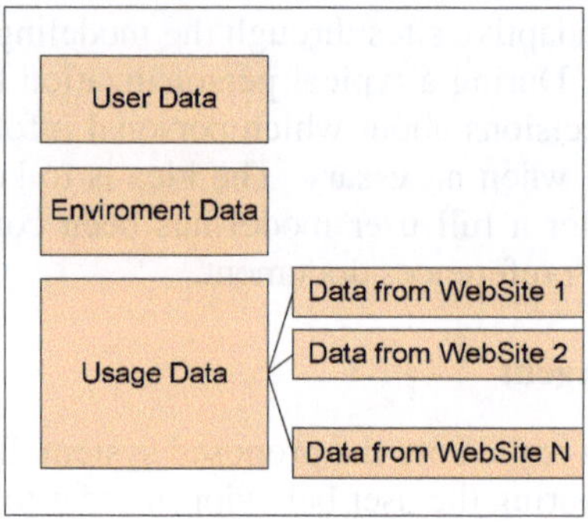

Fig. 2 Main documents that constitute user profiles

5.1.4 User Profiles

Profiles are sets of user data organized in one or multiple XML files. Users may have multiple profiles stored on their client machine. Some of this information is populated by the users and some by user agents and other by service providers. Typically user data is filled manually by the users themselves, environment data by their agents and usage data by the service provider of each site. Any information populated by the service provider requires user authorization first. Typically, usage data are included in the profile document in the form of XML references. Any information included in the user profile is accessible by the user at any time. The language used to describe user information follows a standard format. Figure 2, depicts how user profiles are organized. The sum of the user profiles stored on the client machine constitutes the user model.

5.1.5 User Data Request Document

User data requests are special types of queries produced by service providers and sent to the user modeling agent. Every time service providers need information in order to generate a personalized web page, they request data that reside at the active user profile on the client side. Queries of such kind follow a specific syntax. Service providers can also query profiles for information submitted by other service providers, those for instance that concern usage data of other adaptive sites. This is possible because data contained in the user model(s) follow a specific syntax and the interpretation of those data is publically available. By specifying rules on the privacy policy document users may deny to provide answers to part or the whole of these service requests. The service provider then adjusts the generation of the web page according to the data returned. Any information included into a user profile is accessible by the user at any time.

5.1.6 Privacy Preferences Document

The privacy preferences document is a special type of document in which the user specifies the privacy constraints that should be applied when the information is sent toward the service provider. With the privacy rules defined in this document the user is certain that only strictly specified sensitive information will be transmitted to the corresponding service provider. Initially, general privacy rules can be defined by the user in the privacy preferences document. After that the rules contained in it will be dynamically updated as the user interacts with new sites.

5.1.7 User Data Response Document

A user data response document is a special type of document produced by the user modeling agent and sent toward the service provider. The data that constitutes the document is retrieved from the active user profile based on the provider's requests and privacy preference documents. The data chosen to remain confidential based on the privacy preferences will simply not be added to the response document, thus leaving the corresponding fields empty. We assume that the request document will not contain any fields so sensitive that the denial of their disclosure will result in revealing information about the user.

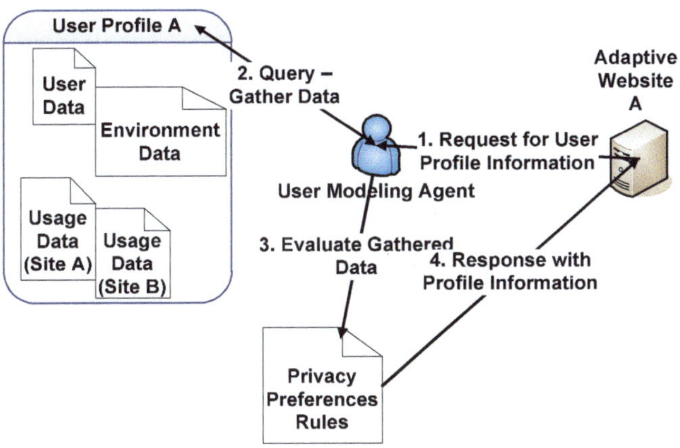

Fig. 3 Overview of the architecture components

5.2 Architecture Components Interaction

In this section we describe how the previously introduced components cooperate in order for a personalization process to be accomplished. There are three steps in this process: (a) the initialization phase, (b) the negotiation phase, and (c) the personalization phase. The architecture components interaction process is presented in Figure 3.

During the initialization phase the user modeling agent prompts the user to create a new user profile, by filling a set of personal information or alternatively by selecting an existing one. As already mentioned it is possible that many user profiles are hosted on the client device at the same time. Typically, at this point, the user fills simple personal information (user data) and preferences. Simultaneously, the agent scans the client for environmental characteristics and updates the user profile accordingly. Then the client may visit the adaptive website. At this point a new usage data entry is created in the user profile, if one does not already exist (e.g., when the user visits the site for the first time). Note that such actions can be completed without the user interaction if a rule already exists in the privacy preferences document.

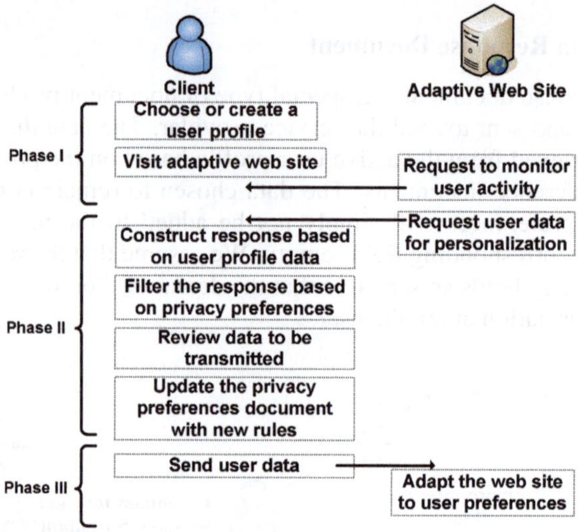

Fig. 4 Architecture component interaction

The negotiation phase is a multi-step process that results in the transmission of user information to the service provider. Initially, the service provider constructs a request (following a specific syntax) of certain user data for a corresponding personalization activity. The requested data may be user data, environmental data, usage data on the site (gathered from previous interactions with the system), or even usage data from any other service provider it considers relevant. This is possible since the data required by each website for personalization is publically known. Thus, service providers maintain lists of relevant information by other providers and check them accordingly. During this process the user agent gathers the corresponding data contained in the user profile to construct a response. Next, the user agent filters the response according to the rules contained in the privacy preferences document(s). Before sending the response toward the service provider the user agent may present an overview of the profile information to be sent to the user. This is done, in order to make last minute changes or to deny sending the profile information for the current session. Finally, the privacy preferences document is updated with new rules, if any.

During the personalization phase, the service provider receives the personalization information the client sent, which is either all the information it requested for, or a subset of them. With the information provided by the user, it adjusts the personalization experience and generates the final web page.

5.3 A Real Life Scenario

In this section a simple, real life scenario is presented in order to examine the operation of the proposed architecture in more detail.

A user wants to visit an online bookstore. Before any casual interaction with the website takes place, the appropriate profile is adopted, i.e., the one the user employs when dealing with e-commerce sites. This profile contains confidential information such as credit card information, real name and shipping address, as well as usage data from various sites (coming from previous interactions with them).

For the first personalization activity to be executed the website needs to know the user's real name and a set of other information relevant to a typical user behaviour on the site, e.g., the favourite book category of the user. As a result the welcome page of the bookstore greets the user by the username and makes some book recommendations according to user preferences. The corresponding request data document is presented in Table 1.

The user agent gathers the data requested from the information contained in the selected user profile. Subsequently, it filters the information to be sent according to the rules contained in the privacy preferences document. The user happens to have a high level of trust for this specific site so the rules are permissive. The user agent presents the information to be sent. So personalization action ends successfully. The transaction completes and the user leaves the site. Since no cookie was accepted from the user's browser the bookstore site will have no recollection of user in subsequent sessions.

Sometime later the user decides to visit a new auction website for the first time. In order to personalize the content of its first page the particular website requires user personal data such as the real name of the user, native language and relative usage data (from other sites) like the user's last purchases. The corresponding request data document is given in Table 2.

The user agent creates the response document as it did in the previous case. Since it is the first time that the user visits this website, there is no relevant privacy rule in the privacy preference document, therefore, by default, none of this information is concealed but the data about to be sent must be shown to the user for evaluation. The user agent presents the data to the user, who in turn decides not to disclose the real name record. A new privacy rule is created and the privacy preferences document is updated. Table 3 presents the resulting privacy preferences document. The information is sent to the auction website, which adjusts its appearance accordingly. Thus, it welcomes the user with a general welcome message (since no real name was provided), translates part of its webpage and recommends some products based on user's past purchases record coming from other sites. An aspect that should be noted here is that the product recommendation was achieved from the very first time the user visited the site.

6 Evaluation

In works [31], [32] the end-user requirements that must be fulfilled by adaptive systems that respect user privacy are recognized as:

- Purpose specification
- Openness
- Simple and appropriate controls
- Limited data retention
- Pseudonymous interaction
- Decentralized control

As it will be explained further down our architecture corresponds to all the aforementioned aspects.

Purpose specification is achieved in two ways. First, as part of the architecture, every site is bound to have a publicly accessible document in which it will be described what kind of personalization activity the site is capable of, what kind of information it requires for each activity and in what way this will benefit the user. Second, every request executed by the adaptive service specifies what the personalization action will be. This is achieved by a dedicated field of the request document (designating the purpose of the request).

The end-user is able to control if personal data is collected or not. Moreover, since the user profiles are stored on their side in simple XML documents, it becomes easy to inspect and manually modify the information contained in each one of them, at any time.

The architecture recognizes the user modeling agent as one of its basic components. Its role is to communicate and negotiate with the service provider, automate procedures and present the information exchanged in a human-readable manner. The existence of this component must guarantee that the user will have a straightforward and smooth control over the system. It is also certain that after some time required for the system to start recognizing patterns in user behavior, the user interaction will be limited to a minimum.

The architecture model assumes that the service provider will not store personal information that has been disclosed by the user. Also, the data gathered by the adaptive sites must be instantly deleted after the end of the session. A mechanism for assuring that these rules will be adhered is not assumed by the architecture. Since the personalization process will always be done on-the-fly, at every time the user visits the site, storing user information would be a meaningless procedure and would constitute suspicious behavior from the part of the service provider. A similar behavior might be regulated by international privacy laws. Moreover, the user always holds the right for denying disclosure of any information considered sensitive. In this way data retention is always limited.

Openness is fulfilled since every time a user visits an adaptive site, a request is created by the service provider for allowing the user behavior to be monitored.

Table 1 The data request document for the bookstore site.

```
<Request-Data>
    <Purpose>
            <Type>
                Recommendation
            </Type>
            <Type>
                Personal Message
            </Type>
            <Requesting_Page>
                http://www.bookstore.com/welcome.html
            </Requesting_Page>
    </Purpose>
    <Sender>
          http://www.bookstore.com/
    </Sender>
    <Data>
          <First_Name/>
          <Last_Name/>
          <Most_Searched_Category/>
          <Last_10_Pruchased/>
    </Data>
</Request-Data>
```

Table 2 Data request document for the auction site

```
<Request-Data>
    <Purpose>
          <Type>
                Recommendation
          </Type>
          <Type>
                Personal Message
          </Type>
          <Type>
                Translation
          </Type>
          <Requesting_Page>
                http://www.auctionwebsite1.com/welcome.html
          </Requesting_Page>
    </Purpose>
    <Sender>
          http://www.bookstore.com/
    </Sender>
    <Data>
          <First_Name/>
          <Last_Name/>
```

Table 3 The final privacy preferences document

```
<Privacy-Preferences>
    <Deny_All>
          <Service>
                http://www.website1.com
          </Service>
          <Service>
                http://www.website2.com
          </Service>
    </Deny_All>
    <Deny>
          <Data>
                <First_Name/>
                <Last_Name/>
                <Service>
                      http://www.auctionwebsite1.com
                </Service>
          </Data>
    </Deny>
</Privacy-Preferences>
```

Users can create and host multiple user profiles. This is done deliberately because: (a) multiple physical entities might use the user's device to access a website, (b) users may act on behalf of others (and not according to their own preferences) at a given time, (c) users may wish to interact with a specific site with a different identity. That is, users may employ any profile from the ones stored in their device. This feature constitutes a mechanism for providing pseudonymous user interaction with the adaptive websites.

The overall architecture of the system is distributed. Data are stored and processed on the client side and integrated user profiles are built by the interaction with various adaptive sites.

Apart from the attributes mentioned earlier more advantages can be recognized when adopting this approach. These are summarized as:

- It is simple to implement and maintain in comparison to other approaches on the field.
- It allows portability of the user profiles, since all information is stored and organized in XML files.
- It can be lightweight since no real process of data is performed on the client side apart from query execution and filtering.
- It allows inter-site collaboration for the creation of robust user profiles in sorter time.
- It allows the scaling of the personalization effect according to the user privacy preferences.

On the downside, the architecture:

- Requires increased user interaction at least when users visit sites for the first time and/or privacy preference rules do not exist beforehand.
- Service providers must not store usage data they gather about clients, but no technical mechanism is provided to enforce that.
- Clients take the responsibility for storing and managing their data in a secure way.
- Service providers must comply with the standard in order other sites can use their usage data in a uniform way.

References

[1] Kolias, C., Kolias, V., Anagnostopoulos, I., Kambourakis, G., Kayafas, E.: Enhancing User Privacy in Adaptive Web Sites with Client-Side User Profiles. In: Third International Workshop on Semantic Media Adaptation and Personalization, SMAP 2008, December 15-16, pp. 170–176 (2008)

[2] Kobsa, A., Koenemann, J., Pohl, W.: Personalised hypermedia presentation techniques for improving online customer relationships. The Knowledge Engineering Review 16(2), 111–155 (2001)

[3] Brusilovsky, P., Maybury, M.T.: From adaptive hypermedia to the adaptive web. Communications of the ACM 45(5) (May 2002)

[4] Kobsa, A.: Personalized hypermedia and international privacy. Communications of the ACM 45(5) (May 2002)
[5] Teltzrow, M., Kobsa, A.: Impacts of User Privacy Preferences on Personalized Systems: A Comparative Study. In: Designing Personalized User Experiences in eCommerce, pp. 315–332. Kluwer Academic (2004)
[6] W3C, `http://www.w3.org/` (accessed November 27, 2011)
[7] W3C, A P3P Preference Exchange Language 1.0 (APPEL1.0), `http://www.w3.org/TR/P3P-preferences/` (accessed November 27, 2011)
[8] W3C, The Platform for Privacy Preferences 1.0 (P3P1.0), `http://www.w3.org/TR/P3P/` (accessed November 27, 2011)
[9] AT&T Privacy Bird, `http://www.privacybird.org/` (accessed November 27, 2011)
[10] Guha, S., Reznichenko, A., Tang, K., Haddadi, H., Francis, P.: Serving Ads from localhost for Performance, Privacy, and Profit. In: Proceedings of Hot Topics in Networking (November 2009)
[11] Preibusch, S.: Implementing Privacy Negotiations in E-Commerce. In: Zhou, X., Li, J., Shen, H.T., Kitsuregawa, M., Zhang, Y. (eds.) APWeb 2006. LNCS, vol. 3841, pp. 604–615. Springer, Heidelberg (2006)
[12] TRUSTe, `http://www.truste.org/` (accessed November 27, 2011)
[13] Bleichenbacher, D., Gabber, E., Gibbons, P.B., Matias, Y., Mayer, A.: On secure and pseudonymous client relationships with multiple servers. In: Proceedings of the 3rd USENIX Electronic Commerce Workshop, pp. 99–108 (1998)
[14] Gabber, E., Gibbons, P., Matias, Y., Mayer, A.: How to make personalized web browsing simple, secure, and anonymous. In: Proceedings of the Conference on Financial Cryptography. Springer, New York (1997)
[15] Kobsa, A., Schreck, J.: Privacy through pseudonymity in user-adaptive systems. ACM Transactions on Internet Technology (TOIT) 3(2), 149–183 (2003)
[16] Finin, T., Weber, J.: Draft specification of the KQML agent-communication language. Tech. Rep. (1993), `http://www.cs.umbc.edu/kqml/kqmlspec/spec.html` (accessed November 27, 2011)
[17] Ishitani, L., Almeida, V., Meira Jr., W.: Masks: Bringing Anonymity and Personalization Together. IEEE Security and Privacy 1(3), 18–23 (2003)
[18] Samarati, P.: Protecting respondents identities in microdata release. IEEE Transactions on Knowledge and Data Engineering 13(6), 1010–1027 (2001)
[19] Gedik, B., Liu, L.: Protecting Location Privacy with Personalized k-Anonymity: Architecture and Algorithms. IEEE Transactions on Mobile Computing 7(1), 1–18 (2008)
[20] Stenneth, L., Yu, P.S., Wolfson, O.: Mobile systems location privacy: "MobiPriv" a robust k anonymous system. In: 2010 IEEE 6th International Conference on Wireless and Mobile Computing, Networking and Communications (WiMob), October 11-13, pp. 54–63 (2010)
[21] You, T., Peng, W., Lee, W.-C.: Protecting moving trajectories using dummies. In: International Workshop on Privacy-Aware Location-Based Mobile Services (2007)
[22] Ghinita, G., Kalnis, P., Khoshgozaran, A., Shahabi, C., Tan, K.-L.: Private queries in location based services: anonymizers are not necessary. In: Proceedings of the 2008 ACM SIGMOD International Conference on Management of Data, Vancouver, Canada, June 09-12 (2008)

[23] Shin, H., Vaidya, J., Atluri, V.: A profile anonymization model for privacy in a personalized location based service environment mobile data management. In: 9th International Conference on MDM 2008, April 27-30, pp. 73–80 (2008)

[24] Xu, Y., Wang, K., Yang, G., Fu, A.W.C.: Online anonymity for personalized web services. In: Proceeding of the 18th ACM Conference on Information and Knowledge Management, Hong Kong, China, November 02-06 (2009)

[25] Shen, Y., Liu, Y., Zhang, Y.: Personalized-Granular k-Anonymity. In: International Conference on Information Engineering and Computer Science, ICIECS 2009, December 19-20, pp. 1–4 (2009)

[26] Toubiana, V., Narayanan, A., Boneh, D., Nissenbaum, H., Barocas, S.: Adnostic: Privacy preserving targeted advertising. In: NDSS (2010)

[27] Fredrikson, M., Livshits, B.: RePriv: Re-imagining content personalization and in-browser privacy. In: 2011 IEEE Symposium on Security and Privacy (SP), May 22-25, pp. 131–146 (2011)

[28] Chen, T., Han, W.-L., Wang, H.-D., Zhou, Y.-X., Xu, B., Zang, B.-Y.: Content Recommendation system based on private dynamic user profile. In: 2007 International Conference on Machine Learning and Cybernetics, August 19-22, vol. 4, pp. 2112–2118 (2007)

[29] Shen, X., Tan, B., Zhai, C.: Implicit user modeling for personalized search. In: CIKM 2005: Proceedings of the 14th ACM International Conference on Information and Knowledge Management, pp. 824–831. ACM (2005)

[30] Shankar, U., Karlof, C.: Doppelganger: Better browser privacy without the bother. In: Proceedings of the 13th ACM Conference on Computer and Communications Security (CCS 2006) (2006)

[31] Brar, A., Kay, J.: Privacy and Security in Ubiquitous Personalized Applications. In: Proc. User Modelling Workshop on Privacy-Enhanced Personalization, Edinburgh, UK, July 25 (2005)

[32] Langheinrich, M.: A privacy awareness system for ubiquitous computing environments. In: Borriello, G., Holmquist, L.E. (eds.) UbiComp 2002. LNCS, vol. 2498, p. 237. Springer, Heidelberg (2002)

Chapter 4: Multimedia and Open Standards

In the context of adaptation and personal delivery of Multimedia/Hypermedia content, one of the main challenges lies in the standardization protocols and mechanisms that need involving in such a diverse and constantly evolving information space. More and more people use different types of client-side technologies, which lead to different context aware requirements. In addition to this heterogeneous environment, the amount of information that satisfy the personal preferences of a user is added/modified in gigantic rates and with different semantic formats due to the new possibilities of resource annotation. In one hand, new opportunities for personalization on the content level are emerged, but on the other hand, several interoperability issues derive for bridging the gap between personalization, adaptation and open standards in Multimedia/Hypermedia technologies. This chapter aims to discover such issues as well as to propose some dominant schemes in this field.

Article 4.1

Title: The XML and Semantic Web Worlds: Technologies, Interoperability and Integration: A survey of the State of the Art

Authors: Nikos Bikakis, Chrisa Tsinaraki, Nektarios Gioldasis, Ioannis Stavrakantonakis, and Stavros Christodoulakis

In the first contribution of this chapter, Bikakis et al. besides an excellent survey they provide in respect to several international XML-based standards (e.g., Dublin Core, MPEG-7, METS, TEI, IEEE LOM, etc.), they discuss the significant interoperability and integration problems raised towards bridging the gap between Semantic Web standardization and XML. In addition, the authors outline the latest efforts derived from W3C, including the latest working drafts and recommendations (e.g., OWL 2, SPARQL 1.1, XML Schema 1.1, etc.). Finally, they end up with an introduction of two dominant frameworks (SPARQL2XQuery and XS2OWL) that seem to be capable enough for bridging the gap, as well as to creating an interoperable environment between current XML technologies and tomorrow semantic-oriented approaches.

Article 4.2

Title: Personalized Multimedia Web Services in Peer to Peer Networks Using MPEG-7 and MPEG-21 Standards

Authors: Emmanouil Skondras, Angelos Michalas, Malamati Louta, and George Kouzas

In the second article of this chapter, the authors (Skondras et al.) firstly provide an overview of the available standards and technologies used for personalization in client/server-based Multimedia Web Services, and mostly around the common MPEG standardization. Then the authors, by using several well-known XML-based and semantic-oriented programming languages (MPQF, SPARQL, OWL) introduce a prototype semantic P2P architecture which delivers personalized audio information. In parallel, the authors discuss on the available standards and technologies that support multimedia retrieval, recommendation and personalization services. Towards that, they end up with a proposal of an embedded mechanism capable of combining user preferences with resource adaptation metadata by receiving feedback from previous user's actions according to the client/server approach.

The XML and Semantic Web Worlds: Technologies, Interoperability and Integration: A Survey of the State of the Art

Nikos Bikakis[1], Chrisa Tsinaraki[2], Nektarios Gioldasis[2], Ioannis Stavrakantonakis[2], and Stavros Christodoulakis[2]

[1] National Technical University of Athens, Greece & Institute for the Management of Information Systems, "Athena" Research Center, Greece
bikakis@dblab.ntua.gr
[2] Lab. of Distributed Multimedia Information Systems & Applications, Technical University of Crete, Greece
{chrisa, nektarios, gstavrak, stavros}@ced.tuc.gr

Abstract. In the context of the emergent *Web of Data*, a large number of organizations, institutes and companies (e.g., *DBpedia*, *ACM*, *IEEE*, *IBM*, *NASA*, *BBC*, etc.) adopt the *Linked Data* practices and publish their data utilizing Semantic Web (SW) technologies. On the other hand, the dominant standard for information exchange in the Web today is XML. Many international standards (e.g., *Dublin Core*, *MPEG-7*, *METS*, *TEI*, *IEEE LOM*, *etc.*) have been expressed in XML Schema resulting to a large number of XML datasets. The SW and XML worlds and their developed infrastructures are based on different data models, semantics and query languages. Thus, it is crucial to provide interoperability and integration mechanisms to bridge the gap between the SW and XML worlds.

In this chapter, we give an overview and a comparison of the technologies and the standards adopted by the XML and SW worlds. In addition, we outline the latest efforts from the W3C groups, including the latest working drafts and recommendations (e.g., OWL 2, SPARQL 1.1, XML Schema 1.1, etc.). Moreover, we present a survey of the research approaches which aim to provide interoperability and integration between the XML and SW worlds. Finally, we present the SPARQL2XQuery and XS2OWL Frameworks, which bridge the gap and create an interoperable environment between the two worlds. These Frameworks provide mechanisms for: (a) Query translation (SPARQL to XQuery translation); (b) Mapping specification and generation (Ontology to XML Schema mapping); and (c) Schema transformation (XML Schema to OWL transformation).

1 Introduction

In the context of the emerging *Web of Data*, a large number of organizations, institutes and companies (e.g., *DBpedia*, *Geonames*, *PubMed*, *DBLP*, *ACM*, *IEEE*,

I.E. Anagnostopoulos et al. (Eds.): Semantic Hyper/Multimedia Adaptation, SCI 418, pp. 319–360.
springerlink.com

IBM, *NASA*, *BBC*, *NSF*, etc.) adopt the *Linked Data* practices; they use the Semantic Web (SW) technologies, publish their data and offer SPARQL endpoints (i.e., SPARQL-based search services) over it. On the other hand, the dominant standard for information exchange in the Web today is XML. In addition, many international standards (e.g., *MPEG-7, MPEG-21*, *VRA Core*, *METS*, *TEI*, *IEEE LOM*, etc.) have been expressed in XML Schema, resulting to a large number of XML datasets.

In the Web of Data, the applications and services have to coexist and interoperate with the existing applications that access legacy systems like, for instance, the very large XML-based audiovisual digital libraries of the digital video broadcasters (e.g., *BBC TV-Anytime Service*, etc.), the XML-based digital libraries of the cultural heritage institutions, etc.

Since the SW (*OWL/RDF/SPARQL*) and XML (*XML Schema/XML/XQuery*) worlds have different data models, different semantics and use different query languages to access them, it is crucial to develop tools and methodologies that will allow bridging the gap between them. In addition, it is unrealistic to expect that all the legacy data (e.g., Relational, XML, etc.) will be converted to SW data. Thus, it is crucial to provide interoperability and integration mechanisms that will allow the SW users to access external heterogeneous data sources from their own working environment. It is also important to offer SPARQL endpoints over legacy data in the Linked Data context.

In this chapter, we deal with the mechanisms that allow the exploitation of the legacy data, in the Web of Data. In particular, in the first part of the chapter (Section 2), we try to present and compare the XML and SW worlds. In particular, we outline and compare the technologies and the standards adopted in the two worlds. In addition, we present the latest efforts from the W3C groups, including the latest working drafts and recommendations (e.g., XML Schema 1.1, OWL 2, SPARQL 1.1, XQuery and XPath Full Text 1.0, XQuery Update Facility 1.0, etc.).

In the second part (Sections 3 and 4), we present a survey of the existing approaches that deal with the interoperability and integration issues between the XML and SW worlds.

Finally, in the third part (Sections 5–8), we outline the SPARQL2XQuery and XS2OWL Frameworks that have been developed in order to provide an interoperable environment between the SW and the XML worlds.

2 XML World vs. Semantic Web World — A Comparison

In this section we outline the XML and SW worlds, we present the adopted technologies and we compare their basic characteristics. Throughout this comparison, we distinguish three levels: (a) The *data level*; (b) The *schema level*; and (c) The *query level*. Table 1 provides an overview of the current W3C standards adopted in each level in the XML and SW worlds.

Table 1 Overview of the W3C Standards currently adopted in the XML and SW Worlds

Level	XML World	Semantic Web
Data	XML	RDF
Schema	XML Schema	RDF Schema – OWL
Query	XQuery – XPath	SPARQL

At the data level, the *Extensible Markup Language* (*XML*) [1] is the data representation language in the XML world, while the *Resource Description Framework* (*RDF*) [2] is used to represent the SW data.

At the schema level, the *XML Schema* [3] is used to describe the structure of the XML data. Currently, the XML Schema 1.0 is a W3C recommendation and the XML Schema 1.1 [4][5] is under development and has reached the W3C working draft level of standardization. Regarding the SW, the *RDF Schema* (*RDFS*) [6] and the *Web Ontology Language* (*OWL*) [7] are exploited to describe the structure and the semantics of the RDF data. Recently, a new version of OWL, OWL 2.0 [8] has become a W3C recommendation.

Finally, at the query level, the *XML Path Language* (*XPath*) [9] and the *XML Query Language* (*XQuery*) [10] are employed for querying XML data. In the SW world, the *SPARQL Protocol and RDF Query Language* (*SPARQL*) [12] is the standard query language for RDF data. Currently, the W3C SPARQL working group [14] is working on the extension of the SPARQL query language in several aspects, resulting in SPARQL 1.1. SPARQL 1.1 includes several components like the *SPARQL 1.1 Query*, *Update*, *Protocol*, *Service Description*, *Property Paths*, *Entailment Regimes*, *Uniform HTTP Protocol for Managing RDF Graphs*, and *Federation Extensions*.

In the rest of this section, we outline the models adopted in the XML and SW worlds at the data level (Section 2.1) and we present and compare their schema (Section 2.2) and query (Section 2.3) level languages.

2.1 Data Level

In this section we present in brief the models adopted at the data level in the XML and SW worlds. Section 2.1.1 describes the XML model that is adopted in the XML world, while Section 2.1.2 outlines the RDF model that is used in the SW world.

2.1.1 XML

The *Extensible Markup Language* (*XML*) [1] is a general-purpose markup language, designed to describe structured documents. XML is based on tags like

HTML, however, XML it does not have a fixed set of tags, but allows users to define their own tags. In addition, unlike HTML, the XML tags have no specific semantics. An XML document consists of plain text and markup, in the form of tags and may be represented as an ordered labeled tree. An XML document may contain the following types of nodes: `document nodes`, `elements`, `attributes`, `text nodes`, `comments`, `processing instructions`, and `namespaces`.

2.1.2 RDF

The *Resource Description Framework* (*RDF*) [2] is a general purpose language for representing information about resources on the Web. To some extent, RDF is a lightweight ontology language. RDF has a very simple and flexible data model, based on the central concept of the `RDF statement`. RDF statements are triples (`subject`, `predicate`, `object`) consisting of the resource (the `subject`) being described, a property (the `predicate`), and a property value (the `object`). In particular, the subject can either be an `IRI` or a `Blank node`. Every predicate must be an `IRI` and the object can be an `IRI`, `Blank node` or `RDF Literal`. A collection of RDF statements (or else `RDF triples`) can be intuitively understood as a directed labeled graph, where the resources are nodes and the statements are arcs (from the subject node to the object node) connecting the nodes. Finally a set of RDF triples is called `RDF Dataset` or `RDF Graph`.

2.2 *Schema Level*

In this section we present the schema languages adopted in the XML and SW worlds. Section 2.2.1 outlines the XML Schema language, used in the XML world. The SW languages are then presented; in particular, RDF Schema is presented in Section 2.2.2 and OWL language is presented in Section 2.2.3. Finally, a comparison between the schema languages adopted by the XML SW worlds is presented in Section 2.2.4.

2.2.1 XML Schema

The *XML Schema* [3] is a schema definition language that has been developed by the W3C and has been expressed in XML syntax. XML Schema is intended to describe the structure and constrain the content of documents written in XML by providing rich structuring capabilities. XML Schema can specify the exact element hierarchy and specify various types of constraints placed on the XML data define options (e.g., limits on the number of occurrences of an element, ranges of the values of elements or attributes, etc.).

In particular, an XML document is composed of `elements`, with the root element delimiting the beginning and the end of the document. Each XML Schema element belongs to an XML Schema type, specified in the `type` attribute.

The elements are distinguished into `complex` and `simple elements`, depending on the kind (simple or complex) of the types they belong to. The XML Schema `simple types` are usually defined as restrictions of the basic datatypes provided by XML Schema (i.e., `xs:string`, `xs:integer`, etc.). Moreover, the XML Schema `complex types` represent classes of XML constructs that have common features, represented by their elements and attributes.

Regarding the document structure, XML Schema, support rich structuring capabilities. The XML Schema elements may either have a predefined order as is specified in the XML Schema element `sequence` or be unordered as is specified in the XML Schema elements `choice` and `all`. The XML Schema `sequence`, `choice` and `all` elements may be nested. The minimum and maximum number of occurrences of the elements, choices and sequences are specified, respectively, in the `minOccurs` and `maxOccurs` attributes (absent `minOccurs` and/or `maxOccurs` correspond to values of 1). Reusable complex structures, combining `sequence`, `choice` and `all` elements may be defined as model `groups`. Finally, the reuse of element definitions is supported by the `substitutionGroup` attribute, which states that the current element is a specialization of another element.

The XML Schema `attributes` describe element features with values of simple type and may form `attribute groups` comprised of attributes that should be used simultaneously.

Default and fixed values may be specified for both attributes and simple type elements, in the `default` and `fixed` attributes respectively. Inheritance is supported for both simple and complex types, and the base types are referenced in the `base` attribute of the type definitions. All the XML Schema constructs may have textual annotations, specified in their `annotation` element.

The top-level XML Schema constructs (`attributes`, `elements`, `simple` and `complex types`, `attribute` and `model groups`) have unique `names` (specified in their `name` attribute). The nested elements and attributes have unique names in the context of the complex types in which they are defined, while the nested (complex and simple) types are unnamed. All the XML Schema constructs may have unique identifiers (specified in their `id` attribute). The top-level constructs may be referenced by other constructs using the `ref` attribute.

XML Schema 1.1. Currently, a new version of XML Schema, XML Schema 1.1 [4][5] is under development and has reached the W3C working draft level of standardization. XML Schema 1.1 is backwards compatible with XML Schema 1.0 and introduces several new features and mechanisms. The most important among them are discussed in the following paragraphs.

The new `assert` element is used to make assertions about element and attribute values, specify relationships and enforce constraints on elements and/or attributes above and beyond the constraints specified in their declarations.

The `alternative` element allows an element to have as type a member of a set of types.

The `override` element replaces, in XML Schema 1.1, the deprecated XML Schema 1.0 `redefine` element.

The `all` element has been enhanced in XML Schema 1.1 to allow elements with multiple occurrences. The `substitutionGroup` element has also been modified in XML Schema 1.1 and allows an element to be substituted by multiple elements. Finally, the `any` and `anyAttribute` elements have been enriched with additional attributes that allow to indicate the extension elements or attributes not allowed in an element.

Regarding the datatypes, XML Schema 1.1 introduces several new features: (a) The new `error` datatype is used to trigger an error; (b) The `anyAtomicType` is introduced, which represents the union of the value spaces of all the primitive types; (c) The XML Schema 1.1 `dateTimeStamp` datatype is introduced, which is identical to the `dateTime`, except from that it requires the time zone to be specified; and (d) The XML Schema 1.1 `yearMonthDuration` and `dayTimeDuration` datatypes are introduced, which are constrained versions of the `duration` datatype, and several others.

Finally, several facets are introduced, like, for instance, the `assertion` facet, which is used to constrain a `simpleType`; the `explicitTimezone` facet which is used with `date` datatypes to specify whether the time zone is required; and the `minScale` and `maxScale` facets which are used with the XML Schema 1.1 `precisionDecimal` datatype, in order to constrain the size of the exponent.

2.2.2 RDF Schema

The *RDF Schema* (*RDFS*) [3] is an extension of RDF designed to describe, using a set of reserved words called the RDFS vocabulary, resources and/or relationships between resources. It provides constructs for the description of types of objects (`classes`), type hierarchies (`subclasses`), properties that represent object features (`properties`) and property hierarchies (`subproperty`).

In particular, a `Class` in RDFS corresponds to the generic concept of a type or category, somewhat like the notion of a class in object-oriented languages, and is defined using the construct `rdfs:Class`. The resources that belong to a class are called its `instances`. An `instance` of a class is a resource having an `rdf:type` property whose value is the specific class. Moreover, a resource may be an instance of more than one classes. Classes can be organized in a hierarchical fashion using the construct `rdfs:subClassOf`. A property in RDFS is used to characterize a class or a set of classes and is defined using the construct `rdf:Property`. The `rdfs:domain` construct is used to indicate that a particular property applies to a designated class, and the `rdfs:range` construct is used to indicate the values of a particular property. In a similar way with the class hierarchies, RDFS provides the `rdfs:subPropertyOf` construct for the definition of property hierarchies.

2.2.3 OWL

The *Web Ontology Language* (*OWL*) [7][8] is the standard language for defining and instantiating Web ontologies. OWL and RDFS have several similarities. OWL is defined as a vocabulary like RDF, however OWL has richer semantics.

An OWL `Class` is defined using the construct `owl:Class` and represents a set of individuals with common properties. All the OWL classes are considered to be subclasses of the class `owl:Thing` and superclasses of the class `owl:Nothing`. Moreover, OWL provides additional constructors for class definition, including the basic set operations, union, intersection and complement that are implemented, respectively, by the constructs `owl:unionOf`, `owl:intersectionOf`, and `owl:complementOf` and several other constructors like, for example, `owl:oneOf`, `owl:equivalentClass`, etc. Regarding the individuals, OWL allows to specify two individuals to be identical or different through the `owl:sameAs` and `owl:differentFrom` constructs. Unlike RDF Schema, OWL distinguishes a property whose range is a datatype value from a property whose range is a set of resources. Thus, the OWL `Datatype properties` are relations between class individuals and XML schema datatypes and are defined using the construct `owl:DatatypeProperty`; The OWL `Object properties` are relations between class individuals and are defined using the `owl:ObjectProperty` construct. Finally, two properties may be stated to be equivalent, using the construct `owl:equivalentProperty`.

OWL 2. The *OWL 2 Web Ontology Language* (*OWL 2*) [8] is a W3C recommendation since the October of 2009. OWL 2 has a very similar overall structure with OWL 1 and is backwards compatible with it, while it introduces a plethora of new features.

Some of these new features are commonly referred as *syntactic sugar*, since these features do not change the expressiveness or the semantics of OWL 1; they have been introduced in order to make some common statements easier to be constructed like, for instance, the `owl:AllDisjointClasses` and `owl:disjointUnionOf` that are used to specify the disjoint classes more easily than using the `owl:disjointWith` and `owl:unionOf` OWL 1 statements.

In addition, a large number of the new OWL 2 features enhance the language expressiveness by adding restrictions and new characteristics over the OWL properties. Among them there are the qualified property cardinality restrictions that are specified using the new constructs `owl:minQualifiedCardinality`, `owl:maxQualifiedCardinality` and `owl:qualifiedCardinality`. Moreover, reflexive, irreflexive, and asymmetric properties are supported using, respectively, the `owl:ReflexiveProperty`, `owl:IrreflexiveProperty` and `owl:AsymmetricProperty` constructs. In addition, a set of classes can be specified to be mutually disjoint using the `owl:propertyDisjointWith` and `owl:AllDisjointProperties` constructs. Furthermore, the new property owl:hasSelf has been introduced to allow relating a class to itself. OWL 2 also allows stating that an individual should not hold certain values for certain properties using the new owl:NegativeObjectPropertyAssertion and owl:NegativeDataPropertyAssertion statements. Finally, a very useful, new feature allows a property to be defined in terms of a chain of object properties, using the construct owl:propertyChainAxiom.

The newly introduced owl:hasKey construct can be used to define that the instances of a class is identified by a set of datatype or object properties, thus providing an identity constraint mechanism.

OWL 1 is highly dependent on XML Schema regarding both the built-in and the user-defined datatypes. Using OWL 2, the users can integrate in their ontologies the datatype definitions, using the XML Schema datatypes and constraints. Moreover, OWL 2 has introduced two new built-in types: owl:real and owl:rational. Finally, OWL 2 allows new datatypes to be defined as the complement of existing datatypes, using the owl:datatypeComplementOf statement.

OWL 2 provides top and bottom object and datatype properties analogous to the owl:Thing and owl:Nothing classes for representing the universal and the empty class. In particular, OWL 2 provides the owl:topObjectProperty and owl:bottomObjectProperty properties, corresponding to the universal and the empty object property respectively, as well as the owl:topDataProperty and owl:bottomDataProperty properties, corresponding to the universal and the empty datatype property respectively.

2.2.4 Schema Level Comparison

In this section we provide a brief comparison between the schema definition languages adopted by the XML the SW worlds. Our comparison is based on a set of generic characteristics over the schema languages presented in the previous sections.

Table 2 presents an overview of this comparison. For each language we present: (a) Its Model Type; (b) Its Concrete Syntax, that is, how language elements are represented; (c) The Basic Constructs of its model; (d) The language Semantics; (e) The Identity Constraint mechanism supported by the schema definition languages; and (f) An overview of the User-Defined Datatypes mechanism (if such a mechanism is supported).

It can be observed from Table 2 that the XML Schema provides a flexible mechanism to support identity constraints as well as rich capabilities for defining user datatypes. In contrast, none of the RDF Schema and OWL 1 languages supports identity constraints or user-defined datatypes. These limitations have been overcome by OWL 2, which introduces an identity constraint mechanism using the hasKey construct and supports user defined datatypes.

Table 2 Comparison of the Schema Definition Languages in terms of a set of generic Characteristics

	Schema Definition Languages			
Characteristic	**XML Schema**	**RDF/S**	**OWL 1**	**OWL 2**
Model Type	Hierarchical	Direct Label Graph	Direct Label Graph	Direct Label Graph
Concrete Syntax	XML	RDF/XML, N3, N-Triples, Turtle	RDF/XML, N3, N-Triples, Turtle	OWL/XML, Functional, Manchester, RDF/XML, N3, N-Triples, Turtle
Basic Constructs	Simple type, Complex type, Attribute, Element, Attribute group, Sequence Choice Annotation Extension, Restriction, Unique, Key, Keyref, Substitution Group, *Alternative*⁺, *Assert*⁺, *Override*⁺, *Redefine*⁺, etc.	Statement, Class, Property, Resource, type, subject, predicate, object, subClassOf, subPropertyOf, domain, range, Datatype, Literal, Bag, Seq, List, Alt, BlankNode	Class, Datatype Property, Object Property, Individual, Thing, Nothing, equivalentClass, intersectionOf, unionOf, complementOf, disjointWith, minCardinality, sameAs, oneof, hasValue, TransitiveProperty, FunctionalProperty, etc	InverseOf, NegativePropertyAssertion, propertyChainAxiom, minQualifiedCardinality, hasSelf, AsymmetricProperty, AllDisjointClasses, disjoint, ReflexiveProperty, maxQualifiedCardinality, UnionOf, etc.
Semantics	Informal	Model Theory	Model Theory, RDF Graphs	Model Theory, Extension of *SROIQ DL*
Identity Constraints	Unique, Key, Keyref	—	—	hasKey
User-Defined Datatypes	minInclusive, minExclusive, maxLength, length, totalDigits, etc.	—	—	xsd:minInclusive, xsd:minExclusive, owl:onDatatype, owl:withRestrictions, etc.

Note. The ⁺ indicates XML Schema 1.1 constructs.

2.3 *Query Level*

In this section we present the query languages used in the XML and SW worlds. Section 2.3.1 outlines the XPath, XQuery 1.0 and XQuery 1.1 languages, used in the XML world. The SPARQL query language that is adopted in the SW world is presented in Section 2.3.2. Finally, a comparison between the query languages adopted by the XML and SW worlds is presented in Section 2.3.3.

2.3 1 XPath and XQuery

Path expressions play a significant role in XML query evaluation, since the path expressions are used to traverse the tree representations of the XML documents and select a node or a set of nodes. The *XML Path Language* (*XPath*) [9] is a W3C recommendation that specifies a path language capable of describing path expressions on the tree data model of XML. The XPath language is essentially a subset of the XQuery language, which is the W3C standard for querying XML data; thus, in the rest of this chapter we will only refer to XQuery language.

The *XML Query Language* (*XQuery*) [10] is based on a tree-structured model of the XML document content. XQuery exploits XPath expressions to address specific parts of the XML documents. The basic structure of the most XQuery queries is the `FLWOR` expression. `FLWOR` stands for the `For`, `Let`, `Where`, `Order By` and `Return` XQuery clauses. `FLWOR`s, unlike path expressions, allow the users to manipulate, transform, and sort the query results.

The `For` and `Let` clauses generate a list of tuples preserving document order, with each tuple consisting of one or more bound variables. In particular, the `For` clause sets up an iteration over the tuples in the tuple list. The `Let` clause is used to set the value of a variable. However, unlike the `For` clause, it does not set up an iteration. The optional `Where` clause serves as a filter for the tuples generated by the `For` and `Let` clauses. The optional `Order By` clause is used to order the results; if no `Order by` clause exists, the order of the tuple list is determined by the `For` and `Let` clauses and by the document order. Finally, every XQuery expression has a `Return` clause that always comes last. The `Return` clause specifies the XML nodes that are included in the results and probably how they are formatted. In addition, XQuery supports conditional expressions based on the keywords `if - else if - else`.

A key feature of XQuery is the large number of built-in functions and operators provided (over 100) [11], covering a broad range of functionality That includes functions for manipulating strings, dates, combine sequences of elements, etc. Moreover, XQuery supports user-defined functions, defined either in the query itself, or in an external library. Both built-in and user-defined functions can be called from almost any place in a query.

2.3.2 SPARQL

The *SPARQL Protocol and RDF Query Language* (*SPARQL*) [15] is a W3C recommendation and it is today the standard query language for RDF data. The evaluation of SPARQL queries is based on graph pattern matching. A *Graph Pattern (GP)* is defined recursively and contains *Triple patterns* and SPARQL operators. The operators of the SPARQL algebra which can be applied on *graph patterns* are: `AND` (i.e., conjunction), `UNION` (i.e., disjunction), `OPTIONAL` (i.e., optional patterns, like left outer join) and `FILTER` (i.e., restriction). The *Triple patterns* are just like *RDF triples* except that each of the subject, predicate and object parts may be a variable. A sequence of conjunctive triple patterns is called *Basic Graph Pattern (BGP)*. The SPARQL `Where` clause consists of a *Graph Pattern* (*GP*).

SPARQL allows four query forms: `Select`, `Ask`, `Construct` and `Describe`. The `Select` query form returns a solution sequence, i.e., a sequence of variables and their bindings. The `Ask` query form returns a Boolean value (`yes` or `no`), indicating whether a query pattern matches or not. The `Construct` query form returns an RDF graph structured according to the *graph template* of the query. Finally, the `Describe` query form returns an RDF graph which provides a "description" of the matching resources. Thus, based on the query forms, the SPARQL query results may be *RDF Graphs*, *SPARQL solution sequences* and *Boolean* values.

SPARQL provides various solution sequence modifiers that can be applied on the initial solution sequence in order to create another, user desired, sequence. The supported SPARQL solutions sequence modifiers are: `Distinct`, `Reduced`, `Limit`, `Offset` and `Order By`. While the `Distinct` modifier ensures that duplicate solutions are eliminated from the solution set, the `Reduced` modifier simply allows them to be reduced. The `Limit` modifier puts an upper bound on the number of solutions returned. Moreover, the `Offset` modifier returns the solutions starting after the specified number of solutions. Finally, the `Order By` modifier, establishes the order of a solution sequence.

SPARQL 1.1. The SPARQL 1.1 is the result of the W3C SPARQL working group [14] on the extension of the SPARQL query language in several aspects. SPARQL 1.1 includes the components *SPARQL 1.1 Query*, *Update*, *Protocol*, *Service Description*, *Property Paths*, *Entailment Regimes*, *Uniform HTTP Protocol for Managing RDF Graphs*, and *Federation Extensions*.

The SPARQL 1.1 tries to eliminate the main limitations of the current SPARQL version like aggregate functions, nested queries, update operations, paths, and several other issues.

The aggregate functions that are supported from almost all the query languages will also be included in the SPARQL 1.1 Query [16]. In particular, SPARQL 1.1 will support the following aggregate functions: `COUNT`, `SUM`, `MIN/MAX`, `AVG`, `GROUP_CONCAT`, and `SAMPLE`.

In addition, nested queries, which are very important in cases where the result from one query is used as an input to another query, will be supported by SPARQL 1.1. Moreover, in order to implement negation, SPARQL 1.1 has adopted the new operators `NOT EXISTS` and `MINUS`.

In the current SPARQL version, a `SELECT` query may project only variables. SPARQL 1.1 allows the `SELECT` queries to project any SPARQL expression: Using the keyword `AS` in the `SELECT` clause, the result of a SPARQL expression is bound to the new variable specified by `AS` and this new variable is going to be projected.

In several cases, in order to find a node, it is necessary to express queries that use fixed-length paths to traverse the RDF graph. The SPARQL 1.1 Property

Paths [18] extends the current basic graph patterns in order to support the expression of path patterns.

SPARQL 1.0 can be used only as a retrieval query language, since it does not support data management operators. The SPARQL Update 1.1 [17] includes several features for graph management. The `INSERT` and `INSERT DATA` operations are exploited to insert new triples in RDF graphs. In addition, the `DELETE` and `DELETE DATA` operations are used to delete triples from an RDF graph. The `LOAD` and `CLEAR` operations perform a group of update operations as a single action. In particular, the `LOAD` operation copies all the triples of a graph into the target graph, while the `CLEAR` operation deletes all the statements from the specified graph. Finally, in order to create a new RDF graph in an RDF graphs repository or to delete an RDF graph from an RDF graphs repository the `CREATE` and `DROP` operations are introduced.

2.3.3 Query Level Comparison

In this section we present a comparison between the query languages adopted in the XML and SW worlds. Our comparison has been based on a set of generic characteristics (Table 3) and on a set of language features (Table 4).

In Table 3 we give an overview of our comparison over a set of generic characteristics. For each query language, we present: (a) Its `Language Type`; (b) Its `Input Data Model`, that is, the data format accessed by the language; (c) The `Basic Elements` of the language; (d) The method exploited by the query language to evaluate the queries (`Evaluation Method`); (e) The `Evaluation Clause`, that is, the clauses of the query language that are used to describe the evaluation settings; (f) The algebra operators defined at each language (`Evaluation Operators`); The language `Semantics`; (g) The `Query Types` supported by the language; (h) The `Result Form Declaration`, that is, how and where the user can specify the form of the results inside a query; (i) The available result forms for each language (`Results Form`); (k) The `Results Modification`, that is, how the return results can be modified; (l) The mechanisms provided for defining evaluation `Restrictions`; (m) The built-in `Operations & Functions` provided by each query language; and (n) The `Logic` adopted by the language operations and functions.

It can be observed from Table 3 that a large number of differences can be identified. These differences are mainly due to: (a) The different data models types (i.e., tree – graph); (b) The functional style of the XQuery language; (c) The different assumptions made (i.e., the *Closed World Assumption* in the XML world vs. the *Open World Assumption* in the SW world); (d) The logic adopted (XQuery adopts Boolean logic while SPARQL adopts three-valued logic), and several others.

Table 3 Comparison of the Query Languages in terms of a set of generic Characteristics

	Query Languages	
Characteristic	**XQuery**	**SPARQL**
Language Type	Functional, Declarative	Declarative
Input Data Model	XML	RDF
Basic Elements	XPaths expressions, For, Let, Where, Return , Order by, If - Else If – Else	Select, Construct, Ask, Describe, Union, Optional, Filter, Limit, Offset, Reduced, Distinct, Order by
Evaluation Method	Tree Traverse, Tree Matching	Graph Matching
Evaluation Clause	Everywhere in the Query	Where Clause
Evaluation Operators	No Standards Operators. Operators can be Implemented Using "Simple" Operators and Clauses (e.g., =, *If–Else If*, etc.)	Union, And, Optional, Filter
Semantics	Closed World Assumption (*CWA*)	Open World Assumption (*OWA*)
Query Types	Not Standard - Combination of FLOWR Expressions	Select, Construct, Ask, Describe
Result Form Declaration	Return Clause Syntax	Query Forms (i.e., Select, Ask, etc.)
Result Form	Flexible	Sequence of Variables Bindings, RDF Graph, Boolean
Result Modifications	Functions	Solution Sequence Modifiers
Restrictions	XPath Predicates, Where Clause Conditions, *If–Else If*	Triple patterns with constant parts, Filter
Operations & Functions	&&, ||, !, =, !=, >, <, >=, <=, +, -, *, /, abs, floor, concat, substring, string-length, lower-case, starts-with, matches, replace, exists, distinct-values, insert-before, empty, count, avg, max, min, sum, etc.	&&, ||, !, =, !=, >, <, >=, <=, +, -, *, /, bound, lang, regex, isIRI, isBlank, isLiteral, str, datatype, sameTerm, langMatches
Logic	Boolean Logic (True / False)	Three-valued Logic (True / False / Error)

Table 4 presents a comparison between the XQuery, the SPARQL 1.0 and the SPARQL 1.1 query languages over several features. For each feature we indicate if the languages fully support it (✓), partially support it (❖) or do not support it (✗). Table 4 is based on the W3C specifications and working drafts and does not consider possible languages extension proposed from other parties.

It can be observed from Table 4 that the XQuery language supports almost all the features, except from the `Support Schema` feature, which is partially supported by the XQuery language. Since the XQuery language can partially exploit schema information, it supports only path expressions, while can not support type-based queries. Note that the `Full Text Search` and `Update` features have been recently introduced as W3C recommendations (see [13] and [12] respectively).

We can observe from Table 4, the SPARQL 1.1 (which is currently a W3C working draft) covers almost all the features which are not supported by the current SPARQL version (SPARQL 1.0). However, the `Full Text Search` feature is not supported by SPARQL 1.1; several SPARQL implementations (e.g., Jena, Sesame, etc.), though, support full text search. In addition, neither the `User-Defined Functions` nor the `Rich Functions` features are supported by the SPARQL 1.1 which particularly supports about 20 build-in functions. Finally, the SPARQL 1.1 does not support the `Flexible Result Form` and partially supports the `Restructure Result` features; SPARQL can restructure the results of `CONSTRUCT` queries, however these results can be only RDF graphs.

Table 4 Comparison of the XQuery and SPARQL Query Languages

	Query Languages		
Feature	**XQuery**	**SPARQL 1.0**	**SPARQL 1.1**
Paths – Reg. Expr.	✓	✗	✓ [18]
Full Text Search	✓ [13]	✗	✗
Nested Queries	✓	✗	✓
Aggregate Functions	✓	✗	✓
Restructure Result	✓	❖	❖
Flexible Result Form	✓	✗	✗
Support Schema	❖	✓	✓
Negation	✓	✗	✓
User-Defined Functions	✓	✗	✗
Rich Functions	✓	✗	✗
Update	✓ [12]	✗	✓ [17]

Legend: ✓ Supported ✗ Not Supported ❖ Partially

3 Motivating Example — Use Cases

In this section, we outline two "real-word" scenarios in order to illustrate the need for bridging the XML and the SW world. In our examples, three hypothetically autonomous partners are involved: `Digital Library X` (Fig. 1) which is an audiovisual digital library that belongs to an institution or an organization, as well as two video on demand servers (Fig. 2), `VoD Server A` and `VoD Server B`. Each partner provides different content and has adopted different technologies to represent and manage their data.

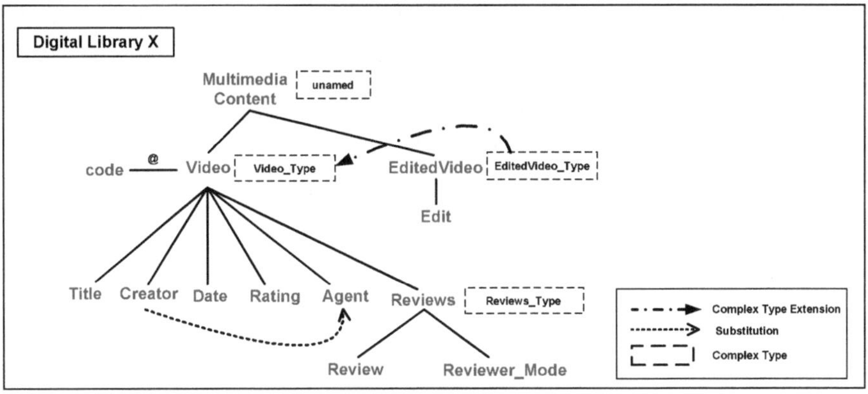

Fig. 1 An excerpt of the XML Schema of Digital Library X describing Audiovisual Material

In particular, `Digital Library X` has adopted XML-related technologies (i.e., XML, XML Schema, and XQuery) and its contents are described in XML syntax, while both servers have chosen SW technologies (i.e., RDF/S, OWL, and SPARQL) for their content. In addition, `Digital Library X` provides information for audiovisual material from several domains, while the video on demand servers only for specific types of movies (i.e., `New Greek Movie` and `Newly Released and Highly Rated`).

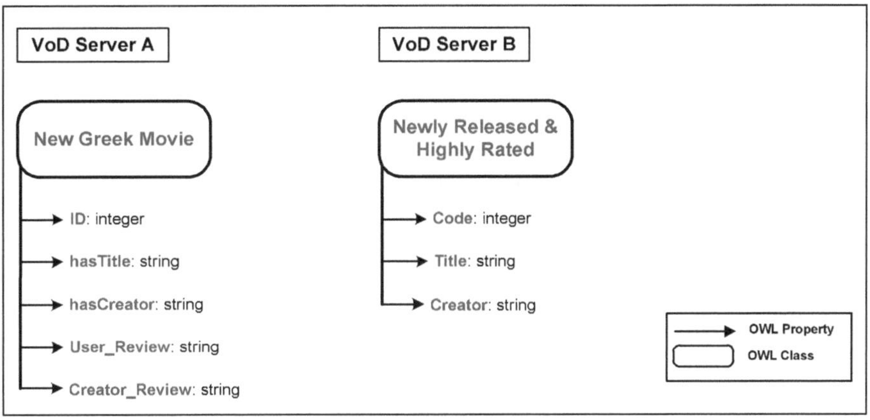

Fig. 2 The Ontologies of `VoD Server A` and `VoD Server B` describing movies

1st Scenario: *Querying XML data based on an automatically generated ontology*. Consider that `Digital Library X` wants to publish their data in the Web of Data using SW technologies, a scenario which is very common nowadays.

In the Linked Data era, a large number of organizations, institutes and companies (e.g., *DBpedia*, *ACM*, *IEEE*, *IBM*, *NASA*, *DBLP*, *BBC*, *NSF*, *Geonames*, *PubMed*, etc.) publish their data utilizing SW standards and technologies. In particular, they offer SPARQL endpoints (i.e., SPARQL-based search services) over their data. In this case, a schema transformation and a query translation mechanism are required. Using the schema transformation mechanism, the XML Schema of `Digital Library X` will be transformed to an ontology. Then, the query translation mechanism will be used to translate the SPARQL queries posed over the generated ontology to XQuery queries over the XML data.

As a special case of this scenario, consider the case, in which `Digital Library X` wants to publish their data in the Web of Data, however, unlike the 1st scenario, `Digital Library X` wants to use existing, well accepted vocabularies (e.g., Friend of a Friend (*FOAF*)[1], Dublin Core (*DC*)[2], etc.).

2nd Scenario: *Querying XML data based on existing ontologies*. Consider Web of Data users and/or applications, who express their queries or have implemented their APIs over the ontologies of `VoD Server A` and/or `VoD Server B` using the SPARQL query language. These users and applications should be able to have access to `Digital Library X`, without being required to adjust their working environment (e.g., query language, schema, API, etc.).

In these cases, a mapping model and a query translation mechanism are required. In such a case, an expert specifies the mappings between the ontologies of `VoD Server A` and `VoD Server B` and the XML Schema of `Digital Library X`. These mappings are, then, exploited by the query translation mechanism in order to translate the SPARQL queries posed over the ontology of `VoD Server A` and `VoD Server B` to XQuery queries to be evaluated over the XML data of `Digital Library X`.

To sum up, in the Linked Data era, publishing legacy data and offering SPARQL endpoints has become a major research challenge. Although, several systems (e.g., *D2R Server* [76], *OpenLink Virtuoso* [78], *TopBraid Composer*[3], etc.) offer SPARQL endpoints[4] over relational data, to best of our knowledge there is no system supporting XML data (except from the SPARQL2XQuery Framework presented here).

4 Bridging the Semantic Web and XML Worlds — A Survey

In order to overcome the heterogeneity among the information systems, a large number of *data integration* [23] and *data exchange systems* [24] have been proposed in the literature. The *data integration systems* provide mechanisms for querying heterogeneous sources in a uniform way based on a global schema. The

---

[1] `http://www.foaf-project.org/`
[2] `http://dublincore.org/schemas/rdfs/`
[3] `http://www.topquadrant.com/products/TB_Composer.html`
[4] Virtual SPARQL endpoints (i.e., with no need to transform the relational data to RDF data).

data exchange systems (also known as *data transformation/translation systems*) restructure the data from the sources according to a global schema. In recent research works, semantics are exploited to bridge the heterogeneity gap among the information systems and provide semantic integration and interoperability [55][56].

In the context of XML, the first research efforts have attempted to provide interoperability and integration between the relational and XML worlds (e.g., [57][58][59][60][61][62][63]). In addition, several approaches focus on data integration and exchange over heterogeneous XML sources (e.g., [64][65][66][67] [68][69][70][71][72]).

Regarding the XML and SW worlds, numerous approaches for transforming XML Schema to ontologies and/or XML data to RDF and vice versa have been proposed in the literature. Moreover, some other recent approaches combine SW and XML technologies in order to transform XML data to RDF and vice versa.

In the rest of this section we present the existing approaches that deal with the interoperability and integration issues between the XML and SW worlds (Section 4.1). The recent approaches are described and compared in Section 4.2. Finally, a discussion about the drawbacks and the limitations of the current approaches is presented in Section 4.3.

4.1 Existing Approaches — An Overview

In this section, we present the literature related to interoperability and integration issues between the SW and XML worlds. Table 5 and present the proposed systems in terms of the environment characteristics and the supported operations. These systems have been distinguished into *data integration systems* (Table 5) and *data exchange systems* (Table 6) and are presented in a chronological order.

Table 5 provides an overview of the *data integration* systems in terms of the `Environment Characteristics` and the supported `Operations`. In particular, the system described in each row is specified in the first column (`System`), the environment characteristics are shown in columns 2–4 and the supported operations are shown in columns 5–6. The environment characteristics include the `Data Models` of the underlying data sources (2nd column), the involved `Schema Definition Languages` (3rd column) and the supported `Query Languages` (4th column). The operations include the `Query Translation` operation (5th column) and the `Schema Transformation` operation (6th column). If a schema transformation operation is supported the value is the operation description; if the method does not support schema transformation, the value is `"no"`.

Table 6 provides an overview of the *data exchange* systems and is structured in a similar way with Table 5. The system described in each row is specified in the first column (`System`), the `Environment Characteristics` are shown in columns 2–3 and the supported `Operations` used are shown in columns 4–5. The environment characteristics include the `Data Models` of the underlying data sources (2nd column) and the involved `Schema Definition Languages` (3rd column). The operations include the `Schema Transformation` operation (4th column), the indication for the `Use of an Existing Ontology` (4th column) and the `Data Transformation` mechanism (5th column). If a schema transformation operation is supported, the value of the third column is the operation description; if the method does not support schema transformation, the value is `"no"`. If the value of the fourth column is `"yes"`, the method supports mappings between XML Schemas and existing ontologies and, as a consequence the XML data are transformed according to the mapped ontologies. Finally, if a data transformation mechanism is provided, the fifth column has its description as value and the value `"no"` if the system does not provide a data transformation mechanism.

It can be observed from Table 5 that the *data integration systems* are older, thus they do not support the currently standard technologies (i.e., XML Schema, OWL, RDF, SPARQL, etc.). Notice also, that, although the *data exchange systems* shown in Table 6 are more recent, they do not support an integration scenario neither they provide query translation mechanism. Instead, they focus on data and schema transformation, exploring how the RDF data can be transformed in XML syntax and/or how the XML Schemas can be expressed as ontologies and vice versa. In the next section, we describe the latest efforts; most of them focus on combining the XML and the SW technologies in order to transform the underlying data.

4.2 Recent Approaches

In this section, we present the latest approaches related to the support of interoperability and integration between the XML and SW worlds. The former approaches utilize the current W3C standards technologies (e.g., XML Schema, RDFS, OWL, XQuery, SPARQL, etc.). With the *DTD2OWL* [54] system be an exception, since it is focus on transforming DTD schemas (instead of XML Schemas) to OWL ontologies. The most of the latest efforts focus on combining the XML and the SW technologies in order to provide an interoperable environment. In particular, they merge SPARQL, XQuery, XPath and XSLT features to transform XML data to RDF and vice versa.

The *Semantic Annotations for WSDL* (*SAWSDL*) W3C Working Group [25] uses XSLT to convert XML data into RDF and a combination of SPARQL and XSLT for the inverse transformation. Additionally, the *Gleaning Resource Descriptions from Dialects of Languages* (*GRDDL*) W3C Working Group [26] uses XSLT to extract RDF data from XML.

XSPARQL [31] combines SPARQL and XQuery in order to achieve the transformation from XML to RDF and back. In the XML to RDF scenario, XSPARQL uses a combination of XQuery expressions and SPARQL `Construct` queries. The XQuery expressions are used to access XML data and the SPARQL `Construct` queries are used to convert the accessed XML data into RDF. In the

Table 5 Overview of the Data Integration Systems in the SW and XML Worlds

Data Integration Systems					
System	Environment Characteristics			Operations	
	Data Model	Schema Definition Language	Query Language	Query Translation	Schema Transformation
STYX (2002) [32][33]	XML	DTD / Graph	OQL / XQuery	OQL → XQuery	no
ICS–FORTH SWIM (2003) [34][35][36]	Relational / XML	DTD / Relational / RDF Schema	SQL / XQuery / RQL	RQL → SQL & RQL → XQUERY	no
PEPSINT (2004) [37][38][39][40]	XML	XML Schema / RDF Schema	XQuery / RDQL	RDQL → XQuery	XML Schema → RDF Schema
Lehti & Fankhauser (2004) [41]	XML	XML Schema / OWL	XQuery / SWQL	SWQL → XQuery	XML Schema → OWL
SPARQL2XQuery [90][91]	XML	XML Schema / OWL	XQuery / SPARQL	SPARQL → XQuery	XML Schema → OWL (Xs2Owl)

Table 6 Overview of the Data Exchange Systems in the SW and XML Worlds

Data Exchange Systems					
System	Environment Characteristics		Operations		
	Data Model	Schema Definition Language	Schema Transformation	Use Existing Ontology	Data Transformation
Klein (2002) [42]	XML / RDF	XML Schema / RDF Schema	no	no	XML → RDF
WEESA (2004) [43]	XML / RDF	XML Schema / OWL	no	yes	XML → RDF
Ferdinand et al. (2004) [44]	XML / RDF	XML Schema / OWL–DL	XML Schema → OWL–DL	no	XML → RDF
Garcia & Celma (2005) [45]	XML / RDF	XML Schema / OWL–FULL	XML Schema → OWL–FULL	no	XML → RDF
Bohring & Auer (2005) [46]	XML / RDF	XML Schema / OWL–DL	XML Schema → OWL–DL	no	XML → RDF
Gloze (2006) [47]	XML / RDF	XML Schema / OWL	no	no	XML ↔ RDF
JXML2OWL (2006) [48]	XML / RDF	XML Schema / OWL	no	yes	XML → RDF
XS2OWL (2007) [20][21]	XML / RDF	XML Schema / OWL–DL	XML Schema → OWL–DL	no	XML ↔ RDF
GRDDL (2007) [25]	XML / RDF	not specified	no	no	XML ↔ RDF [5]
SAWSDL (2007) [26]	XML / RDF	not specified	no	no	XML ↔ RDF [5]
Thuy et al. (2007 & 2008) [49][50]	XML / RDF	DTD / OWL–DL	DTD → OWL–DL [5]	no	XML → RDF [5]
Janus (2008) [51]	XML / RDF	XML Schema / OWL–DL	XML Schema → OWL–DL	no	no
XSPARQL (2008) [31]	XML / RDF	not specified	no	no	XML ↔ RDF [5]

[5] The transformation is performed in a semi-automatic way that requires user intervention.

Table 6 *(continued)*

Data Exchange Systems					
System	Environment Characteristics		Operations		
	Data Model	Schema Definition Language	Schema Transformation	Use Existing Ontology	Data Transformation
Droop et al. (2007 & 2008) [28][29][30]	XML / RDF	not specified	no	no	XML → RDF [5]
Cruz & Nicolle (2008) [52]	XML / RDF	XML Schema / OWL	no	yes	XML → RDF
XSLT+SPARQL (2008) [53]	XML / RDF	not specified	no	no	RDF → XML
DTD2OWL (2009) [54]	XML / RDF	DTD / OWL–DL	DTD → OWL–DL	no	XML → RDF
TopBraid Composer (Maestro Edition) – TopQuadrant [3] (Commercial Product)	XML / RDF	not specified / OWL	XML → OWL	no	XML ↔ RDF [5]
XS2OWL 2.0 (2011) [22]	XML / RDF	XML Schema 1.1 / OWL 2	XML Schema 1.1 → OWL 2	no	XML ↔ RDF

RDF to XML scenario, XSPARQL uses a combination of SPARQL and XQuery clauses: The SPARQL clauses are used to access RDF data, and the XQuery clauses are used to format the results in XML syntax.

In *XSLT+SPARQL* [53] the XSLT language is extended in order to embed SPARQL `SELECT` and `ASK` queries. The SPARQL queries are evaluated over RDF data and the results are transformed to XML using XSLT expressions.

In other approaches, SPARQL queries are embedded into XQuery and XSLT queries [27]. In [28][29][30], XPath expressions are embedded in SPARQL queries. The former approaches try to process XML and RDF data in parallel and benefit from the combination of the SPARQL, XQuery, XPath and XSLT language characteristics. Also, a method that transforms XML data to RDF and translates XPath queries into SPARQL has been proposed in [28][29][30].

4.3 Drawbacks and Limitations — A Discussion

In this section we present a discussion over the existing approaches related to the support of interoperability and integration between the XML and SW worlds and highlight their main drawbacks and limitations.

As already mentioned, the existing data integration systems (Table 5) are quite old and do not support the current standard technologies (e.g., XML Schema, OWL, RDF, SPARQL, etc.). On the other hand, the data exchange systems (Table 6) are more recent. However, they neither support integration scenarios nor they provide query translation mechanisms. Instead, they focus on data transformation. Finally, none of the existing systems can offer SPARQL endpoints (i.e., SPARQL-based search services) over XML data (except from the SPARQL2XQuery Framework presented later in this chapter).

The main drawback of the latest approaches ([25][26][27][28][29][30][31][53]) is that there is no way to express XML retrieval queries using the SPARQL query language.

In particular, the users of these systems are forced to: (a) be familiar with both the XML and SW models and languages; (b) be aware of both ontologies and XML Schemas in order to express their queries; and (c) be aware of the syntax and the semantics of each of the above approaches in order to express their queries. In addition, each of these approaches has adopted its own syntax and semantics by modifying and/or the merging the standard technologies. These modifications may also result in compatibility problems. A consequence of the assumptions made by these approaches is that they have been evaluated over only a few kilobytes of data.

The major limitation of the existing data exchange systems, which provide schema transformation mechanisms, is that they do not support the XML Schema identity constraints (i.e., `key`, `keyref`, `unique`). In addition, none of them supports the XML Schema user-defined simple datatypes. Finally, none of the existing approaches considers the new constructs introduced by XML Schema 1.1.

These limitations have been overcome by the XS2OWL Framework [20][22], which is presented later in this chapter (Section 6).

The XS2OWL Framework belongs to the data exchange systems category. It provides a transformation model for the automatic and accurate expression of the XML Schema semantics in OWL syntax. In addition, it allows the transformation of XML data in RDF format and vice versa. The current version of the XS2OWL Transformation Model [22] exploits the OWL 2 semantics, in order to achieve a more accurate representation of some XML Schema constructs. For instance, the XML Schema identity constraints (i.e., `key`, `keyref`, `unique`) can now be accurately represented in OWL 2 syntax (which was not feasible with OWL 1.0), thus overcoming the most important limitation of the previous version of XS2OWL Transformation Model. In addition, this version supports the new XML constructs introduced by XML Schema 1.1 [4]. To the best of our knowledge, this is the first work that fully captures the XML Schema semantics and support the XML Schema 1.1 constructs.

Regarding the data integration systems and in particular the query translation operation, a major limitation in the existing literature is that there do not exist approaches addressing the SPARQL to XQuery translation.

The SPARQL to SQL translation has been extensively studied and several systems and approaches have been proposed in the literature both for integrating relational data with the SW (e.g., [73][74][75][76][77][78][79]) and for accessing RDF data stored in relational databases (e.g., [80][81][82][83]). However, despite the significant research efforts on the SPARQL to SQL translation, to the best of our knowledge there is no work addressing the SPARQL to XQuery translation. Given the very high importance of XML and the related standards in the web, this is a major limitation in the existing research.

In the Linked Data era, publishing legacy data and offering SPARQL endpoints has become a major research challenge. Although, several systems (e.g., *D2R Server* [76], *OpenLink Virtuoso* [78], *TopBraid Composer*[1]) offer SPARQL endpoints over relational data, to best of our knowledge there is no system supporting XML data.

The SPARQL2XQuery Framework [90][91] presented in this chapter (Section 5), provides a formal model for the expression of mappings from OWL to XML Schema and a generic method for SPARQL to XQuery translation. The SPARQL2XQuery Framework supports both manual and automatic mapping specification. In addition, it has been integrated with the XS2OWL Framework, thus facilitating the automatic mapping generation and maintenance. Moreover, the SPARQL2XQuery Framework has been evaluated over large datasets.

Compared to the latest approaches ([25][26][27][28][29][30][31][53]), in the SPARQL2XQuery Framework working scenarios the users (a) work only on SW technologies; (b) are not expected to know the underlying XML Schema or even the existence of XML data; and (c) they express their queries only in standard (i.e., without modifications) SPARQL syntax.

5 The SPARQL2XQuery Framework — An Overview

In this section, we present an overview of the SPARQL2XQuery Framework [90] [91] that has been developed to provide an interoperable environment between the SW (*OWL/RDF/SPARQL*) and XML (*XML Schema/XML/XQuery*) worlds.

In particular, the SPARQL2XQuery Framework offers an interoperable environment where SPARQL queries are automatically translated to XQuery queries, in order to access XML data across the Web. The SPARQL2XQuery Framework provides a mapping model for the expression of OWL to XML Schema mappings and a method for SPARQL to XQuery translation. To this end, our Framework supports both manual and automatic mapping specification between ontologies and XML Schemas. In order to support the automatic mapping specification scenario, the SPARQL2XQuery has been integrated with our XS2OWL Framework which generates OWL ontologies that fully capture the XML Schema semantics. The system architecture is presented in Fig. 3.

The SPARQL2XQuery Framework provides an essential component for the Linked Data environment that allows setting SPARQL endpoints over existing XML data, as well as a fundamental part of ontology-based integration frameworks involving XML sources.

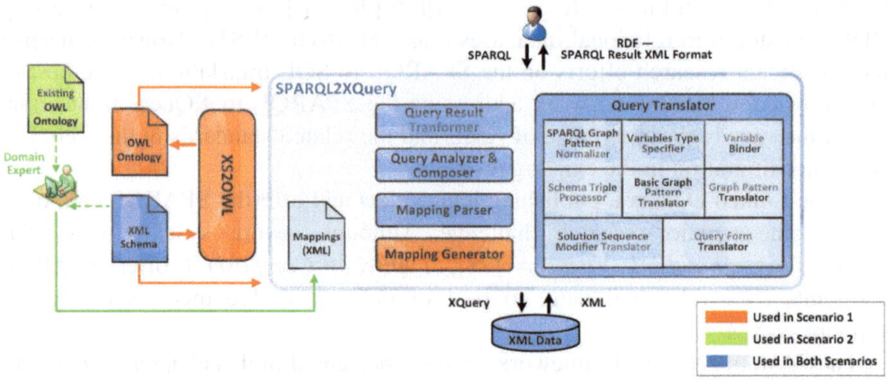

Fig. 3 System Architecture. In the first scenario, the XS2OWL Framework is used to generate an OWL ontology from the XML Schema. The mappings are automatically specified and stored. In the second scenario, a domain expert specifies the mappings between an existing ontology and the XML Schema. In both scenarios, SPARQL queries are processed and translated into XQuery queries for accessing the XML data. The results are transformed in the preferred format and returned to the user.

As shown in Fig. 3, our working scenarios involve existing XML data that follow one or more XML Schemas. Moreover, the SPARQL2XQuery Framework supports two different scenarios:

1st Scenario: Querying XML data based on automatically generated OWL ontologies. In that case, the following steps take place:

(a) Using the XS2OWL Framework, the XML Schema is expressed as an OWL ontology.
(b) The Mappings Generator component, taking as input the XML Schema and the generated ontology; automatically generates, maintains and stores the mappings between them in XML format.
(c) The SPARQL queries posed over the generated ontology are translated by the Query Translator component to XQuery expressions.
(d) The query results are transformed by the Query Result Transformer component into the desired format (SPARQL Query Result XML Format [19] or RDF).

In this case, the SPARQL2XQuery Framework can be utilized as a fundamental component of *hybrid ontology–based integration* [55] frameworks (e.g., [88]), where the schemas of the XML data sources are represented as OWL ontologies and these ontologies are further mapped to a global ontology.

2nd Scenario: Querying XML data based on existing OWL ontologies. In this case the following steps take place:

(a) Existing OWL ontologies are manually mapped by a domain expert to the XML Schema.
(b) The SPARQL queries posed over the existing ontology are translated to XQuery expressions.
(c) The query results are transformed into the desired format.

In both scenarios, the systems and the users that pose SPARQL queries over the ontology are not expected to know the underlying XML Schemas or even the existence of XML data. They express their queries only in standard SPARQL, in terms of the ontology that they are aware of, and they are able to retrieve XML data.

Our Framework provides the following operations:

1. **Schema Transformation.** Every XML Schema can be automatically transformed into OWL ontology, using the XS2OWL Framework.
2. **Mappings Generation.** The mappings between the OWL representations of the XML Schemas and the XML Schemas and their OWL representations can be either automatically detected or manually specified and stored as XML documents.
3. **Query Translation.** Every SPARQL query that is posed over the OWL representation of the XML Schemas (first scenario) or over the existing ontology (second scenario), is translated into an XQuery query that can be answered from the XML data.
4. **Query Result Transformation.** The query results are transformed in the preferred format.

5. **XML – RDF Transformation.** Transformation of XML data in RDF syntax and vice versa.

Finally, the SPARQL2XQuery Framework is going to be integrated in an ontology-based mediator framework [87][88][89] that we are developing now and is going to provide semantic interoperability and integration support between distributed heterogeneous sources using the standard SW and XML technologies.

6 The XS2OWL Framework — An Overview

In this section we describe the schema transformation process exploited in the first scenario supported by the SPARQL2XQuery Framework in order to express the XML Schema semantics in OWL syntax. This is accomplished by the XS2OWL Framework [20][22], which implements the XS2OWL *Transformation Model* . The XS2OWL Transformation Model allows the accurate representation of the XML Schema constructs in OWL syntax without any loss of the XML Schema semantics.

Here we present, an extended and updated version of the XS2OWL Transformation Model, XS2OWL 2.0 [22] exploits the OWL 2 semantics, in order to achieve a more accurate representation of some XML Schema constructs. For instance, the XML Schema identity constraints (i.e., `key`, `keyref`, `unique`) can now be accurately represented in OWL 2 syntax (which was not feasible with OWL 1.0), thus overcoming the most important limitation of the previous version of XS2OWL Transformation Model. Additionally, this version supports the new XML constructs introduced by XML Schema 1.1. To the best of our knowledge, this is the first work that fully captures the XML Schema semantics and support the XML Schema 1.1 constructs.

As shown in Fig. 4, the XS2OWL Framework takes as input an XML Schema *XS* and generates: a) An OWL *Schema ontology* O_S that captures the XML Schema semantics; and (b) A *Backwards Compatibility ontology* O_{BC} which keeps the correspondences between the O_S constructs and the *XS* constructs, and systematically capture the semantics of the XML Schema constructs that cannot be directly captured in O_S (since they cannot be represented by OWL semantics).

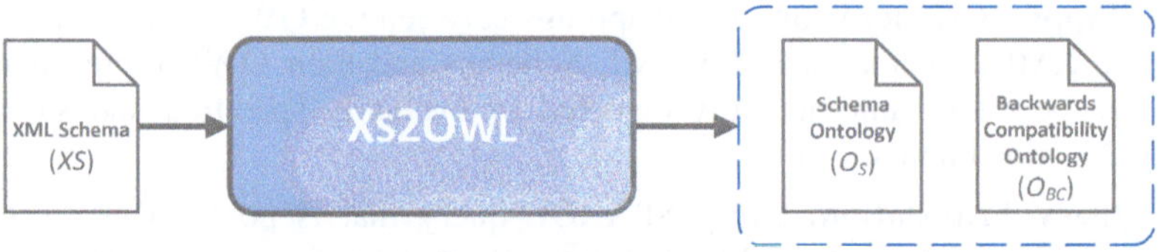

Fig. 4 The XS2OWL Framework

The *OWL Schema Ontology* O_S, which directly captures the XML Schema semantics, is exploited in the first scenario supported by the SPARQL2XQuery Framework. In particular, O_S is utilized by the users while forming the SPARQL queries. In addition, the SPARQL2XQuery Framework processes O_S and XS and generates a list of mappings between the constructs of O_S and XS (details in Section 7).

The ontological infrastructure generated by the XS2OWL Framework, additionally supports the transformation of XML data into RDF format and vice versa [21]. For transforming XML data to RDF, O_S can be exploited to transform XML documents structured according to XS into RDF descriptions structured according to O_S. However, for the inverse process (i.e., transforming RDF documents to XML) both O_S and O_{BC} should be used, since the XML Schema semantics that cannot be captured in O_S are required. For example, the accurate order of the XML sequence elements should be preserved; but this information cannot be captured in O_S.

A complete listing of the correspondences between the XML Schema constructs and the OWL constructs, as they are specified in the XS2OWL Transformation Model, is presented in Table 7.

Table 7 Correspondences between the XML Schema Constructs in OWL Syntax, according to the XS2OWL Transformation Model

XML Schema Construct	OWL Construct
Complex Type	Class
Simple Datatype	Datatype Definition
Element	(Datatype or Object) Property
Attribute	Datatype Property
Sequence	Unnamed Class – Intersection
Choice	Unnamed Class – Union
Annotation	Comment
Extension, Restriction	subClassOf axiom
Unique (Identity Constraint)	HasKey axiom *
Key (Identity Constraint)	HasKey axiom – ExactCardinality axiom *
Keyref (Identity Constraint)	Object Property Range *
Substitution Group	SubPropertyOf axioms
Alternative +	In the Backwards Compatibility Ontology
Assert +	In the Backwards Compatibility Ontology
Override, Redefine +	In the Backwards Compatibility Ontology
Error +	Datatype

Note. The + indicates the new XML Schema constructs introduced by the XML Schema 1.1 specification. The * indicates the OWL 2 constructs.

6.1 XML Schema Transformation Example

In this section we present a concrete example that demonstrates the expression of the semantics of the XML Schema in OWL syntax using the XS2OWL Framework.

```
<xs:schema xmlns:xs="http://www.w3.org/2001/XMLSchema">

<xs:complexType name="Video_Type">
    <xs:group ref="videoGroup"/>
    <xs:attribute name="code" type="xs:integer"/>
</xs:complexType>

      <xs:complexType name="EditedVideo_Type">
<xs:complexContent>
    <xs:extension base="Video_Type">
      <xs:sequence>
      <xs:element name="Edit" type="xs:string" minOccurs="0" maxOccurs="unbounded"/>
      </xs:sequence>
    </xs:extension>
  </xs:complexContent>
</xs:complexType>

<xs:complexType name="Reviews_Type">
  <xs:sequence>
    <xs:element name="Review" type="xs:string"/>
    <xs:element name="Reviewer_Mode" type="xs:string"/>
  </xs:sequence>
</xs:complexType>

<xs:element name="MultimediaContent">
  <xs:complexType>
    <xs:sequence>
      <xs:element name="Video" type="Video_Type" minOccurs="0" maxOccurs="unbounded"/>
      <xs:element name="EditedVideo" type="EditedVideo_Type" minOccurs="0" maxOccurs="unbounded"/>
    </xs:sequence>
  </xs:complexType>
</xs:element>

<xs:group name="videoGroup">
  <xs:sequence>
      <xs:element name="Title" type="xs:string"/>
      <xs:element ref="Creator" minOccurs="1" maxOccurs="unbounded"/>
      <xs:element name="Date" type="xs:date" />
      <xs:element name="Rating" type="xs:float" />
      <xs:element name="Reviews" type="Reviews_Type" minOccurs="0" maxOccurs="unbounded"/>
  </xs:sequence>
</xs:group>

      <xs:element name=" Creator" type="xs:string"/>
      <xs:element name="Agent" substitutionGroup="Creator" type="xs:string">

</xs:schema>
```

Fig. 5. XML Schema Document describing Audiovisual Material

Fig. 5 presents the XML Schema document that corresponds to the XML Schema tree representation of Fig. 1.

The XML Schema of Fig. 5 has the root element `MultimediaContent`, which is a list of videos and edited videos that has been implemented as an XML Schema sequence that may contain any number of `Video` elements of type `Video_Type` and any number of `EditedVideo` elements of type `EditedVideo_Type`. The complex type `Video_Type` represents videos and contains the model group `videoGroup` and the `code` attribute, of integer type, which represents the video identification code.

The `videoGroup` model group is a reusable sequence of elements, including: (a) The `Title` element, of string type, which represents the title of a video; (b) The referenced top-level `Creator` element, of string type, which occurs at least once and represents the creator of a video; (c) The `Date` element, of type `xs:date`, which represents the date of the video creation; (d) The `Rating` element, of float type, which represents the average rating score of a video; and (e) The `Reviews` element, of the `Reviews_Type` type, which represents the reviews of a video.

The complex type `EditedVideo_Type` extends the complex type `Video_Type` and represents edited videos. In addition to the elements and attributes defined in the context of `Video_Type`, the complex type `EditedVideo_Type` has the `Edit` element, of string type, which specifies zero or more edits (e.g., cuts, filters, etc.) that have been applied on the edited video.

The complex type `Reviews_Type` is a list of reviews, implemented as an XML Schema sequence that may contain any number of `Review` and `Reviewer_Mode` elements of string type. These elements represent the review text and the type of the reviewer (e.g., simple user, creator, etc.).

Finally, the top-level element `Agent` is an element that may substitute the `Creator` element, as is specified in its `substitutionGroup` attribute.

Using the XS2OWL Framework, the XML Schema of Fig. 5 is expressed in OWL syntax. The constructs and the semantics of the *Schema ontology* are presented in Table 8 and Table 9. In particular:

- Information about the classes is provided in Table 8, including the class `rdf:ID`, the `name` of the corresponding XML Schema complex type and its superclass(es) (`rdfs:subClassOf`) name(s).
- Information about the datatype properties (*DTP*) and the object properties (*OP*) is provided in Table 9, including the property `rdf:ID`, the `name` of the corresponding XML Schema element or attribute, the property type (*DTP* for the datatype properties or *OP* for the object properties), the property domain(s) (`rdfs:domain`) and the property range(s) (`rdfs:range`).

Table 8 Representation of the XML Schema Complex Types in the Schema Ontology (O_S)

XML Schema Complex Types	Ontology Classes	
	rdf:ID	**rdfs:subClassOf**
Video_Type	Video_Type	owl:Thing
EditedVideo_Type	EditedVideo_Type	Video_Type
Reviews_Type	Reviews_Type	owl:Thing
MultimediaContent (*unnamed complex type*)	NS_ MultimediaContent_UNType	owl:Thing

Table 9 Representation of the XML Schema Elements and Attributes in the Schema Ontology (O_S)

XML Schema Elements & Attributes	Ontology Properties			
	Type	**rdf:ID**	**rdfs:domain**	**rdfs:range**
Title	DTP	Title_videoGroup__xs_string	Video_Type	xs:string
Creator	DTP	Creator__xs_string	Video_Type	xs:string
Date	DTP	Date_videoGroup__xs_date	Video_Type	xs:date
Rating	DTP	Rating_videoGroup__xs_float	Video_Type	xs:float
Agent	DTP	Agent__xs_string	Video_Type	xs:string
code	DTP	code__xs_integer	Video_Type	xs:integer
Reviews	OP	Reviews_videoGroup__Reviews_Type	Video_Type	Reviews_Type
Review	DTP	Review__xs_string	Reviews_Type	xs:string
Reviewer_Mode	DTP	Reviewer_Mode__xs_string	Reviews_Type	xs:string
Edit	DTP	Edit__xs_string	EditedVideo_Type	xs:string
Video	OP	Video_Video_Type	NS_ MultimediaContent _UNType	Video_Type
EditedVideo	OP	EditedVideo__EditedVideo_Type	NS_ MultimediaContent _UNType	EditedVideo_Ty pe
MultimediaContent	OP	MultimediaContent__NS_ MultimediaContent_UNType	owl:Thing	NS_ MultimediaCo ntent_UNType

In addition, the constructs of the Backwards Compatibility ontology generated by the XS2OWL for the XML Schema of Fig. 5 are presented in Table 10. Notice that the XML Schema constructs are represented in the Backwards Compatibility ontology as individuals of the Backwards Compatibility ontology classes. For every XML Schema construct Table 10 presents: (a) the Class of the corresponding individual in O_{BC}; (b) the unique `rdf:ID` of the individual (3rd column); and (c) the `rdf:ID` of the corresponding OWL construct in the Schema ontology (4th column).

Table 10 Representation of the `Persons` XML Schema Constructs in the Backwards Compatibility Ontology (O_{BC})

XML Schema Constructs	Backwards Compatibility Individuals		
	Class	**rdf:ID**	**Schema Ontology Construct rdf:ID**
Title	ElementInfoType	Title_videoGroup__xs_string__ei	Title_videoGroup__xs_string
Title	DatatypePropertyInfoType	Title_videoGroup__xs_string	Title_videoGroup__xs_string
Date	ElementInfoType	Date_videoGroup__xs_date__ei	Date_videoGroup__xs_date
Date	DatatypePropertyInfoType	Date_videoGroup__xs_date	Date_videoGroup__xs_date
Rating	ElementInfoType	Rating_videoGroup__xs_float__ei	Rating_videoGroup__xs_float
Rating	DatatypePropertyInfoType	Rating_videoGroup__xs_float	Rating_videoGroup__xs_float
Reviews	ElementInfoType	Reviews_videoGroup__Reviews_Type__ei	Reviews_videoGroup__Reviews_Type
Creator	ElementInfoType	Creator__xs_string__ei	Creator__xs_string
Creator	DatatypePropertyInfoType	Creator__xs_string	Creator__xs_string
Agent	ElementInfoType	Agent__xs_string__ei	Agent__xs_string
Agent	DatatypePropertyInfoType	Agent__xs_string	Agent__xs_string
Review	ElementInfoType	Reviews_Type_Review__xs_string__ei	Review__xs_string
Review	DatatypePropertyInfoType	Reviews_Type_Review__xs_string	Review__xs_string
Reviewer_Mode	ElementInfoType	Reviews_Type_Reviewer_Mode__xs_string __ei	Reviewer_Mode__xs_string
Reviewer_Mode	DatatypePropertyInfoType	Reviews_Type_Reviewer_Mode__xs_string	Reviewer_Mode__xs_string
MultimediaContent	ElementInfoType	MultimediaContent__NS_MultimediaContent_UNType__ei	MultimediaContent__NS_MultimediaContent_UNType
MultimediaContent	DatatypePropertyInfoType	MultimediaContent__NS_MultimediaContent_UNType	MultimediaContent__NS_MultimediaContent_UNType
Code	DatatypePropertyInfoType	Video_Type_code__xs_integer	code__xs_integer
Edit	ElementInfoType	EditedVideo_Type_Edit__xs_string__ei	Edit__xs_string
Edit	DatatypePropertyInfoType	EditedVideo_Type_Edit__xs_string	Edit__xs_string
Reviews_Type	ComplexTypeInfoType	Reviews_Type	Reviews_Type
EditedVideo_Type	ComplexTypeInfoType	EditedVideo_Type	EditedVideo_Type
Video_Type	ComplexTypeInfoType	Video_Type	Video_Type

The XML Schema of Fig. 5 and the schema ontology O_S generated by XS2OWL are depicted in Fig. 6. The correspondences between the XML Schema and the generated ontology are represented with dashed grey lines.

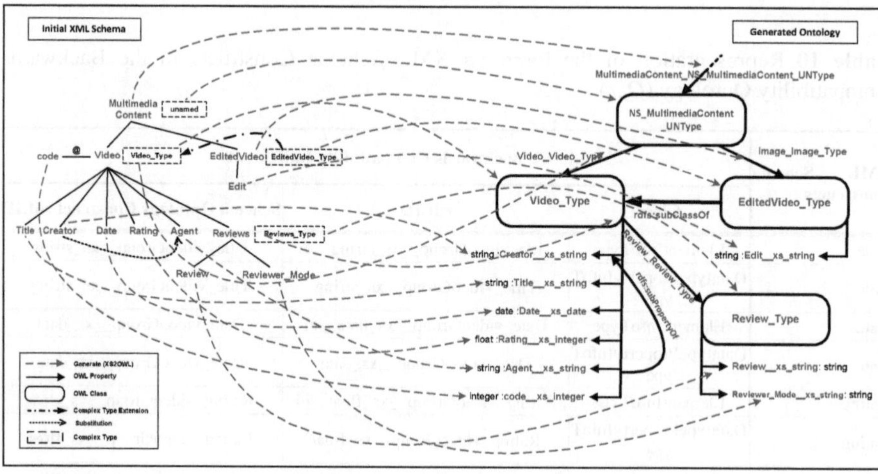

Fig. 6 The XML Schema of Fig. 5 and the Schema Ontology O_S generated by XS2OWL with their correspondences drawn in dashed grey lines

7 The SPARQL2XQuery Framework — Mapping Model

In this section we outline the mapping model adopted by the SPARQL2XQuery Framework in the context of SPARQL to XQuery translation, in order to allow the expression of mappings between OWL ontologies and XML Schemas.

At the *Schema level* (OWL Ontology/XML Schema), associations between the ontology constructs (i.e., classes, properties, etc.) and the XML Schema constructs (i.e., elements, complex types, etc.) are obtained. Moreover, at the *Data level*, the XML data follow the XML Schema. As a result, the XML Schema construct occurrences in the XML data can be identified and addressed using a set of XPath expressions (*XPath Set*). Thus, based on the correspondences between the ontology and the XML Schema, the ontology constructs are associated with the corresponding XPath expressions (that point to the corresponding XML nodes at the XML data level). Consequently, a *mapping* μ in the context of the SPARQL to XQuery translation is the association of an OWL construct with *XPath Sets*.

In the first scenario supported by the SPARQL2XQuery Framework the mappings are automatically generated. In this case, the XS2OWL Framework is exploited for expressing the semantics of an XML Schema *XS* in OWL syntax, then the mappings between the automatically generated OWL ontology O_S and *XS* are automatically detected, generated and maintained by the SPARQL2XQuery Framework.

The mapping generation is carried out by the *Mappings Generator* component, which takes as input *XS* and O_S. The Mappings Generator component parses the input files and obtains a set ***M*** of mappings, expressed in XML syntax, between all the constructs of O_S and the XPath sets that address all the corresponding XML nodes in the documents that follow *XS*.

In the second scenario, an XML Schema must be mapped with existing ontologies; in this case, the mappings must be manually specified by an expert. Thus, in some cases the XML Schema and the ontology are partially mapped.

7.1 Mapping Examples

In this section we provide two mapping examples based on both the scenarios supported by the SPARQL2XQuery Framework. In the first example, we present a manual mapping specification case (2nd Scenario), where the VoD server ontologies (Fig. 2) are manually mapped by an expert to the XML Schema of the `Digital Library X` (Fig. 1). In the second example, we outline the automatic mapping generation between the XML Schema of the `Digital Library X` and its corresponding OWL ontology that is automatically generated by the XS2OWL Framework (Fig. 6) (1st Scenario). In this case, the SPARQL2XQuery Framework, automatically generates, maintains and stores the mappings between the XML Schema of the digital library and the generated ontology in XML format.

Example 1. *Manual Mapping Specification*
In Fig. 7, excerpts of the VoD server ontologies have been (partially) mapped to the XML Schema of the `Digital Library X` (2nd Scenario). The mappings are presented with dashed grey lines.

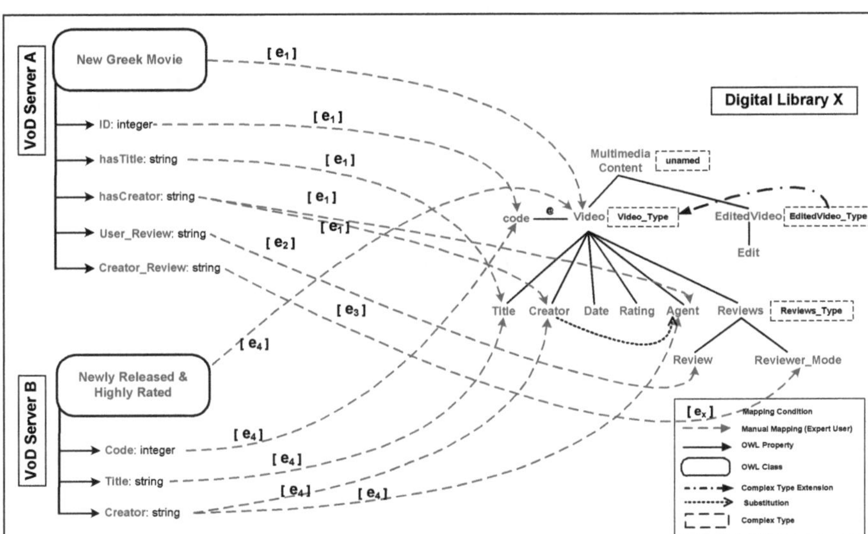

Fig. 7 The Existing ontologies of `VoD Server A & B` are manually mapped with the XML Schema of `Digital Library X`

It can be observed from Fig. 7 that several mappings have been identified. For instance, the class `New Greek Movies` from the `VoD Server A` can be mapped to the XML Schema element `Movie`, under the condition e_1. e_1 is a condition that holds for the movies with a `code` attribute that starts with `"960"` (i.e., Greek code) and a `Date` element equal to `"2011"`. Such a mapping in the context of SPARQL to XQuery translation is of the following form:

μ_1: New Greek Move ≡ { */MultimediaContent/ Video[./Date = 2011 and starts-with(. @code, "960")]*}

In a similar way, the properties `ID` and `hasTitle` can be mapped to the `code` attribute and the `Title` element respectively under the e_1 condition. Such mappings are of the following form:

μ_2: ID≡ { */MultimediaContent/ Video[./Date= 2011 and starts-with(. @code, "960")]/ @code*}

μ_3: hasTitle ≡ { */MultimediaContent/ Video[./Date= 2011 and starts-with(. @code, "960")]/ Title* }

In addition, the property `User_Review` can be mapped to the `Review` element under the conjunction (i.e., *and*) of the e_1 condition with a condition e_2 that holds for the reviews with a `Reviewer_Mode` element equal to `"User"`. Also, the `Creator_Review` property can be mapped to the `Review` element under the conjunction of the e_1 condition with a condition e_3 that holds for the reviews with a `Reviewer_Mode` element equal to `"Creator"`. Such mappings are of the following form:

μ_4: User_Review ≡ { */MultimediaContent/ Video[./Date= 2011 and starts-with(.@code, "960") and ./Reviews/ Reviewer_Mode = "User"]/ Review* }

μ_4: Creator_Review ≡ { */MultimediaContent/ Video[./Date= 2011 and starts-with(.@code, "960") and ./Reviews/ Reviewer_Mode = "Creator"]/ Review* }

In the `VoD Server B`, the class `Newly Released & Highly Rated` can be mapped to the XML Schema element `Video`, under the condition e_4. e_4 is a condition that holds for videos with a `Date` element equal to `"2011"` and a `Rating` element greater than `"7.5"`.

μ_5: Newly Released & Highly Rated ≡ { */MultimediaContent/ Video[./Date = 2011 and./Rating> 7.5]* }

Similarly, the property `Creator` can be mapped to the union of the `Creator` and `Agent` elements under the condition e_4.

μ_5: Creator ≡ { */MultimediaContent/ Video[./Date = 2011 and./Rating> 7.5]/ Creatpor, /MultimediaContent/ Video[./Date = 2011 and./Rating> 7.5]/ Agent* } ■

Example 2. *Automatic Mapping Generation*

Consider the XML Schema of the `Digital Library X` and the corresponding ontology that has been automatically generated by XS2OWL (Fig. 6). In this case, the *Mappings Generator* component automatically generates the mappings between all the ontology constructs (classes and properties) and the XPath Sets that address the corresponding XML nodes. The generated mappings are listed below.

Generated Mappings

Classes:

μ_1: NS_ MultimediaContent _UNType ≡ { */MultimediaContent* }
μ_2: Video_Type ≡ { */MultimediaContent/ Video* }
μ_3: EditedVideo_Type ≡ { */MultimediaContent/ EditedVideo* }
μ_4: Reviews_Type ≡ { */MultimediaContent/ Video/ Reviews, /MultimediaContent/ EditedVideo/ Reviews* }

Datatypes Properties:

μ_5: code__xs_integer ≡ { */MultimediaContent/ Video/ @code , /MultimediaContent/ EditedVideo/ @code* }
μ_6: Creator__xs_string ≡ { */MultimediaContent/ Video/ Creator , /MultimediaContent/ EditedVideo/ Creator* }
μ_7: Agent__xs_string ≡ { */MultimediaContent/ Video/ Agent , /MultimediaContent/ EditedVideo/ Agent* }
μ_8: Title_videoGroup__xs_string ≡ { */MultimediaContent/ Video/ Title , /MultimediaContent/ EditedVideo/ Title* }
μ_9: Date_videoGroup__xs_date ≡ { */MultimediaContent/ Video/ Date, /MultimediaContent/ EditedVideo/ Date* }
μ_{10}: Rating_videoGroup__xs_float ≡ { */MultimediaContent/ Video/ Rating , /MultimediaContent/ EditedVideo/ Rating* }
μ_{11}: Review__xs_string { */MultimediaContent / Video/ Review , /MultimediaContent/ EditedVideo/ Review*}
μ_{12}: Reviewer_Mode__xs_string ≡ { */MultimediaContent/ Video/ Review_Mode,*
/MultimediaContent/ EditedVideo/ Review_Mode }
μ_{13}: Edit__xs_ string ≡ { */MultimediaContent/ EditedVideo/ Edit* }

Object Properties:

μ_{14}: MultimediaContent __NS_ MultimediaContent _UNType ≡ { */MultimediaContent* }
μ_{15}: Video__Video_Type ≡ { */MultimediaContent/ Video* }
μ_{16}: EditedVideo__EditedVideo_Type ≡ { */MultimediaContent/ EditedVideo* }
μ_{17}: Reviews_videoGroup__Reviews_Type ≡ { */MultimediaContent/ Video/ Reviews ,*
/MultimediaContent/ EditedVideo/ Reviews }

■

8 The SPARQL2XQuery Framework — Query Translation

In this section, we present an overview of the SPARQL to XQuery query translation process, which is performed by the *Query Translator* component. The *Query Translator* takes as input a SPARQL query and the mappings between an ontology and an XML Schema and translates the SPARQL query to semantically correspondent XQuery expressions w.r.t. the mappings. To the best of our knowledge, this is the first work addressing this issue.

The query translation process is based on a generic method and a set of algorithms for translating SPARQL queries to XQuery expressions under strict compliance with the SPARQL semantics. The translation covers all the syntax variations of the SPARQL grammar [15]; as a consequence, it can handle every SPARQL query. In addition, the translation process is generic and scenario independent, since the mappings are represented in an abstract formal form as *XPath Sets*. The mappings may be automatically generated (in the first scenario supported by the XS2OWL Framework) or manually specified by a mapping process carried out by an expert (in the second scenario supported by the XS2OWL Framework).

The *Query Translator* component comprises of the following sub-components:

– The *SPARQL Graph Pattern Normalizer*, that rewrites a Graph Pattern (*GP*) (i.e., the SPARQL `Where` clause) to an equivalent normal form, resulting, in a simpler and more efficient translation process.

- The *Variable Type Specifier*, that identifies the types of the variables, in order to detect any conflict arising from the syntax provided by the user, as well as to identify the form of the results for each variable.
- The *Schema Triple Processor* that processes *Schema Triples* (triples referring to the ontology structure and/or semantics) and binds the appropriate XPaths to the SPARQL variables contained in the Schema Triples.
- The *Variable Binder*, that is used for the assignment of the appropriate XPaths to the SPARQL variables, thus enabling the translation of *GPs* to XQuery expressions.
- The *Basic Graph Pattern Translator*, that performs the translation of *Basic Graph Patterns* (*BGP*) into XQuery expressions.
- The *Graph Pattern Translator*, that translates *GPs* into XQuery expressions. The concept of a *GP* is defined recursively. The *Basic Graph Pattern Translator* sub-component translates the basic components of a *GP* (i.e., *BGPs*) into XQuery expressions, which however have to be properly associated in the context of the *GP* by applying the SPARQL algebra operators (i.e., `AND`, `OPT`, `UNION` and `FILTER`) among them, using XQuery expressions
- The *Solution Sequence Modifier Translator*, that translates the SPARQL solution sequence modifiers using XQuery clauses and built-in functions.
- The *Query Form Translator*, that is responsible for building the appropriate result structure. SPARQL supports four query forms (i.e., `Select`, `Ask`, `Construct` and `Describe`). According to the query form, the structure of the final result is different (i.e., an RDF graph, a Result sequence or Boolean value).

8.1 Query Translation Example

Consider a query posed over the ontology of Fig. 6. The query, expressed in both natural language and in SPARQL syntax, is presented below.

Natural Language Query

For the instances of the Video_Type class, return the code and the Title, if the creator is "Johnson John", the title includes the string "Music" and the rating is higher than "5". The query must return at most 50 result items ordered by the rating value in descending order and by the id value in ascending order, skipping the first 10 items."

SPARQL Query

```
PREFIX ns:  <http://example.com/ns#>
PREFIX rdf: <http://www.w3.org/1999/02/22-rdf-syntax-ns#>
SELECT ?id ?title
WHERE {  ?video   rdf:type                              ns:Video_Type .
         ?video   ns:Creator__xs_string                 "Johnson John" .
         ?video   ns: code__xs_integer                  ?id .
         ?video   ns: Title_videoGroup__xs_string ?     title .
         ?video   ns: Rating_videoGroup__xs_float ?     rating .
         FILTER ( regex( ?title , "Music") && ?rating > 5 )
} ORDER BY  DESC ( ?rating )  ASC ( ?id )
LIMIT 50  OFFSET 10
```

The SPARQL2XQuery Framework takes as input the mappings of Example 2 and translates the above SPARQL query to the XQuery query presented below.

Translated XQuery Query

```
let $doc := collection("http://www.music.tuc.gr/mediaXMLDB/...")
let $Modified_Results :=(
    let $Results :=(
       for $video  in  $doc/MultimediaContent/Video[./Creator = "Johnson John" ]
       for $id     in  $video/@code
       for $title  in  $video/Title[matches( . , "Music" )]
       let $rating : =  $video/Rating[. > 5 ]
       return( <Result> <id>{ string($id) }</id> , <title>{ string($title) }</title> </Result> )
    )
   return ( let $Ordered_Results :=(
                          for $iter in $Results
                            order by $iter/rating descending empty least , $iter/id empty least
                           return($iter) )
            return ( $Ordered_Results[ position( ) > 10 and position( ) <= 60 ]) )
)
return ( <Results>{ $Modified_Results }</Results> )
```

9 Conclusions

In this chapter we have described the mechanisms that allow the exploitation of the legacy data, in the Web of Data. In particular, in the first part of the chapter (Section 2), we have presented and compared the XML and SW worlds including the technologies and the standards adopted in the two worlds.

In the second part (Sections 3 and 4) we have present a survey of the existing approaches that deal with the interoperability and integration issues between the XML and SW worlds.

Finally, in the third part (Sections 5–8), we have made a brief presentation of the SPARQL2XQuery and XS2OWL Frameworks that have been developed to provide an interoperable environment between the SW and the XML worlds.

References

[1] Bray, T., et al. (eds.): Extensible Markup Language (XML) 1.1. W3C Recommendation (September 2006), http://www.w3.org/TR/xml11/

[2] Manola, F., Milles, E. (eds.): RDF Primer. W3C Recommandation (February 2004), http://www.w3.org/TR/rdf-primer

[3] Fallside D. (ed.): XML Schema Part 0: Primer. W3C Recommendation (October 2004), http://www.w3.org/TR/xmlschema-0/

[4] Gao, S., Sperberg-McQueen, C.M., Thompson, H.S. (eds.): W3C XML Schema Definition Language (XSD) 1.1 Part 1: Structures, W3C Working Draft, December 3 (2009), http://www.w3.org/TR/xmlschema11-1/

[5] Peterson, D., Gao, S., Malhotra, A., et al. (eds.): W3C XML Schema Definition Language (XSD) 1.1 Part 2: Datatypes, W3C Working Draft, December 3 (2009), http://www.w3.org/TR/xmlschema11-2/

[6] Brickley, D., Guha, R.V. (eds.): RDF Vocabulary Description Language 1.0: RDF Schema. W3C Recommendation (Febuary 2004), http://www.w3.org/TR/rdf-schema

[7] McGuinness, D.L., van Harmelen, F. (eds.): OWL Web Ontology Language: Overview. W3C Recommendation, Febuary 10 (2004), http://www.w3.org/TR/owl-features

[8] Motik, B., Schneider, P.F.P., Parsia, B. (eds.): OWL 2 Web Ontology Language: Structural Specification and Functional-Style Syntax, W3C Recommendation, October 27 (2009), http://www.w3.org/TR/owl2-syntax/

[9] Berglund, A., et al. (eds.): XML Path Language (XPath) 2.0. W3C Recommendation (January 2007), http://www.w3.org/TR/xpath20/

[10] Siméon, J., Chamberlin, D. (eds.): XQuery 1.0: an XML Query Language. W3C Recommendation (January 2007), http://www.w3.org/TR/xquery/

[11] Malhotra, A., et al. (eds.): XQuery 1.0 and XPath 2.0 Functions and Operators W3C Recommendation, 2nd edn. (December 2010), http://www.w3.org/TR/xpath-functions/

[12] Robie, J., Chamberlin, D., et al. (eds.): XQuery Update Facility 1.0 W3C Recommendation (March 2011), http://www.w3.org/TR/xquery-update-10/

[13] Case, P., Dyck, M., Holstege, M.: XQuery and XPath Full Text 1.0, W3C Recommendation (March 2011), http://www.w3.org/TR/xpath-full-text-10/

[14] W3C SPARQL Working Group, http://www.w3.org/2009/sparql/wiki/Main_Page

[15] Prud'hommeaux, E., Seaborne, A. (eds.): SPARQL Query Language for RDF. W3C Recommendation (January 2008), http://www.w3.org/TR/rdf-sparql-query/

[16] Harris, S., Seaborne, A. (eds.): SPARQL 1.1 Query Language W3C Working Draft (October 2010), http://www.w3.org/TR/sparql11-query/

[17] Schenk, S., Gearon, P., Passant, A. (eds.): SPARQL 1.1 Update W3C Working Draft (October 2010), http://www.w3.org/TR/sparql11-update/

[18] Seaborne, A. (ed.): SPARQL 1.1 Property Paths W3C Working Draft (January 2010), http://www.w3.org/TR/sparql11-property-paths/

[19] Beckett, D. (ed.): SPARQL Query Results XML Format. W3C Recommendation (January 2008), http://www.w3.org/TR/rdf-sparql-XMLres/

[20] Tsinaraki, C., Christodoulakis, S.: Interoperability of XML Schema Applications with OWL Domain Knowledge and Semantic Web Tools. In: Proc. of ODBASE 2007 (2007)

[21] Tsinaraki, C., Christodoulakis, S.: Support for Interoperability between OWL based and XML Schema based Applications. In: the Proc. of DELOS Conference II (2007)

[22] Stavrakantonakis, I., Tsinaraki, C., Bikakis, N., Gioldasis, N., Christodoulakis, S.: SPARQL2XQuery 2.0: Supporting Semantic-based Queries over XML Data. In: The Proc. of SMAP (2010)

[23] Lenzerini, M.: Data Integration: A Theoretical Perspective. In: Proc. of PODS 2002 (2002)

[24] Fagin, R., Kolaitis, P., Miller, R., Popa, L.: Data exchange: semantics and query answering. Theor. Comput. Sci (TCS) 336(1) (2005)

[25] Farrell, J., Lausen, H. (eds.): Semantic Annotations for WSDL and XML Schema. W3C Recommendation, W3C (August 2007), http://www.w3.org/TR/sawsdl/

[26] Connolly, D. (ed.): Gleaning Resource Descriptions from Dialects of Languages W3C Recommendation, W3C (September 2007), http://www.w3.org/TR/grddl/

[27] Groppe, S., Groppe, J., Linnemann, V., Kukulenz, D., Hoeller, N., Reinke, C.: Embedding SPARQL into XQuery/XSLT. In: Proc. of ACM SAC 2008 (2008)

[28] Droop, M., Flarer, M., Groppe, J., Groppe, S., Linnemann, V., Pinggera, J., Santner, F., Schier, M., Schöpf, F., Staffler, H., Zugal, S.: Embedding XPATH Queries into SPARQL Queries. In: Proc. of the 10th International Conference on Enterprise Information Systems 2008 (2008)

[29] Droop, M., Flarer, M., Groppe, J., Groppe, S., Linnemann, V., Pinggera, J., Santner, F., Schier, M., Schöpf, F., Staffler, H., Zugal, S.: Bringing the XML and Semantic Web Worlds Closer: Transforming XML into RDF and Embedding XPath into SPARQL. In: Filipe, J., Cordeiro, J. (eds.) Enterprise Information Systems. LNBIP, vol. 19, pp. 31–45. Springer, Heidelberg (2009)

[30] Droop, M., Flarer, M., Groppe, J., Groppe, S., Linnemann, V., Pinggera, J., Santner, F., Schier, M., Schöpf, F., Staffler, H., Zugal, S.: Translating xPath queries into SPARQL queries. In: Meersman, R., Tari, Z. (eds.) OTM-WS 2007, Part I. LNCS, vol. 4805, pp. 9–10. Springer, Heidelberg (2007)

[31] Akhtar, W., Kopecký, J., Krennwallner, T., Polleres, A.: XSPARQL: Traveling between the XML and RDF Worlds – and Avoiding the XSLT Pilgrimage. In: Bechhofer, S., Hauswirth, M., Hoffmann, J., Koubarakis, M. (eds.) ESWC 2008. LNCS, vol. 5021, pp. 432–447. Springer, Heidelberg (2008)

[32] Amann, B., Beeri, C., Fundulaki, I., Scholl, M.: Querying XML Sources Using an Ontology-Based Mediator. In: Meersman, R., et al. (eds.) CoopIS 2002, DOA 2002, and ODBASE 2002. LNCS, vol. 2519, Springer, Heidelberg (2002)

[33] Amann, B., Beeri, C., Fundulaki, I., Scholl, M.O.: Ontology-based integration of XML web resources. In: Horrocks, I., Hendler, J. (eds.) ISWC 2002. LNCS, vol. 2342, p. 117. Springer, Heidelberg (2002)

[34] Christophides, V., Karvounarakis, G., Magkanaraki, A., Plexousakis, D., Tannen, V.: The ICS-FORTH Semantic Web Integration Middleware (SWIM). IEEE Data Eng. Bull. 26(4), 11–18 (2003)

[35] Christophides, V., Karvounarakis, G., Koffina, I., Kokkinidis, G., Magkanaraki, A., Plexousakis, D., Serfiotis, G., Tannen, V.: The ICS-FORTH SWIM: A Powerful Semantic Web Integration Middleware. In: Proc. of SWDB 2003 (2003)
[36] Koffina, I., Serfiotis, G., Christophides, V., Tannen, V.: Mediating RDF/S Queries to Relational and XML Sources. Int. J. Semantic Web Inf. Syst (IJSWIS) 2(4), 68–91 (2006)
[37] Xiao, H., Cruz, I.: Integrating and Exchanging XML Data Using Ontologies. In: Spaccapietra, S., Aberer, K., Cudré-Mauroux, P. (eds.) Journal on Data Semantics VI. LNCS, vol. 4090, pp. 67–89. Springer, Heidelberg (2006)
[38] Xiao, H., Cruz, I.: RDF-based metadata management in peer-to- peer systems. In: Proc. of 2nd IST MMGPS Workshop 2004 (2004)
[39] Cruz, I., Xiao, H., Hsu, F.: Peer-to-Peer Semantic Integration of XML and RDF Data Sources. In: Moro, G., Bergamaschi, S., Aberer, K. (eds.) AP2PC 2004. LNCS (LNAI), vol. 3601, pp. 108–119. Springer, Heidelberg (2005)
[40] Cruz, I., Huiyong, X., Hsu, F.: An Ontology-Based Framework for XML Semantic Integration. In: Proc of IDEAS 2004 (2004)
[41] Lehti, P., Fankhauser, P.: XML data integration with OWL: Experiences and challenges. In: Proc. of 3rd International Symposium on Applications and the Internet, SAINT 2004 (2004)
[42] Klein, M.: Interpreting XML via an RDF schema. In: Proc. of Second International Workshop on Electronicy Business Hubs, WEBH 2002 (2002)
[43] Reif, G., Jazayeri, M., Gall, H.: Towards semantic Web Engineering: WEESA-mapping XML schema to ontologies. In: Proc. of WWW 2004 Workshop on Application Design, Development, and Implementation Issues in the Semantic Web (2004)
[44] Ferdinand, M., Zirpins, C., Trastour, D.: Lifting XML Schema to OWL. In: Koch, N., Fraternali, P., Wirsing, M. (eds.) ICWE 2004. LNCS, vol. 3140, pp. 354–358. Springer, Heidelberg (2004)
[45] Garcia, R., Celma, O.: Semantic integration and retrieval of multimedia meta-data. In: Proc. of Fifth International Workshop on Knowledge Markup and Semantic Annotation (2005)
[46] Bohring, H., Auer, S.: Mapping XML to OWL Ontologies. In: Leipziger Informatik-Tage 2005, pp. 147–156 (2005)
[47] Battle, S.: Gloze: XML to RDF and back again. In: 2006 Jena User Conference (2006)
[48] Rodrigues, T., Rosa, P., Cardoso, J.: Mapping XML to Exiting OWL ontologies. In: International Conference WWW Internet 2006, Murcia, Spain, October 5-8 (2006)
[49] Thuy, P.T.T., Lee, Y.K., Lee, S., Jeong, B.S.: Transforming Valid XML Documents into RDF via RDF Schema. In: International Conference on Next Generation Web Services Practices. IEEE (October 2007)
[50] Thuy, P.T.T., Lee, Y.K., Lee, S., Jeong, B.S.: Exploiting XML Schema for Interpreting XML Documents as RDF. In: International Conference on Services Computing SCC 2008 (2008)
[51] Bedini, I., Gardarin, G., Nguyen, B.: Deriving Ontologies from XML Schema. In: Proc. EDA 2008, vol. B-4, pp. 3–17 (2008)
[52] Cruz, C., Nicolle, C.: Ontology Enrichment and Automatic Population from XML Data. In: Proc. of 4th Int. Workshop on Ontology-based Techniques, ODBIS 2008 (2008)

[53] Berrueta, D., Labra, J.E., Herman, I.: XSLT+SPARQL: Scripting the Semantic Web with SPARQL embedded into XSLT stylesheets. In: Proc. of 4th Workshop on Scripting for the Semantic Web 2008 (2008)
[54] Thuy, P.T.T., Lee, Y.K., Lee, S.: DTD2OWL: automatic transforming XML documents into OWL ontology. In: Proc. of Int. Conf. Interaction Sciences 2009 (2009)
[55] Wache, H., Voegele, T., Visser, U., Stuckenschmidt, H., Schuster, G., Neumann, H., Huebner, S.: Ontology-based integration of information - A survey of existing approaches. In: Proc. of Workshop on Ontologies and Information Sharing 2001 (2001)
[56] Noy, N.F.: Semantic Integration: A Survey Of Ontology-Based Approaches. SIGMOD Record 33(4) (2004)
[57] Miller, R.J., Haas, L.M., Hernández, M.: Schema Mapping as Query Discovery. In: Proc. of VLDB 2000 (2000)
[58] Popa, L., Velegrakis, Y., Miller, R.J., Hernández, M.A., Fagin, R.: Translating Web Data. In: Proc. of VLDB 2002 (2000)
[59] Yu, C., Popa, L.: Constraint-based XML Query Rewriting for Data Integration. In: Proc. of SIGMOD 2004 (2004)
[60] DeHaan, D., Toman, D., Consens, M., Özsu, T.: A Comprehensive XQuery to SQL Translation using Dynamic Interval Encoding. In: Proc. of SIGMOD 2003 (2003)
[61] Thiran, P., Estievenart, F., Hainaut, J.L., Houben, G.J.: A Generic Framework for Extracting XML Data from Legacy Databases. J. Web Eng. (JWE) 4(3) (2005)
[62] Kappel, G., Kapsammer, E., Retschitzegger, W.: Integrating XML and Relational Database Systems. World Wide Web (WWWJ) 7(4) (2004)
[63] Krishnamurthy, R., Kaushik, R., Naughton, J.F.: XML-to-SQL Query Translation Literature: The State of the Art and Open Problems. In: Bellahsène, Z., Chaudhri, A.B., Rahm, E., Rys, M., Unland, R. (eds.) XSym 2003. LNCS, vol. 2824, pp. 1–18. Springer, Heidelberg (2003)
[64] Deutsch, A., Tannen, V.: Reformulation of XML Queries and Constraints. In: Calvanese, D., Lenzerini, M., Motwani, R. (eds.) ICDT 2003. LNCS, vol. 2572, pp. 225–238. Springer, Heidelberg (2002)
[65] Bohannon, P., Fan, W., Flaster, M., Narayan, P.: Information Preserving XML Schema Embedding. In: Proc. of VLDB 2005 (2005)
[66] Halevy, A., Ives, Z., Mork, P., Tatarinov, I.: Piazza: Data Management Infrastructure for Semantic Web Applications. In: Proc. WWW 2003 (2003)
[67] Halevy, A.Y., Ives, Z.G., Suciu, D., Tatarinov, I.: Schema Mediation in Peer Data Management Systems. In: Proc. of ICDE (2003)
[68] Tatarinov, I., Halevy, A.: Efficient Query Reformulation in Peer Data Management Systems. In: Proc. of SIGMOD 2004 (2004)
[69] Ives, Z., Halevy, A., Mork, P., Tatarinov, I.: Piazza: mediation and integration infrastructure for Semantic Web data. J. Web Sem. (WS) 1(2) (2004)
[70] Arenas, M., Libkin, L.: XMLdata Exchange: Consistency and Query Answering. Journal of ACM (JACM) 55(2) (2008)
[71] Aguilera, V., Cluet, S., Milo, T., Veltri, P., Vodislav, D.: Views in a large-scale XML repository. VLDB Journal 11(3) (2002)
[72] Bonifati, A., Chang, E.Q., Ho, T., Lakshmanan, L.V.S., Pottinger, R., Chung, Y.: Schema Mapping and Query Translation in Heterogeneous P2P XML Databases. VLDB Journal 19(2) (2010)
[73] Sahoo, S., Halb, W., Hellmann, S., Idehen, K., Thibodeau, T., Auer, S., Sequeda, J., Ezzat, A.: A Survey of Current Approaches for Mapping of Relational Databases to RDF. Technical Report. RDB2RDF W3C Working Group (2009)

[74] Das, S., Sundara, S., Cyganiak, R. (eds.): R2RML: RDB to RDF Mapping Language, W3C Working Draft, October 28 (2010), http://www.w3.org/TR/r2rml/
[75] Sequeda, J., Depena, R., Miranker, D.: Ultrawrap: Using SQL Views for RDB2RDF. In: Proc. of ISWC (2009)
[76] Bizer, C., Cyganiak, R.: D2R Server - Publishing Relational Databases on the Semantic Web. In: Proc. of ISWC (2006)
[77] Chen, H., Wu, Z., Wang, H., Mao, Y.: RDF/RDFS-based Relational Database Integration. In: Proc. of ICDE (2006)
[78] Blakeley, C.: Mapping Relational Data to RDF with Virtuoso's RDF Views. OpenLink Software (2007)
[79] Zemke, F.: Converting SPARQL to SQL, Technical Report (October 2006), http://lists.w3.org/Archives/Public/public-rdf-dawg/2006OctDec/att-0058/sparql-to-sql.pdf
[80] Krishnamoorthy, K., Kumar, R., Dua, S.R.: Converting SPARQL Queries to SQL Queries. Microsoft Corporation patent application, http://www.faqs.org/patents/app/20090138437#ixzz0jE2i19bW
[81] Ma, L., Wang, C., Lu, J., Cao, F., Pan, Y., Yu, Y.: Effective and Efficient Semantic Web Data Management over DB2. In: Proc. of SIGMOD (2008)
[82] Chebotko, A., Lub, S., Fotouhib, F.: Semantics preserving SPARQL-to-SQL translation. Data & Knowl. Eng. (DKE) 68(10), 973–1000 (2009)
[83] Elliott, B., Cheng, E., Thomas-Ogbuji, C., Ozsoyoglu, Z.M.: A Complete Translation from SPARQL into Efficient SQL. In: Proc. of International Database Engineering & Applications Symposium IDEAS (2009)
[84] Cyganiak, R.: A relational algebra for SPARQL. Hewlett-Packard Laboratories Technical Report (2005), http://www.hpl.hp.com/techreports/2005/HPL-2005-170.html
[85] Klein, M., Fensel, D., Harmelen, F., Horrocks, I.: The relation between ontologies and XML Schemata. In: Proc. of the Workshop on Applications of Ontologies and Problem-Solving Methods (2000)
[86] Perez, J., Arenas, M., Gutierrez, C.: Semantics and Complexity of SPARQL. ACM Trans. Database Syst (TODS) 34(3) (2009)
[87] Makris, K., Bikakis, N., Gioldasis, N., Tsinaraki, C., Christodoulakis, S.: Towards a Mediator Based on OWL and SPARQL. In: Lytras, M.D., Damiani, E., Carroll, J.M., Tennyson, R.D., Avison, D., Naeve, A., Dale, A., Lefrere, P., Tan, F., Sipior, J., Vossen, G. (eds.) WSKS 2009. LNCS, vol. 5736, pp. 326–335. Springer, Heidelberg (2009)
[88] Makris, K., Gioldasis, N., Bikakis, N., Christodoulakis, S.: Ontology Mapping and SPARQL Rewriting for Querying Federated RDF Data Sources. In: Proc. of ODBASE (2010)
[89] Makris, K., Gioldasis, N., Bikakis, N., Christodoulakis, S.: SPARQL Rewriting for Query Mediation over Mapped Ontologies. Technical Report 2010 MUSIC/TUC (2010), http://www.music.tuc.gr/reports/SPARQLREWRITING.PDF
[90] Bikakis, N., Gioldasis, N., Tsinaraki, C., Christodoulakis, S.: Querying XML Data with SPARQL. In: Bhowmick, S.S., Küng, J., Wagner, R. (eds.) DEXA 2009. LNCS, vol. 5690, pp. 372–381. Springer, Heidelberg (2009)
[91] ikakis N., Tsinaraki C., Gioldasis N., Christodoulakis S.: The SPARQL2XQuery Framework. Technical Report, http://www.music.tuc.gr/reports/SPARQL2XQUERY.PDF

Personalized Multimedia Web Services in Peer to Peer Networks Using MPEG-7 and MPEG-21 Standards

Emmanouil Skondras[1], Angelos Michalas[1], Malamati Louta [2], and George Kouzas [3]

[1] Dpt of Informatics & Computer Technology, Technological Educational Institute of Western Macedonia, 52100 Kastoria, Greece
manolisskondras@yahoo.gr, amichalas@kastoria.teikoz.gr
[2] Dpt of Informatics and Telecommunications Engineering, University of Western Macedonia, 50100, Kozani, Greece
louta@uowm.gr
[3] Dpt of Financial and management engineering, University of the Aegean, 82100, Chios, Greece
georgekouzas@gmail.com

Abstract. Multimedia information has been increased in the recent years while new content delivery services enhanced with personalization functionalities are provided to users. Several standards are proposed for the representation and retrieval of multimedia content. This paper makes an overview of the available standards and technologies. Furthermore a prototype semantic P2P architecture is presented which delivers personalized audio information. The metadata which support personalization are separated in two categories: the metadata describing user preferences stored at each user and the resource adaptation metadata stored at the P2P network's web services. The multimedia models MPEG-21 and MPEG-7 are used to describe metadata information and the Web Ontology Language (OWL) to produce and manipulate ontological descriptions. SPARQL is used for querying the OWL ontologies. The MPEG Query Format (MPQF) is also used, providing a well-known framework for applying queries to the metadata and to the ontologies.

1 Introduction

Nowadays, the volume of multimedia data is increasing rapidly in many information channels. Network infrastructures enable multimedia information to be easily transferred to users. The delivery of multimedia services is a common task. Several systems are emerging using various retrieval methods and algorithms to provide multimedia content. However more users tend to require information retrieval services which include high quality features such as semantic description and personalization of information.

I.E. Anagnostopoulos et al. (Eds.): Semantic Hyper/Multimedia Adaptation, SCI 418, pp. 361–383.
springerlink.com

A number of standards have been developed providing personalized multimedia content delivery services. MPEG-7 and MPEG-21 provide schemes which define efficiently multimedia metadata and user preferences. In particular user preferences include choices for specific content types, filtering and browsing modes as well as usage environment parameters. OWL provides a framework for the representation of semantic information. It offers a semantic understanding of metadata structures used for personalization. Also SPARQL and MPQF are powerful tools for querying RDF repositories and OWL ontologies.

In this paper the available standards and technologies that support multimedia retrieval services are presented. Additionally a P2P prototype application that delivers personalized audio information to users is described presenting thus an innovative approach for standards' usage. The personalization process is automated and decentralized. The information which describes the user preferences is created and stored at the client. The P2P side information includes audio resources and resource adaptation metadata, minimizing thus, the central storage and computational requirements. This significantly reduces the response time of the system, handling multiple concurrent requests from users during normal operation.

The framework uses MPEG-7 and MPEG-21 for the description of audio content as well as the users' preferences. The metadata information is managed using Web Ontology Language (OWL) ontologies, which provide semantic descriptions of the multimedia content. SPARQL is used for querying the metadata and the relative ontologies. The MPQF provides a well-known framework for applying the queries.

A client may request to listen to an audio track, upload a new audio track or retrieve a catalog of audio tracks that match specific criteria. Additionally the framework is capable to propose audio files which comply with user preferences and/or overall statistics extracted from the aggregation of all users' preferences stored as resource adaptation metadata. Moreover, a mechanism capable of combining user preferences with resource adaptation metadata is defined, which receives feedback from previous users' actions.

The remainder of the paper is organized as follows. In section 2, the related research literature is revisited. Section 3 presents an overview of the standards followed in this study. Section 4 describes the software architecture that supports the prototype application, as well as the software elements and modules required. Finally, section 5 concludes our work and presents possible future extensions and plans.

2 Related Work

The rapid increase in multimedia content has challenged the academic and industrial communities into the development of information retrieval tools enhanced with personalization and adaptation capabilities. An increasing number of these applications use defined standards and well-known query formats to support personalization.

In the work described in [1], the authors propose a personalization process that customizes rich multimedia documents to the needs of an individual reader. Multimedia documents, such as textbooks, reference materials and leisure materials, inherently use techniques making them accessible to people with disabilities, who are incapable of using printed materials. The authors address issues of establishing user personalization profiles, as well as adapting and customizing content, interaction and navigation. Customization of interaction and navigation leads to different user interfaces, as well as different structural content presentation. Customization of content includes insertion of a summary, synchronization of sign language video with text highlighting, selfvoicing capability, alternative support for screen readers, as well as reorganization of layout to accommodate large fonts.

The work described in [2] examines a metadata based approach, supporting the personalization process for knowledge workers who interact with 38 distributed information objects. An architecture supporting the personalization process is described, along with a prototype personalization environment. Its metadata are decentralized, in terms that the information is stored locally at client-side. The authors discuss the advantages, as well as the challenges of the suggested approach.

The authors of the approach presented in [3] introduce a wide view of personalization and user profiles, making the preferences available to a range of services and devices. Behind every instance of personalization is a profile that stores the user preferences, context of use and other information capable to deliver user experiences that describe individual users' needs and preferences. It is based upon the fact that users' needs depend on the context and current situation, (e.g. "At home", "In a Meeting", "In the Car").

In the approach proposed in [4] the user of a multimedia database returns relevance ranking to his retrieval intention for top n data of a retrieval result. Using this feedback information, the framework produces an adjustment data inherent to the user and utilizes it for personalization.

In the work described in [5] a region of interest (ROI) approach for image retrieval is presented. An "attention window" of an image is determined. Consequently, regions of interest are segmented within the attention window and relative luminance features are considered for image's decomposition. The MPEG-7 standard is used providing feature descriptions about the extracted segments. Finally, the similarity between the regions of interest is observed in respect of their MPEG-7 descriptions.

In [6], the design and the implementation of a MPEG-7 based Multimedia Retrieval System for Film Heritage is presented. The multimedia content has been indexed using an Annotation Tool based on MPEG-7 standard. An MPEG-7 Compliant Ontology in OWL DL has been developed to fulfill the requirements of the system. This ontology has been instantiated so that the retrieval process can be handled. This work has been assessed during the validation of the CINeSPACE project, which aims to design and implement a mobile rich media collaborative information exchange platform, accessible through a wide variety of networks (cities WiMax, WANs etc.) for the promotion of Film Heritage.

In the work described in [7], the issues associated with designing a video personalization and summarization system in heterogeneous usage environments are addressed, providing in parallel, a tutorial that introduces MPEG-7 and MPEG-21 within these contexts. The authors introduce a framework for a three-tier summarization system (server, middleware and client). The server maintains the content sources, the MPEG-7 metadata descriptions, the MPEG-21 rights expressions and content adaptability declarations. The client exploits the MPEG-7 user preferences and the MPEG-21 usage environments, in order to retrieve and display the personalized content. The middleware contains the personalization and adaptation engines, which select, adapt, and deliver the summarized rich media content to the user. The system includes MPEG-7 annotation tools, semantic summarization engines, real-time video transcoding and composition tools, application interfaces for PDA devices as well as browser portals.

In [8] a model for integrating semantic user preference descriptions within the MPEG-7/21 standard is presented. The approach preserves the hierarchical structure of the MPEG-21/7 user preference descriptions. The implementation of the model is presented, which allows descriptions of domain ontologies, semantic content descriptions and user preference descriptions in an OWL/RDF environment and also supports automatic conversion of the proposed extensions to MPEG-21/7 descriptions.

The work described in [9] presents an agent based multimedia broadcasting framework using MPEG-21/7 and Foundation for Intelligent Physical Agents (FIPA) standards [10]. A FIPA implementation is used as platform for exchanging user preferences and program information, based on the classical client-server architecture. The user preferences are modeled in respect to the MPEG-21/7 User Preference description scheme.

[11] presents an adaptation model for content personalization by integrating MPEG-7/21 metadata. It uses web services as basic modules. A central web service is used. It selects and monitors a suitable workflow in respect of user preferences, content semantics, network constrains as well as terminal capabilities. Each web service evaluates the MPEG-7/21 description and adapts the multimedia material. Thus, user gets the best possible quality in respect of his terminal specifications.

[12] presents the MP7QL query language. It is a language for querying MPEG-7 descriptions and allows querying every aspect of an MPEG-7 multimedia description. Its design has taken into account the MPEG-7 Query Format Requirements. The queries utilize the user preferences as context, enabling thus personalized multimedia content retrieval. The MP7QL allows the specification of queries that refer to multimedia content satisfying specific criteria (such as "give me the multimedia objects where a goal is scored"), semantic entities (such as "give me the players affiliated to the soccer team Barcelona") and constructs of domain ontologies expressed using MPEG-7 syntax (such as "give me the subclasses of the Player class").

The work presented in [13] pays attention to the semantic retrieval for sports information in World Wide Web. The SPARQL query language is used. It realizes intelligent retrieval according to relations between sports such as "synonymy of", "kind of" and "part of". The process is as follows: Firstly, a sports-ontology is created. Then data are collected from data sources and annotated with the ontology. The search engine completes semantic matching of retrieval conditions through ontology reasoning for user's request and finds out the eligible data.

This paper is an extension of our work described in [14]. A prototype architecture delivering personalized audio information over a P2P network using a combination of available standards and different technologies is presented. The personalization process is distributed since user preferences metadata are created and stored locally at each client while the network includes audio resources and resource adaptation metadata. The audio information promoted to the user is formed taking into account both the user preferences metadata and the usage history metadata of the network. The framework achieves better performance as the computational load and personalization of data are shared among each user and the P2P network. To the best of our knowledge, this study is the first one providing a distributed personalization model over a P2P network which exploits the capabilities of Web Services for handling registration, discovery and content lookup over the network, MPEG 7/21 technologies for managing metadata as well as SPARQL and MPQF standards for the retrieval of multimedia information.

3 Used Standards

This section provides an overview of the technologies and standards employed in the development of the application prototype. These are classified in two categories. The technologies responsible for the storage and delivery of content, including P2P networks [15] and Web Services [16]. The technologies supporting the semantic description and personalization of information, including MPEG-7 [17], MPEG-21 [18], OWL [19], SPARQL [20] and MPQF [21].

3.1 P2P Networks

P2P networks are typically used for enabling peers' interaction. Each peer can act either as a client making requests to other peers or as a server responding to incoming requests. P2P networks form an overlay network at the application layer which is different from the underlying physical network. Data, including stored content in any format are transferred between P2P nodes using the underlying physical infrastructure. A P2P system supports node mobility and fault tolerance operations.

There are two main architectures for P2P networks the centralized and the decentralized. Centralized architectures such as [22] have central management servers which keep information for the content of peers and control peer activities. A central server receives queries from nodes about the location of files in other nodes. Its records are dynamically updated whenever changes occur in the P2P network.

In decentralized architectures, central servers do not exist whereas requests are directed from peers to other peers of the network. Decentralized schemes are classified into three categories, namely structured, unstructured and hybrid.

In structured P2P networks, peers are organized according to specific criteria. The topology of the overlay network and the location of files are organized according to a specific algorithm. A location protocol such as the Distributed hash table (DHT) is used to allocate files to peer nodes. DHT provides lookup functionalities employing key-value pairs to route a query to the peer containing the required resource. The task of updating the key-value mapping is distributed among nodes. DHT networks handle effectively node arrivals, departures and failures. Examples of DHT networks include Pastry [23], CAN [24], Chord [25], and Tapestry [26].

In unstructured P2P networks such as Gnutella [27] and KaZaA [28], the overlay links are randomly formed. Each new node entering the network chooses its peers according to a specific algorithm. If a peer needs to find a file, it queries its neighbors. Commonly, a query is flooded through the network to find peers that share the file. Flooding is a convenient technique for searching popular highly replicated items, however it lacks efficiency in case the requested items are rare. In the later case flooding may produce large traffic loads to the network without guaranteeing that the required files will be found due to possible restrictions on the flood range.

Hybrid architectures [29] combine structured and unstructured searching techniques to improve performance in P2P networks. Requests for popular items are performed using flooding while rare items are searched using DHT techniques. A challenging issue in hybrid P2P networks is to define mechanisms for the characterization of items as popular or rare.

3.2 Web Services

A web service is a software component designed to support interoperable communication for processes over a network. Web services can be implemented using existing languages and platforms extending easily functionalities of applications. Moreover, they complement J2EE, CORBA [30] and other standards allowing easy integration with existing distributed applications. A web service exploits a set of standards [16] including the Simple Object Access Protocol (SOAP), the Web Services Description Language (WSDL) and the Universal Description Discovery and Integration protocol (UDDI).

The Simple Object Access Protocol (SOAP) is an XML based protocol for exchanging structured data with web services. SOAP can be used in combination with a variety of other protocols such as HTTP, SMTP and FTP. The SOAP messaging framework [31] defines the SOAP message construct, guidlines for processing SOAP messages, mechanisms which provide message extensibility features as well as rules for carrying SOAP messages from an underlying protocol. A SOAP message consists of an "Envelope" element which contains a "Header" and a "Body" element. The "Envelope" is the root element of the message. The "Header" contains transmission information. The "Body" contains the message information to be transmitted and a

"Fault" element which is used for error reporting. Tables 1 and 2 present a SOAP request and a SOAP response message respectively.

Table 1 A SOAP request message

```
POST /InAudioTitles HTTP/1.1
Host: www.itisanexample.gr
Content-Type: application/soap+xml; charset=utf-8
Content-Length: nnn
<?xml version="1.0"?>
<soap:Envelope
xmlns:soap="http://www.w3.org/2001/12/soap-envelope"
soap:encodingStyle="http://www.w3.org/2001/12/soap-encoding">
<soap:Body xmlns:examplens="http://www.itisanexample.gr/examplens ">
  < examplens:GetAudioTitles>
    < examplens: AudioAuthor >AuthorA </ examplens:AudioAuthor>
  </ examplens: GetAudioTitles >
</soap:Body>
</soap:Envelope>
```

Table 2. A SOAP response message

```
HTTP/1.1 200 OK
Content-Type: application/soap+xml; charset=utf-8
Content-Length: nnn
<?xml version="1.0"?>
<soap:Envelope xmlns:soap="http://www.w3.org/2001/12/soap-envelope" soap:encodingStyle="http://www.w3.org/2001/12/soap-encoding">
<soap:Body xmlns:mpeg7="http://www.mpeg.org/MPEG7/2000">
  <mpeg7:CreationPreferences>
    <mpeg7:Title mpeg7:preferenceValue="12" xml:lang="en">track1.mp3</mpeg7:Title>
  </mpeg7:CreationPreferences>
</soap:Body>
</soap:Envelope>
```

The Web Services Description Language (WSDL) [32] is a description language in XML which defines web services and access rules for them. WSDL defines services as a set of end points called ports which can exchange messages. Once a client initiates a connection with a web service he reads the WSDL file at the server to find which operations are available. A typical WSDL document contains the following elements:

- Types: Contains data type definitions in XML Schemas.
- Message: Includes the definitions of the data needed to perform an operation.
- Operation: Defines a function provided by the relative web service.
- PortType: Describes the web service's interface where the supported operations are defined.
- Binding: Indicates the transport protocol (SOAP) and the PortType of a service.
- Port: Specifies a connection point to a web service described as a combination of a binding with a network addresses.
- Service: Contains a collection of ports which provide access to service's operations.

The Universal Description Discovery and Integration protocol (UDDI) [16] defines a model for publishing and discovering web services. It is implemented as a central element of the service oriented approach. There are approximately forty well defined SOAP messages according to the UDDI version 2 specifications, providing publication as well as retrieval operations. Each registration consists of five data structures types assisting the management of different information. Each structure defines elements for serving business or technical purposes which are defined as follows:

- BusinessEntity structure: It provides information about the publisher of a service as well as the provided services.
- BusinessService structure: It represents a service classification and is a logical child of a businessEntity structure.
- BindingTemplate structure: It represents technical features about services and is a logical child of a businessService structure.
- TModel structure: It provides abstract reference descriptions of how to interact with the web service as well as service compatibility issues.
- PublisherAssertion structure: It extends businessEntity's functionalities by defining relationships between publishers.

In a typical scenario a service provider hosts a Web service. Additionally he defines a relative service description which publishes either directly to a service requestor or to a UDDI service registry. Subsequently, the service requestor retrieves the service description locally or from the service registry and interacts with the Web service exchanging SOAP messages.

3.3 Multimedia Description Standards

MPEG-7 [17] is a standard developed from Moving Pictures Expert Group (MPEG) for the description of multimedia information. The standard provides a framework for the description of multimedia content encoded in any existing scheme such as MPEG1, MPEG2, and MPEG4. Metadata are stored in XML allowing efficient indexing, searching and filtering of multimedia data. MPEG-7 defines the following elements:

- Description tools, which include Descriptors (D) and Description Schemes (DS). Descriptors define the syntax and the semantics of metadata elements. Description Schemes contain Descriptions, other Description Schemes as well as relationships between them.
- A Description Definition Language (DDL), which is used for defining the syntax of Description Tools and creating new or extending existing Description Schemes.
- System tools which provide mechanisms for multiplexing descriptors and synchronizing descriptions with content defining an open framework for multimedia applications

The MPEG-21 [18] standard defines a framework for effectively managing multimedia resources. MPEG-21 uses the architectural concept of the Digital Item. A Digital Item is a combination of resources (such as videos, audio tracks, images), metadata (such as descriptors, identifiers), and structures describing the relationships between resources. Digital Items are declared using the Digital Item Declaration Language (DIDL). MPEG-21 Digital Item Adaptation (DIA) architecture and the MPEG-7 Multimedia Description Schemes (MDS) for content and service personalization provide a Usage Environment which models user preferences. The Usage Environment Description of the DIA framework contains the following attributes:

- The User Characteristics, which specify user features, including:
- The User Info, where user information is stored.
- The User Preferences, describing the user browsing, filtering and search preferences.
- The Usage History, where the history of user interaction with digital items is presented.
- The Presentation Preferences, which describe user preferences concerning the means of presentation of multimedia information.
- The Accessibility Characteristics, responsible for content adaptation concerning users with auditory or visual impairments.
- The Terminal Capabilities, which describe the technical characteristics of user devices.
- The Natural Environment Characteristics, providing information about the location and time of a user in a particular environment, as well as audio-visual characteristics which may include noise levels and illumination properties of the natural environment.
- The Network Characteristics, which specify the network characteristics parameters including bandwidth utilization, packet delay and packet loss.

3.4 OWL and Query Languages

The RDF Schema (RDFS) [33] provides structures for knowledge representation. It deals with the organization of ontological hierarchies such as classes, relationships and properties. However complex structures or restrictions such as the scope of properties or the cardinality of attributes can not be supported in RDFS. The need of a more powerful ontology language leads us to the Web Ontology Language (OWL).

OWL [34] is a family of knowledge representation languages used for composing ontologies. It is considered as an extention of the RDFS and its specifications have been authorized by the World Wide Web Consortium. Ontologies are described in owl documents by defining classes, properties and individuals. Classes are collection of concepts, attributes are properties of classes and individuals represent the objects of a particular class.

SPARQL is an SQL-like language developed for issuing queries [13] to RDF and OWL repositories. Queries are expressed in triple patterns similar to RDF whereas RDF subjects, predicates and objects could be variables. Additional language features include conjunctive or disjunctive patterns as well as value filters. SPARQL components are described in three specifications. The query language specification

[35] presents the SPARQL language structures. The query results XML specification [36], defines the format of the results returned from SPARQL queries as XML documents. The SARQL protocol [37] defines the framework for sending queries from clients to remote server using HTTP or SOAP messages.

The Mpeg Query Format (MPQF) [21] defines the interaction between clients and multimedia repositories. It specifies the message format of clients' requests and multimedia services responses. In contrast to SPARQL, the MPQF doesn't specify any transfer protocol (such as HTTP) however the SOAP message model can be used. In addition to that an extensive set of error messages is defined. Moreover advanced multimedia retrieval operations are supported via a rich set of defined query types:

- QueryByMedia: Query using a specified image, video, audio or text.
- QueryByDescription: Query using metadata expressed in XML format.
- QueryByFreeText: Query using keywords.
- QueryByFeatureRange: Query using features such as bitrate or media duration.
- SpatialQuery: Query for spatial elements within media objects.
- TemporalQuery: Query for temporal characteristics within media objects.
- QueryByROI: Query for spatial-temporal characteristics within media objects.
- QueryByXQuery: A container which supports XQuery expressions.
- QueryByRelevanceFeedback: Query taking into consideration results from previous searches which are characterized as bad or good examples.
- QueryBySPARQL: Query which embodies a SPARQL query.

Combination of query types is also supported providing advanced multimedia requests. Table 3 presents an example which combines three query types (QueryByMedia, QueryByDescription and QueryBySPARQL) and retrieves images according to the parameters of the relevant condition blocks.

Table 3 A multipart MPQF query

```
<MpeqQuery><Query><Input>
 <QueryCondition>
  <TargetMediaType>image/*</TargetMediaType>
   <Condition xsi:type="AND">
    <Condition xsi:type="QueryByMedia">
     <MediaResource xsi:type="MediaResourceType"><MediaResource>
      <InlineMedia type="image/jpeg"><MediaData64>PeQYmmW4ML8m2iQ3AzMBTbmodr</MediaData64></InlineMedia>
     </MediaResource></MediaResource>
    </Condition>
    <Condition xsi:type="QueryByDescription" matchType="exact">
     <DescriptionResource resourceID="des01"> <AnyDescription xmlns:mpeg7=" http://www.mpeg.org/MPEG7/2000">
       <mpeg7:Mpeg7>
       <mpeg7:MediaFormat><mpeg7:FileFormat mpeg7:href="urn:mpeg:mpeg7:cs:FileFormatCS:2001:3">
         <mpeg7:Name xml:lang="en">JPEG</mpeg7:Name></mpeg7:FileFormat>
       </mpeg7:MediaFormat>
      </mpeg7:Mpeg7>
     </AnyDescription> </DescriptionResource>
    </Condition>
    <Condition xsi:type="QueryBySPARQL">
     <SPARQL>
   PREFIX mpeg7: < http://www.mpeg.org/MPEG7/2000>
   SELECT ?Title WHERE { ?x mpeg7:title ?title . FILTER (?Genre=Natural || ?Genre=Animal ) }
     </SPARQL>
    </Condition>
   </Condition>
  </QueryCondition>
</Input></Query></MpegQuery>
```

Furthermore MPQF supports query management tools which define inquiries for multimedia services that satisfy specific requirements. In this way, a management request searches for multimedia services that satisfy specific requirements. Following, a management response returns the appropriate service descriptions. The standard includes procedures for service discovery, query formats for service capabilities and descriptions of service capabilities.

4 A Prototype Architecture

In this section a P2P prototype application that delivers personalized information is presented (Figure 1). Each peer interacts with a main-web-service, an MPQF parser and a SPARQL parser, also implemented as web services. A Network Access Server (NAS) is used providing the necessary login functionalities to the users and a UDDI service enhances web services' interaction capabilities. UDDI describes each peer's content as well as web services' capabilities.

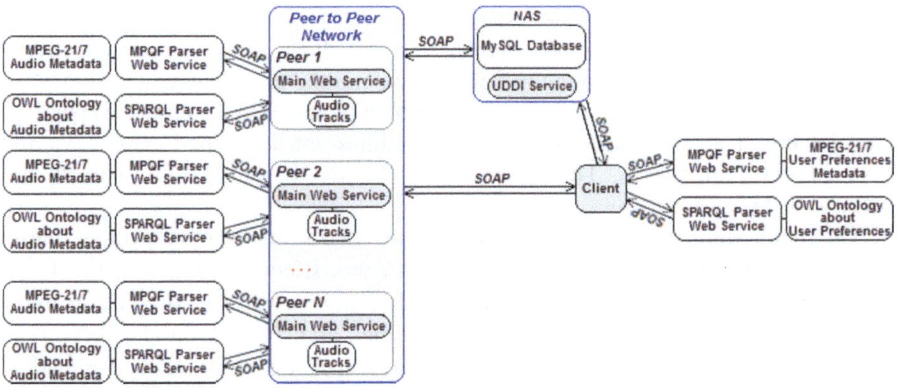

Fig. 1 The basic modules of our architecture

The information contained in the UDDI registry service can be classified in two main categories. The first, provides abstract reference descriptions about the web services and according to the UDDI terminology is referred as tModel (Table 4). The second, defines services implementations as well as detailed technical information about services functionalities. Each peer interacts with a set of web services (main-web-service, MPQF parser and SPARQL parser) and publishes its capabilities to the UDDI service. The client communicates with the UDDI service and retrieves the appropriate information needed to interact effectively with the peers and receive personalized information.

The architecture is decentralized in respect to the information required to achieve personalization. User related preferences are created and stored at each client. Resource adaptation metadata along with the resources are the only to be composed and stored centrally at the P2P network. As an effect, distribution of

Table 4 An example of tModel entry

```
<businessEntity xmlns="urn:uddi-org:api" businessKey="00000000-0000-0000-0000-000000000000">
 <name>Main Web Service</name>
 <description xml:lang="en">The main web service</description>
 <businessServices>
   <businessService  businessKey="00000000-0000-0000-0000-000000000000" serviceKey="11111111-1111-1111-1111-111111111111">
   <name>Promote Audio</name>
   <bindingTemplates>
   <bindingTemplate serviceKey="11111111-1111-1111-1111-111111111111"
     bindingKey="AAAAAAAA-AAAA-AAAA-AAAA-AAAAAAAAAAAA">
   <accessPoint URLType="http">http://peerhost.com/mainwebservice.jws</accessPoint>
    <tModelInstanceDetails>
     <tModelInstanceInfo
     tModelKey="UUID:BBBBBBBB-BBBB-BBBB-BBBB-BBBBBBBBBBBB"/>
    </tModelInstanceDetails>
   </bindingTemplate>
  </bindingTemplates>
 </businessService>
 </businessServices>
 <categoryBag>
 <keyedReference
  tModelKey="UUID:AB12C1DA-AD34-12A1-DACD-A1DAC23C12AC"
  keyName="Multimedia personalization services"
  keyValue="70.24.12.11.08"/>
 </categoryBag>
</businessEntity>
```

both computational load and personalization data is achieved improving framework's scalability. The main-web-service and the client interact with MPQF parser and SPARQL parser web services for MPQF and SPARQL queries parsing, respectively. The web services and client modules are developed using Java and Java Media Framework. The MPEG-21/7, SPARQL and MPQF parsers are mapped to Java classes.

The main-web-service contains the audio tracks and the respective audio metadata using MPEG-7 in an MPEG-21 structure. It communicates with MPQF parser and SPARQL parser web services. The audio tracks are divided in thirty different audio categories (speech, crowd, animal, audio background effects, pop, classical, dance, electronic etc.). Audio metadata include user defined metadata (artist, producer, production year and category), technical oriented metadata (bitrate, sample rate, track duration, upload date and last download date, audio channels, audio format, file size) as well as usage history metadata (track's popularity in respect to all tracks, track's popularity in its category and recommended similar tracks). Table 5 presents a sample of the audio metadata structure.

Each client organizes its metadata using MPEG-21/7 user preferences element of MPEG-21 usage environment. The client's metadata rely on user's preferences (favorite audio categories and top 10 audio tracks in each category). Table 6 presents a sample of the user preferences metadata structure.

Moreover, suitable OWL ontologies for metadata manipulation have been created. The main-web-service extends the OWL ontology presented in [38] for managing the audio metadata. The client uses its personal metadata which contain the semantic description of user preferences, based on the OWL ontology presented in the class diagram of Figure 2.

Table 5 Sample of the audio metadata structure


```
<mpeg21:DIDL xmlns:mpeg21="urn:mpeg:mpeg21:2002:02-mpeg21-NS" xmlns:mpeg7="http://www.mpeg.org/MPEG7/2000">
 <mpeg21:Container>
  <mpeg21:Item>
   <mpeg21:Descriptor>
    <mpeg21:Statement mpeg7:mimeType="text/plain">Metadata about  audio track.</mpeg21:Statement>
    </mpeg21:Descriptor>
     <mpeg21:Component>
      <mpeg21:Resource mpeg7:mimeType="application/xml">
       <mpeg7:Mpeg7>

        <mpeg7:CreationPreferences>
         <mpeg7:Title mpeg7:preferenceValue="12" xml:lang="en">track1.mp3</mpeg7:Title>
        </mpeg7:CreationPreferences>

        <mpeg7:CreationInformation>
         <mpeg7:Creation>
          <mpeg7:Creator>
              <mpeg7:Role mpeg7:href="urn:mpeg:mpeg7:cs:RoleCS:2001:AUTHOR" />
              <mpeg7:Agent xsi:type="PersonType">
               <mpeg7:Name>
                 <mpeg7:GivenName>John</mpeg7:GivenName>
                <mpeg7:FamilyName>Johny</mpeg7:FamilyName>
               </mpeg7:Name>
              </mpeg7:Agent>
          </mpeg7:Creator>
          <mpeg7:Creator>
           <mpeg7:Role mpeg7:href="urn:mpeg:mpeg7:cs:RoleCS:2001:Publisher"/>
           <mpeg7:Agent xsi:type="PersonType">
             <mpeg7:Name>
               <mpeg7:GivenName>George</mpeg7:GivenName>
               <mpeg7:FamilyName>Smith</mpeg7:FamilyName>
             </mpeg7:Name>
            </mpeg7:Agent>
          </mpeg7:Creator>

          <mpeg7:Abstract>
             <mpeg7:FreeTextAnnotation>VeryGood</mpeg7:FreeTextAnnotation>
             <mpeg7:StructuredAnnotation>
               <mpeg7:What>
                 <mpeg7:Name>Music Track</mpeg7:Name>
              </mpeg7:What>
             </mpeg7:StructuredAnnotation>
           </mpeg7:Abstract>

          <mpeg7:CreationCoordinates>

              <mpeg7:CreationDate>
                <mpeg7:TimePoint>2010-05-11</mpeg7:TimePoint>
                <mpeg7:Duration>P7D</mpeg7:Duration>
              </mpeg7:CreationDate>
          </mpeg7:CreationCoordinates>
         </mpeg7:Creation>
        </mpeg7:CreationInformation>

        <mpeg7:ClassificationPreferences>
         <mpeg7:Genre mpeg7:preferenceValue="47"
           mpeg7:href="urn:mpeg:ContentCS:1">
          <mpeg7:Name xml:lang="en">Acappella</mpeg7:Name>
         </mpeg7:Genre>
        </mpeg7:ClassificationPreferences>

        <mpeg7:MediaLocator>
         <mpeg7:MediaUri>tracks/track1.mp3</mpeg7:MediaUri>
        </mpeg7:MediaLocator>
        <mpeg7:MediaTime>
         <mpeg7:MediaTimePoint>T00:00:00F100</mpeg7:MediaTimePoint>
         <mpeg7:MediaDuration>T00:13:07F100</mpeg7:MediaDuration>
        </mpeg7:MediaTime>

        <mpeg7:MediaFormat>
        <mpeg7:Content mpeg7:href="urn:mpeg:mpeg7:cs:ContentCS:2001:2">
         <mpeg7:Name xml:lang="en">audio</mpeg7:Name>
        </mpeg7:Content>
        <mpeg7:Medium mpeg7:href="urn:mpeg:mpeg7:cs:MediumCS:2001:2.1.1 ">
         <mpeg7:Name xml:lang="en">HD</mpeg7:Name>
        </mpeg7:Medium>
        <mpeg7:FileFormat mpeg7:href="urn:mpeg:mpeg7:cs:FileFormatCS:2001:3">
         <mpeg7:Name xml:lang="en">MP3</mpeg7:Name>
        </mpeg7:FileFormat>
        <mpeg7:FileSize>787082</mpeg7:FileSize>
        <mpeg7:BitRate mpeg7:minimum="N/A" mpeg7:average="8000"
        mpeg7:maximum="N/A"></mpeg7:BitRate>
        <mpeg7:AudioCoding>
         <mpeg7:Format mpeg7:href="urn:mpeg:mpeg7:cs:AudioCodingFormatCS:2001:1">
           <mpeg7:Name xml:lang="en">MP3</mpeg7:Name>
         </mpeg7:Format>
         <mpeg7:AudioChannels mpeg7:track="2"></mpeg7:AudioChannels>
         <mpeg7:Sample mpeg7:rate="22050" mpeg7:bitPer="0">
         </mpeg7:Sample>
        </mpeg7:AudioCoding>
       </mpeg7:MediaFormat>
      </mpeg7:Mpeg7>
     </mpeg21:Resource>
    </mpeg21:Component>
   </mpeg21:Item>
  </mpeg21:Container>
 </mpeg21:DIDL>
```

Table 6. Sample of the user preferences metadata structure

```
<mpeg21:DIDL xmlns:mpeg21="urn:mpeg:mpeg21:2002:02-mpeg21-NS">
<mpeg21:Container>
  <mpeg21:Item>
    <mpeg21:Descriptor>
      <mpeg21:Statement mimeType="text/plain">This item is a metadata block about first's preferences.</mpeg21:Statement>
    </mpeg21:Descriptor>
    <mpeg21:Component>
      <mpeg21:Resource mimeType="application/xml">
        <Mpeg7 xmlns="http://www.w3.org/2000/XMLSchema-instance"  type="complete">
          <UserPreferences>
            <UserIdentifier protected="true">
              <UserName>John</UserName>
            </UserIdentifier>
            <UsagePreferences allowAutomaticUpdate="true">
              <FilteringAndSearchPreferences protected="true">
                <ClassificationPreference>
                  <Genre href="urn:mpeg:GenreCS" preferenceValue="42">
                    <Name>Acappella</Name>
                  </Genre>
                          <Genre href="urn:mpeg:GenreCS" preferenceValue="30">
                    <Name>Animal</Name>
                  </Genre>
                  <Genre href="urn:mpeg:GenreCS" preferenceValue="2">
                    <Name>Acoustic</Name>
                  </Genre>
                  <Genre href="urn:mpeg:GenreCS" preferenceValue="34">
                    <Name>Bass</Name>
                  </Genre>
                  <Genre href="urn:mpeg:GenreCS" preferenceValue="75">
                    <Name>Classical</Name>
                  </Genre>
                  <Genre href="urn:mpeg:GenreCS" preferenceValue="40">
                    <Name>HipHop</Name>
                  </Genre>
                  <Genre href="urn:mpeg:GenreCS" preferenceValue="91">
                    <Name>Jazz</Name>
                  </Genre>
                  <Genre href="urn:mpeg:GenreCS" preferenceValue="38">
                    <Name>Pop</Name>
                  </Genre>
                  <Genre href="urn:mpeg:GenreCS" preferenceValue="24">
                    <Name>Speech</Name>
                  </Genre>
…
                  <Genre href="urn:mpeg:GenreCS" preferenceValue="17">
                    <Name>Traditional</Name>
                  </Genre>
                  <Genre href="urn:mpeg:GenreCS" preferenceValue="52">
                    <Name>Other</Name>
                  </Genre>
                </ClassificationPreference>
              </FilteringAndSearchPreferences>
            </UsagePreferences>
          </UserPreferences>
        </Mpeg7>
      </mpeg21:Resource>
    </mpeg21:Component>
  </mpeg21:Item>
</mpeg21:Container>
</mpeg21:DIDL>
```

The client interacts with the P2P network and sends the user's preferences along with the respective credentials. Next, the NAS checks user credentials, establishes a session between the client and the P2P network. Thereafter, the P2P network promotes music tracks to the clients, according to their choices and preferences. A client can upload a new music track, request to listen to a specific music track as well as retrieve a catalog of music tracks that match specific criteria (e.g. belong in a specific music category, comply with user preferences) using SPARQL queries. The aforementioned functionalities are discussed below.

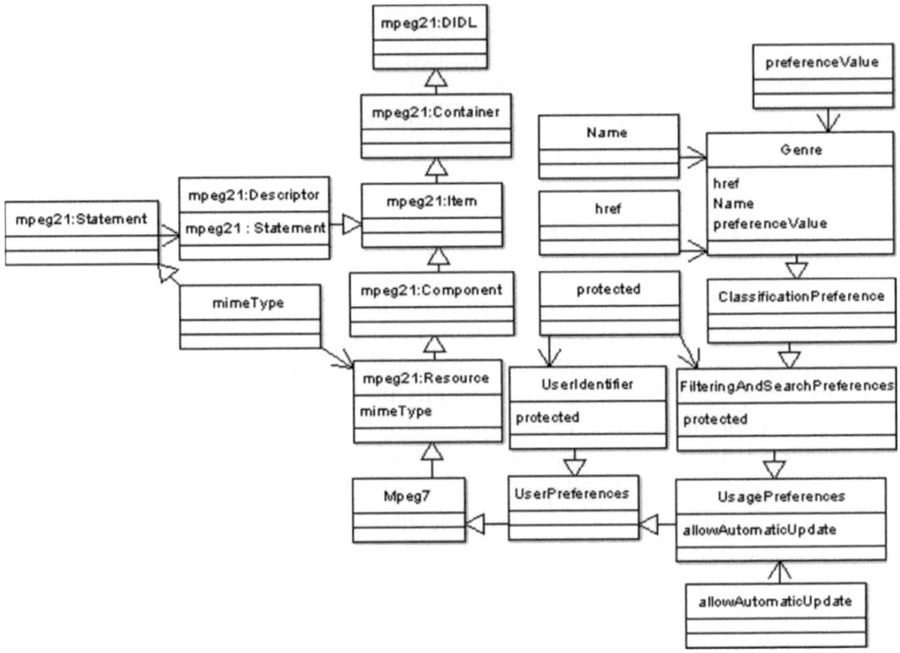

Fig. 2 OWL ontology about user preferences metadata

4.1 Audio Track Upload

User defined metadata of a specific resource are created from the client when a new audio track is uploaded to the P2P network. The client interacts with the UDDI service and finds which peers can receive the new audio file.

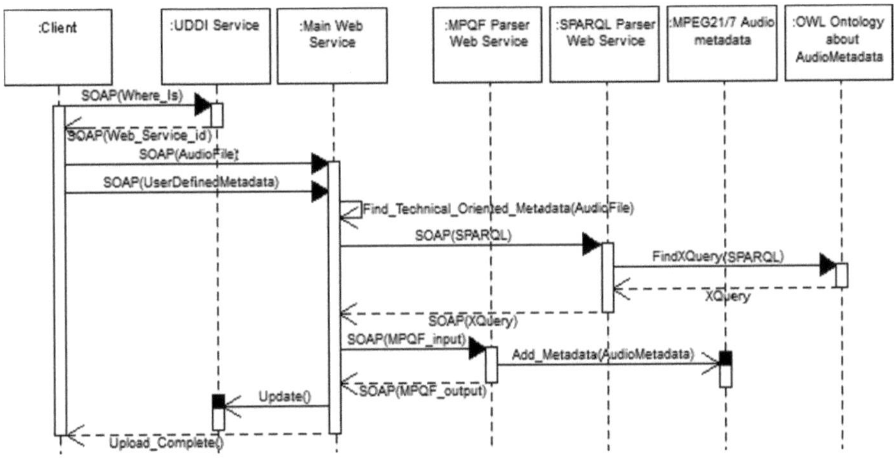

Fig. 3 Adding a new audio file

The main-web-service uses the Java Media Framework to analyze the uploaded audio track and extracts technical oriented metadata. After these actions, the service interacts with the SPARQL/MPQF parsers and inserts all the audio metadata into the metadata file according to the relative standards and to the OWL ontology. The file upload operation is presented in Figure 3.

4.2 Audio Track Request

The client interacts with the UDDI service, finds which peers contain a specific audio track and makes a request to them. Each peer main-web-service, interacts with the SPARQL parser and obtains the metadata structure expressed in OWL in order to satisfy a client request. Subsequently the MPQF parser receives the query together with the metadata structure and returns the requested multimedia info. As a next step, each peer sends a part of the requested audio track. The client retrieves the audio information and updates its preferences. Thereafter, it sends its updated preferences to the main-web-services as feedback information, to enhance future retrievals. The P2P network's web services interact with each other to synchronize the usage history metadata.

Furthermore each peer involved in the track request process proposes audio tracks to the clients, based on their choices and preferences. The list of audio tracks promoted to the user is formed according to an adaptive weighting method of the user preferences metadata stored at each client and the usage history metadata stored at the main-web-service.

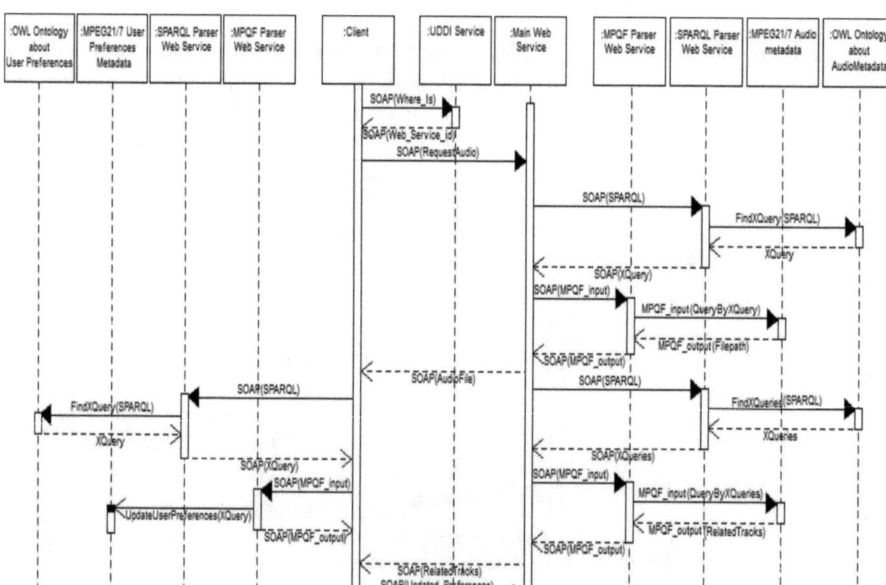

Fig. 4 Web service proposes audio tracks

At each client, the weights of the user preferences metadata w_p and the usage history metadata w_h are updated according to the user actions. When the user requests to listen to an audio file that has been promoted due to the user preferences values, w_p increases while w_h decreases. Adversely, the opposite operation is performed when the audio file has been promoted due to the usage history values. Weight values are updated according to the formula:

$$w_{p/h} = w_{p/h} \pm \frac{pref\_value_i}{\sum_{j=1}^{N} pref\_value_j}$$

where $pref\_value_i$ stands for the preference value of the requested audio file, while N represents the number of all audio files and $w_p + w_h = 1$. The system operation is graphically illustrated in the sequence diagram of Figure 4.

4.3 SPARQL Request

The client retrieves audio catalogues matching specific criteria using SPARQL queries. The client interacts with the UDDI service and retrieves a list of the available peers. Then it makes a respective request containing the SPARQL query to the relevant peers. Each peer main-web-service receives the SPARQL query, interacts with the SPARQL parser and obtains the metadata structure expressed in OWL in order to satisfy the client's request. Consequently the MPQF parser receives the query together with the metadata structure and returns the requested audio list. Finally, the main-web-service sends the extracted list to the client. Figure 5 presents the catalogue retrieval process.

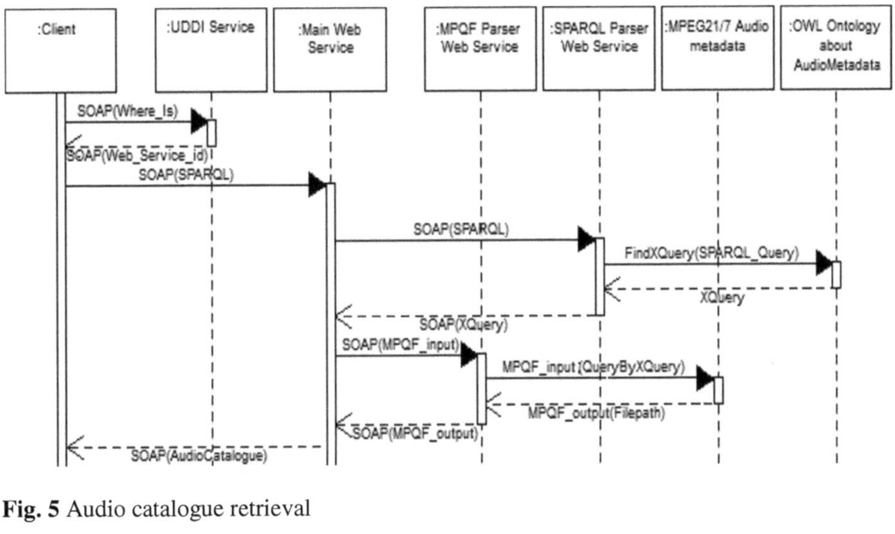

Fig. 5 Audio catalogue retrieval

5 Case Study

This section presents an example of our framework's functionality. At first, the client contacts the P2P network and sends his preferences according to the relative OWL ontology using a SOAP message. As a next step the client may retrieve an audio catalog using the SPARQL query presented in Table 7. The query retrieves a catalog of audio files according to the arguments of the "FILTER" statement. The results are ordered by descendance sequence according to their popularity.

Table 7 The client retrieves an audio catalog using SPARQL

```
PREFIX mpeg7: < http://www.mpeg.org/MPEG7/2000>
SELECT ?Title
WHERE { ?x mpeg7:title ?Title .
FILTER (?Genre=Speech
     && ?Author=Yakub
     && ?Subject=Blood Pressure || ?Subject=Cholesterol
     && ?CreationDate>=2001-01-01 && ?CreationDate<=2004-03-15
     && ?Language=English
     && ?MediaDuration>=3600
     && ?Format=MP3 || ?Format=WAV
     && ?Filesize<=20000000
     )
 }ORDER BY DESC(?preferenceValue)
```

Following, when the catalog is received, the client requests a specific audio track. The P2P network's web services use the relative OWL ontology to manipulate the audio metadata and send the requested track to the client. The client receives a form which plays the requested audio track and presents a list of similar tracks. This list contains promoted audio tracks according to user's preferences stored locally at the client as well as to the usage history stored at the P2P network. The user can request any track from the list. The selection of an audio track results to an update of user preferences metadata at the client and the usage history resource adaptation metadata at the P2P network.

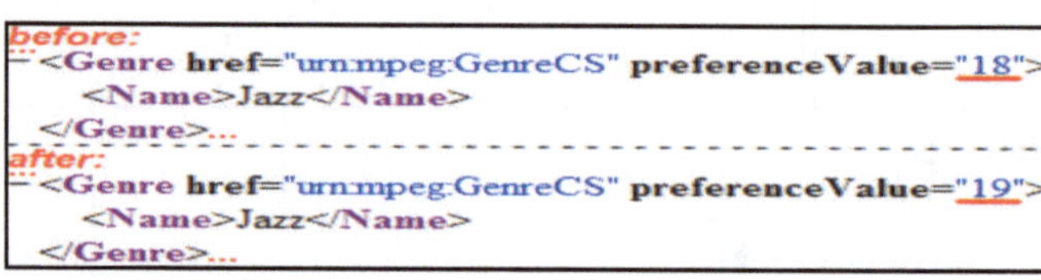

Fig. 6 The relative user preference block before and after the client's request

Figure 6 presents the user preference metadata block before and after the client's request. Accordingly, the 'preferenceValue' concerning audio track's genre is increased.

Similarly, figure 7 presents the CreationPreferences and the Classification Preferences metadata blocks stored at the relative main-web-services of the P2P network before and after client's request. The 'preferenceValue' of the former

```
before:
-<mpeg7:CreationPreferences>
   <mpeg7:Title mpeg7:preferenceValue="12" xml:lang="en">Lullaby.mp3</mpeg7:Title>
 </mpeg7:CreationPreferences>...
-<mpeg7:ClassificationPreferences>
 - <mpeg7:Genre mpeg7:preferenceValue="47" href="urn:mpeg:ContentCS:1">
     <mpeg7:Name xml:lang="en">Jazz</mpeg7:Name>
   </mpeg7:Genre>
 </mpeg7:ClassificationPreferences>...
-------------------------------------------------------------------------
after:
-<mpeg7:CreationPreferences>
   <mpeg7:Title mpeg7:preferenceValue="13" xml:lang="en">Lullaby.mp3</mpeg7:Title>
 </mpeg7:CreationPreferences>...
-<mpeg7:ClassificationPreferences>
 - <mpeg7:Genre mpeg7:preferenceValue="48" href="urn:mpeg:ContentCS:1">
     <mpeg7:Name xml:lang="en">Jazz</mpeg7:Name>
   </mpeg7:Genre>
 </mpeg7:ClassificationPreferences>...
```

Fig. 7 The relative audio metadata blocks before and after the client's request

block shows the number of times the relative audio track has been requested from all users. Respectively, the 'preferenceValue' of the later shows the times the relative genre of tracks has been requested from all users. Both values are increased after the client request.

Subsequently the client selects to listen to audio files promoted by the user preferences values. Figure 8 illustrates the percentage of the promoted audio files based on the user preferences values for 5 consecutive requests. The proposed adaptive weighting method improves the retrieval process by increasing the percentage of the promoted audio files according to user choices.

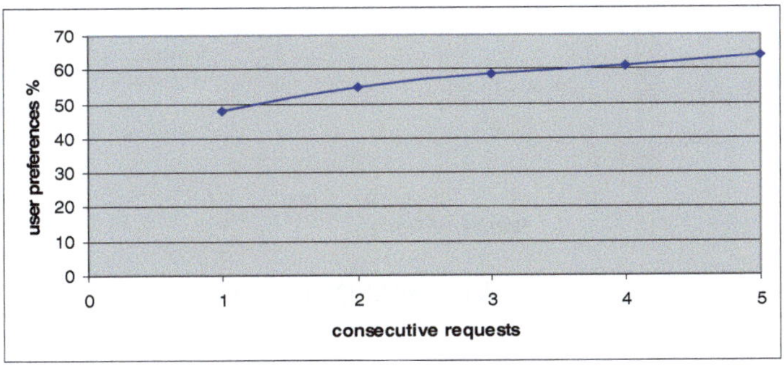

Fig. 8 Audio files promoted to the user based on the user preferences values

Figure 9 presents the response times for audio proposals in respect to a SPARQL or an audio track request. In this scenario each peer contains 20 audio tracks and the relative audio catalogue is extracted according to the mechanisms described in the previous sections. As the peers number increases, the response time increases proportionally as an effect of the load of information process. Additionally, the SPARQL response times are marginally greater compared to the relative response times from audio requests. This is an effect of the complexity of SPARQL requests resulting to higher computational requirements.

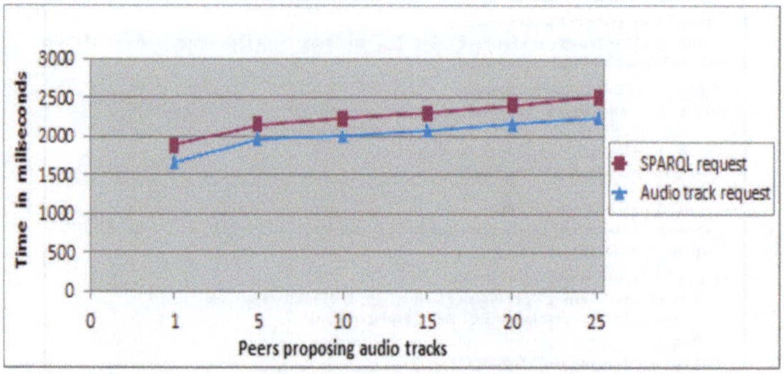

Fig. 9 Response times for audio track and SPARQL request in respect of peers' number

Consequently, Figure 10 presents the response times for audio proposals in respect to the tracks contained in each peer. Similarly the SPARQL response times are greater than the relative audio request times. Respectively, as the tracks number increases the response time increases as well due to the information load processed per peer.

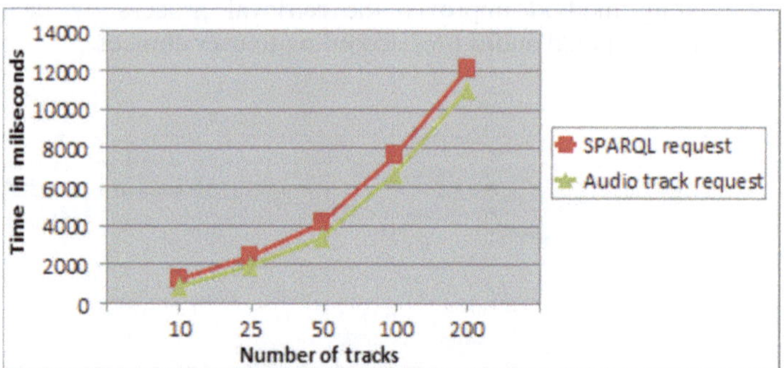

Fig. 10 Response times for audio track and SPARQL request in respect of tracks' number

6 Conclusions

Our approach relies on MPEG-21 and MPEG-7 standards to achieve personalization. MPEG-21 DIDL and DIA are used handling Digital Items declaration and user preferences, respectively. Moreover, the appropriate OWL ontologies are used for managing the metadata. The framework is implemented over web services in a P2P network. It uses SOAP messages for services communication and applies queries to the metadata and to the ontologies using MPQF and SPARQL models. A UDDI service is also used describing each peer's content as well as web services' capabilities and enhancing thus framework's functionalities. The architecture is decentralized improving framework's scalability. Each client organizes its own metadata locally. The P2P network hosts the resource adaptation metadata along with the resources, proposing audio tracks to the clients based on their choices and preferences. A client can also retrieve audio catalogues using SPARQL queries.

The model presented in this paper can be applied to any type of multimedia resources. Additionally applications conforming to MPEG-21 and MPEG-7 may use the metadata produced by our framework.

Future work includes the extension of the MPQF parser's functionalities to support the QueryByMedia query type. Thus, it will give the capability to the web services to propose audio tracks according to other sample audio tracks. For instance, the client will send along with its request an audio track that contains violin and the web services will promote audio tracks that also contain this musical instrument.

References

[1] Petrie, H.L., Weber, G., Fisher, W.: Personalization, interaction, and navigation in rich multimedia documents for print-disabled users. IBM Systems Journal 44(3), 629–635 (2005)

[2] Hicks, D.L., Tochtermann, K.: Personalizing Information Spaces: A Metadata Based Approach. In: International Conference on Dublin Core and Metadata Applications, Tokyo, Japan, October 24-26, pp. 213–220 (2001)

[3] Kovacikova, T., Petersen, F., Pluke, M., Alvarez, V.A., Bartolomeo, G., Frisiello, A., Zetterstrom, E., Cadzow, S.: Personalization and User Profile Standardization. In: Seventh IASTED International Conference on Communication Systems and Networks, Track 629-043, Palma de Mallorca, Spain, September 1-3, pp. 629–43 (2008)

[4] Takayama, T., Sasaki, H., Kuroda, S.: Personalization by Relevance Ranking Feedback in Impression-based Retrieval for Multimedia Database. Journal of Systemics, Cybernetics and Informatics 3(2), 85–89 (2005)

[5] Seo, M., Ko, B., Chung, H., Nam, J.Y.: ROI-Based Medical Image Retrieval Using Human-Perception and MPEG-7 Visual Descriptors. In: Sundaram, H., Naphade, M., Smith, J.R., Rui, Y. (eds.) CIVR 2006. LNCS, vol. 4071, pp. 231–240. Springer, Heidelberg (2006)

[6] Cobos, Y., Sarasua, C., Linaza, M.T., Jimerez, I., Garcia, A.: Retrieving Film Heritage content using an MPEG-7 Compliant Ontology. In: Third International Workshop on Semantic Media Adaptation and Personalization, Prague, Czech Republic, December 15-16, pp. 63–68 (2008)

[7] Tseng, B., Lin, C.Y., Smith, J.: Using MPEG7 and MPEG-21 for Personilizing Video. IEEE Multimedia Journal 11(1), 42–53 (2004)

[8] Tsinaraki, C., Christodoulakis, S.: Semantic User Preference Description in MPEG-7/21. In: Fourth Hellenic Data Management Symposium, pid 6, Athens, Greece, August 25-26 (2005)

[9] Kim, M., Lim, J., Kang, K., Kim, J.: Agent-Based Intelligent Multimedia Broadcasting within MPEG-21 Multimedia Framework. ETRI Journal 26(2) (April 2004)

[10] The Foundation for Intelligent Physical Agents, `http://www.fipa.org/`

[11] Kohncke, B., Balke, W.T.: Preference-driven personalization for flexible digital item adaptation. Multimedia Systems 13(2), 119–130 (2007)

[12] Tsinaraki, C., Christodoulakis, S.: A User Preference Model and a Query Language that allow Semantic Retrieval and Filtering of Multimedia Content. In: SMAP 2006: First International Workshop on Semantic Media Adaptation and Personalization, Athens, Greece, December 4-5, pp. 121–128 (2006)

[13] Zhai, J., Zhou, K.: Semantic Retrieval for Sports Information Based on Ontology and SPARQL. In: ISME 2010: International Conference of Information Science and Management Engineering, Beijing, China, August 7-8, vol. 2, pp. 395–398 (2010)

[14] Skondras, E., Michalas, A., Louta, M., Ioannis, A.: A Personalized Audio Server using MPEG-7 and MPEG-21 standards. In: IEEE 5th International Workshop on Semantic Media Adaptation and Personalization, Limassol, Cyprus, December 9-10, pp. 37–43 (2010)

[15] Oram, A.: Peer to Peer: Harnessing the Power of Disruptive Technologies. O'Reilly & Associates (2001)

[16] Kreger, H.: Web Services Conceptual Architecture. IBM (May 2001)

[17] MPEG-7, ISO/IEC JTC1/SC29/WG11 Coding of Moving Pictures and Audio, http://mpeg.chiariglione.org/standards/mpeg-7/mpeg-7.htm

[18] MPEG-21, ISO/IEC JTC1/SC29/WG11 Coding of Moving Pictures and Audio, http://mpeg.chiariglione.org/standards/mpeg-21/mpeg-21.htm

[19] Web Ontology Language, World Wide Web Consortium, http://www.w3.org/TR/owl-features/

[20] SPARQL Query Language for RDF, http://www.w3.org/TR/rdf-sparql-query/

[21] Döller, M., Tous, R., Gruhne, M., Yoon, K., Sano, M., Burnett, I.S.: The MPEG Query Format: Unifying Access to Multimedia Retrieval Systems. IEEE Multimedia 5(4), 82–95 (2008)

[22] Lee, I., He, Y., Guan, L.: Centralized P2P Streaming with MDC. In: Seventh IEEE Workshop on Multimedia Signal Processing, October 30-November 2, pp. 1–4 (2005)

[23] Rowstron, A., Druschel, P.: Pastry: Scalable, decentralized object location, and routing for large-scale peer-to-peer systems. In: Guerraoui, R. (ed.) Middleware 2001. LNCS, vol. 2218, pp. 329–350. Springer, Heidelberg (2001)

[24] Ratnasamy, S., Francis, P., Handley, M., Karp, R., Shenker, S.: A Scalable Content-addressable Network. In: Proceedings of ACM SIGCOMM, San Diego, CA, pp. 161–172 (August 2001)

[25] Stoica, I., Morris, R., Karger, D., Kaashoek, M.F., Balakrishnan, H.: Chord: A Scalable Peer-to-Peer Lookup Service for Internet Applications. In: Proceedings of ACM SIGCOMM, San Diego, CA, pp. 149–160 (August 2001)

[26] Hildrum, K., Kubiatowicz, J.D., Rao, S., Zhao, B.Y.: Distributed object location in a dynamic network. Theory of Computing Systems 37(3), 405–440 (2004)

[27] Gnutella, http://rfc-gnutella.sourceforge.net/

[28] KazaA. KaZaA, http://www.kazaa.com

[29] Loo, B.T., Huebsch, R., Stoica, I., Hellerstein, J.M.: The Case for a Hybrid P2P Search Infrastructure. In: Voelker, G.M., Shenker, S. (eds.) IPTPS 2004. LNCS, vol. 3279, pp. 141–150. Springer, Heidelberg (2005)

[30] Qusay, M.: Overview of CORBA. In: Distributed Programming with Java, ch. 11, pp. 153–161. Manning Publications (January 2001)

[31] SOAP Version 1.2 Part 1: Messaging Framework, W3C Recommendation, 2nd edn., April 27 (2007), http://www.w3.org/TR/2007/REC-soap12-part1-20070427/
[32] Web Services Description Language (WSDL) 1.1, W3C Note, March 15 (2001), http://www.w3.org/TR/wsdl
[33] RDF Vocabulary Description Language 1.0: RDF Schema, W3C Recommendation, February 10 (2004), http://www.w3.org/TR/rdf-schema/
[34] OWL 2 Web Ontology Language Document Overview, W3C Recommendation, October 27 (2009), http://www.w3.org/TR/owl2-overview/
[35] SPARQL Query Language for RDF, W3C Recommendation, January 15 (2008), http://www.w3.org/TR/rdf-sparql-query/
[36] SPARQL Query Results XML Format, W3C Recommendation, January 15 (2008), http://www.w3.org/TR/rdf-sparql-XMLres/
[37] SPARQL Protocol for RDF, W3C Working Draft, January 14 (2005), ftp://www.w3.org/TR/2005/WD-rdf-sparql-protocol-20050114/
[38] MPEG-7 OWL Ontology, http://metadata.net/mpeg7/mpeg7.owl

Author Index

MIX
Papier aus verantwortungsvollen Quellen
Paper from responsible sources
FSC® C105338

If you have any concerns about our products,
you can contact us on
ProductSafety@springernature.com

In case Publisher is established outside the EU,
the EU authorized representative is:
Springer Nature Customer Service Center GmbH
Europaplatz 3, 69115 Heidelberg, Germany

Printed by Libri Plureos GmbH
in Hamburg, Germany